# SUPERCOLLIDERS AND SUPERDETECTORS

**THE SCIENCE AND CULTURE SERIES — PHYSICS**

Series editor: A. Zichichi, *European Physical Society, Geneva, Switzerland*

1. Perspectives for New Detectors in Future Supercolliders, 1991
2. Data Structures for Particle Physics Experiments: Evolution or Revolution?, 1991
3. Image Processing for Future High-Energy Physics Detectors, 1992
4. GaAs Detectors and Electronics for High-Energy Physics, 1992
5. Supercolliders and Superdetectors, 1993
6. Properties of SUSY Particles, 1993

THE SCIENCE AND CULTURE SERIES — PHYSICS





# SUPERCOLLIDERS AND SUPERDETECTORS

Proceedings of the 19th and 25th Workshops of the INFN Eloisatron Project

Erice, Italy

19 – 26 November 1991

17 – 22 November 1992

Editors

**W. A. Barletta and H. Leutz**

Series Editor

**A. Zichichi**



**World Scientific**
*Singapore • New Jersey • London • Hong Kong*

*Published by*

World Scientific Publishing Co. Pte. Ltd.

P O Box 128, Farrer Road, Singapore 9128

*USA office:* Suite 1B, 1060 Main Street, River Edge, NJ 07661

*UK office:* 73 Lynton Mead, Totteridge, London N20 8DH

**SUPERCOLLIDERS AND SUPERDETECTORS**



ISBN 981-02-1595-9

**Printed in Singapore**

# CONTENTS


Preface ix

**Accelerator Part**

Maximizing the Luminosity of Hadron Supercolliders at 10–100 TeV 3
*W. A. Barletta*

Performance Limitations Beyond LHC/SSC 43
*J. Gareyte*

Strategies for the Eloisatron Based on Beam–Beam and Synchrotron Radiation Considerations 54
*J. Gareyte*

Scaling Laws and Cost Optimization for Hadron Colliders Around 100 TeV 58
*E. Keil*

Design Criteria for the Eloisatron Dipole Magnet 67
*K. Koepke*

Injector for a 100 TeV Collider ("Eloisatron") 80
*K. Schindl*

SSC Beam Dynamics Scaled to the Eloisatron 89
*D. Ritson*

Impedance of Many Holes in a Liner Within a Beam Pipe 95
*R. L. Gluckstern*

Diffusion Phenomenology of Transverse Beam Dynamics and Losses in Hadronic Colliders 102
*A. Gerasimov*

The Effect of Chromatic Decoherence on Transverse Injection Oscillation Damping 117
*G. P. Jackson*

$\mathcal{Z}$-Transform Description of Feedback Acting on a Coupled Beam 131
*R. Talman*

Transverse Feedback in a 100 TeV Storage Ring 151
*G. Lambertson*

PS for LHC—Emittance Conservation at Low Energy 157
*R. Cappi*

Requirements for the Transverse Injection Oscillation Dampers 169
*G. Jackson*

Longitudinal Damper for the SSC 173
*D. Boussard*

Wide-Band Transverse Feedback for Hadron Supercolliders 181
*R. Cappi, G. Lambertson and L. Vos*

The SSC Transverse Feedback Systems 188
*W. Chou and J. Peterson*

Radiation Damage Considerations in a High Luminosity Collider:
The Interaction Region 209
*D. M. Lee*

Results of IR Working Group 219
*D. Ritson*

**Detector Part**

Particle Detection at Eloisatron 227
*H. Leutz*

Central Tracking in High Luminosity Future Colliders 234
*C. D'Ambrosio*

Read-out and Data Acquisition for Tracking Devices at
Future Colliders 254
*T. Gys*

Tracking and Pattern Recognition at High Luminosity Accelerators 270
*U. Gensch*

Calorimetry for High Energy and High Luminosity Hadron Colliders 286
*A. Contin*

Development of a 13.5 T $Nb_3Sn$ Dipole Magnet for Future Particle Accelerators 305
*Q. T. Lin, G. F. Luan, W. J. Shen, L. Y. Wang and A. Zichichi*

Radiation at Hadron Colliders: Scaling to the Highest Energies 311
*D. E. Groom*

Materials in the Radiation Environment of High Luminosity Colliders 327
*H. Schönbacher*

Electronic Components in a High Luminosity Environment 337
*F. Wulf, D. Bräunig, A. Boden and D. Fritsch*

Recommendations of the Detector Group 392

# PREFACE

The present volume is based on the 19th and 25th Workshops of the INFN ELOISATRON (ELN) Project held at the Ettore Majorana Centre for Scientific Culture in Erice, Sicily from 19 to 26 Nov. 1991 and 17 to 22 Nov. 1992 respectively. The 19th workshop addressed "Hadron Supercollider at the Highest Energy and Luminosity; the 25 workshop was titled "Supercolliders and Superdetectors".

The contributions to the workshops address the following main topics of concern to the ELN project : 1) What are the energy and luminosity limitations of proton supercolliders using conventional technology ? 2) What are practical design specifications for the accelerator sub-systems of the ELOISATRON, operating at 100 TeV per beam ? 3) What are the limiting particle detector technologies and how do they constrain the design of the colliders at the highest energies and luminosity.

The 19th and 25th ELOISATRON workshops aimed at establishing the pratical limits of conventional accelerator technology to explore the high energy frontier with the greatest possible dicovery potential. The workshops have confirmed the practicality of a collider with 100 TeV per beam operating with a luminosity of $10^{34}$ $cm^{-2}s^{-1}$. Indeed, it appears that detector technology rather than accelerator technology may set the practical operating luminosities of future proton colliders for energies exceeding 10 TeV per beam.

The papers concerning Accelerator Technology and Beam Physics aim at providing a broad perspective on the beam physics and accelerator technology that will determine the performance of future proton colliders aimed at extending the energy frontier of elementary particle physics. The first four papers discuss the scaling of accelerator technology from that of presently operating machines to a supercollider with beam energies of 100 TeV or greater. These general discussions form the basis for a more detailed examination of the extension of constituent accelerator technologies such as superconducting magnets to 100 TeV. A discussion of the injection chain for ELN shows that the accelerator designer has a wide latitude in choosing the bunch structure in a supercollider without degrading the time-average luminosity. Papers on scaling beam dynamics to extremely high energy and luminosity set the stage for an extensive discussion of the design of feedback control systems that must act to preserve the emittance of the beam in the supercollider in spite of the presence of noise sources and potential beam instabilities. Finally a discussion of radiation damage considerations and of interaction region designs sets for the section on Particle Detectors.

The section on Particle Detectors aimed at recommendations in order to make the machine design compatible with their requirements. The first contribution shows that Higgs cross-sections increase much faster with ELN centre-of-mass energy than the minimum biased events, which are responsible for the radiation dose. The following articles emphasize the interconnections between collider and tracker parameters, e.g. vertex densities, influence of solenoidal magnet fields on radiation dose, the aspects of readout and data acquisition and the requirements for pattern recognition and track finding. Specific options for calorimetric detectors are also examined. The contributions on radiation hardness attempt quantitative assessment of radiation levels, an overview of radiation effects on organic polymers, metals and inorganic materials with particular attention to electronic components in a high luminosity environment. Finally, the detector group recommends some changes in machine parameters, e.g. reduction of beam crossing angle and an energy upgrade of ELN rather than a luminosity increase. To achieve this goal, the development of 13.5 T dipole magnets is considered.

W.A. Barletta
Lawrence Berkeley Laboratory
Berkeley, USA

H. Leutz
CERN, Geneva, Switzerland

# Accelerator Part

# MAXIMIZING THE LUMINOSITY OF HADRON SUPERCOLLIDERS AT 10 - 100 TeV

William A. Barletta

*Lawrence Berkeley Laboratory*
*Berkeley, CA 94720*
*and*
*Department of Physics, University of California Los Angeles*
*405 Hilgard Avenue, Los Angeles CA*

Extending the frontiers of experimental high energy physics requires building accelerators of ever higher particle energies and luminosities. This paper presents a general framework for choosing self-consistent characteristics for hadron colliders at energies of 10 – 100 TeV per beam. It concludes with an examination of the potential of synchrotron–based colliders to explore PeV energies.

## 1. Introduction

As physicists seek to build colliders of ever higher particle energy, they are generally faced with the task of simultaneously increasing the luminosity to compensate for the decrease in cross sections. Extending the physics reach and discovery potential of hadron supercolliders to the maximum extent will require accelerator designers learning how to achieve luminosities much greater than $10^{35}$ cm$^{-2}$ s$^{-1}$ at energies of at least 100 TeV per beam. The application of the presently available accelerator technologies embodied in the designs of the LHC[1.1] (at 8 TeV per beam and $10^{34}$ cm$^{-2}$ s$^{-1}$) and the SSC[1.2] (20 TeV per beam at $10^{33}$ cm$^{-2}$ s$^{-1}$) to an ELOISATRON[1.3] (ELN) operating at 100 TeV per beam would yield a collider with a luminosity of $10^{34}$ cm$^{-2}$ s$^{-1}$. Even higher energies and luminosities are clearly possible.

This paper presents a general context for assessing the performance of future hadron colliders. It begins by illustrating machine trends and by defining key collider characteristics. Descriptions of limiting beam physics and limiting accelerator technologies constitute the next two sections. These considerations have been folded into a simple spreadsheet-based computer code that allows one to perform self-consistent parameter searches to maximize luminosity. Section 5 presents some results from the parameter searches. The paper concludes by sketching the characteristics of an ultimate hadron supercollider, i.e., an Ultra-ELOISATRON (UELN).

## 2. Trends in collider energy and luminosity

The search for understanding the nature of mass and the dynamics underlying the physical universe has led particle physicists to explore phenomena at the frontier of energies. The modern tools for the experimental explorations are colliders with ever higher beam energies and ever higher luminosities. The trends in the performance of present and future hadron colliders are illustrated in figure 1. What considerations determine these trends? What technologies limit collider performance? What are the energy dependences of

trends? What technologies limit collider performance? What are the energy dependences of the limiting phenomena? The answer to these questions lies in an explication of limiting beam physics and limiting technologies.

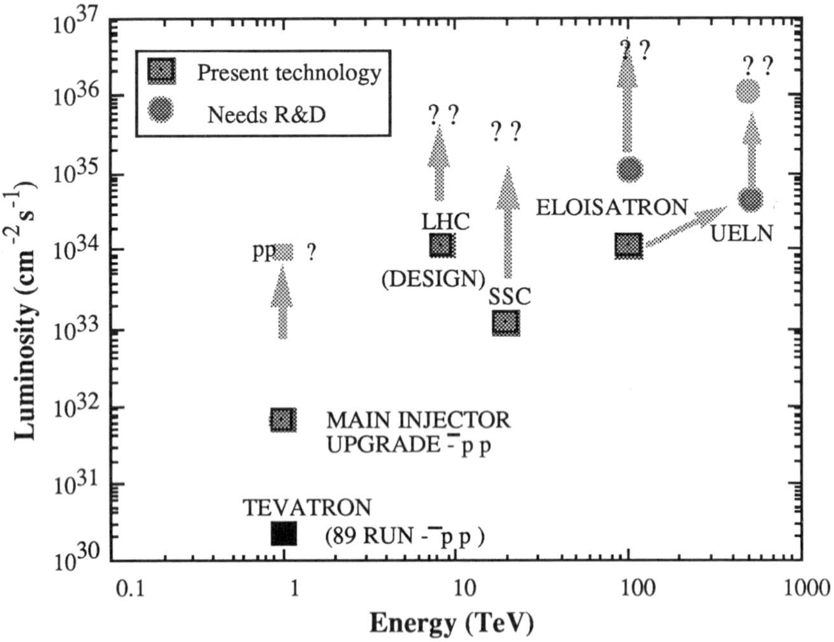


Figure 1. The luminosity goals of present and future hadron colliders

2.1. *Luminosity*

For collisions at a frequency, $f_{coll}$, between bunches with populations $N_1$ and $N_2$ and with equal sizes at the interaction point, the luminosity is given by the well known expression

$$\mathcal{L} = \frac{N_1 N_2 f_{coll}}{4\pi \sigma_x \sigma_y (1 + q^2)^{1/2}} \tag{1}$$

where $\sigma_x$ and $\sigma_y$ are the Gaussian horizontal and vertical radii of the beams. The correction factor, q, for the case of a finite crossing angle, $\alpha$, in the vertical plane is given by

$$q = \frac{\alpha \sigma_z}{2 \sigma_y} . \tag{2}$$

For simplicity assume that both beams have bunches of equal population, $N_B$. Then writing the beam radius in terms of the normalized emittance, $\varepsilon_n$, the usual relativistic factor $\gamma$ and the value of the $\beta$-function at the interaction point, $\beta^*$, one has

$$\mathcal{L} = \frac{N_B^2 \, c \, \gamma}{4\pi \left( \varepsilon_{n,x} \beta_x^* \varepsilon_{n,y} \beta_y^* \right)^{1/2} S_B} \,. \tag{3}$$

In Eq. (3) the collision frequency has been replaced by the c divided by the bunch spacing $S_B$. If the bunch length, $\sigma_z$, is greater than $\beta^*$ the radius of the bunch will not be constant during the collision. In that case one must replace $\sigma_{x,y}$ in Eq.(1) with an effective radius, $\overline{\sigma}$ defined as

$$\overline{\sigma}_i = \sigma_i \left( 1 + \frac{\sigma_z^2}{\beta_i^{*\,2}} \right)^{1/2} . \tag{4}$$

For beta function chosen so that at the interaction point

$$\frac{\varepsilon_x}{\beta_x^*} = \frac{\varepsilon_y}{\beta_y^*} = \frac{\varepsilon}{\beta^*} \,, \tag{5}$$

one can write the luminosity as

$$\mathcal{L} = \frac{N_B^2 \, c \, \gamma}{4\pi \, \varepsilon_n \, \beta^* \, S_B} = \frac{1}{e \, r_p} \left( \frac{N_B \, r_p}{4\pi \, \varepsilon_n} \right) \left( \frac{\gamma I}{\beta^*} \right) \equiv \frac{1}{e \, r_p} \, \xi \left( \frac{\gamma I}{\beta^*} \right) . \tag{6}$$

In Eq. (6) I is the average beam current in the storage ring and $r_p$ is the classical radius of the proton. The luminosity rises naturally with increasing beam energy though not without some "cost" in practical difficulties of machine design. The difficulties of increasing the luminosity even faster than its natural linear growth with energy are associated with increasing the beam current.

### 2.2. *Design strategies*

A recent examination[2.1] of the luminosity limits of the SSC and the LHC (Snowmass, 1990) and earlier studies[2.2, 2.3, 2.4] suggested the following approaches to collider design to maximize the luminosity of colliders:

1) increasing the charge per bunch,
2) increasing the number of bunches,
3) increasing the crossing angle to allow more rapid bunch separation,
4) tilting the bunch with respect to the direction of motion at the interaction point ("crab-crossing")
5) minimizing the $\beta$ function at the interaction point

These strategies and their attendant physics and technology issues are exactly those being pursued in the design studies of very high luminosity B factories at SLAC, CERN, and Cornell and of the $\phi$ factories at UCLA and Laboratorio Nazionale Frascati. Upon closer examination of these issues one finds that the physical phenomenon that underlies nearly all design difficulties of supercolliders in the range of 10 to 100 TeV (and beyond) is the emission of copious synchrotron radiation by the protons. Even the practical problems of controlling the consequences of synchrotron radiation are similar to those for electron storage rings, if one assumes that the vacuum pipe in the hadron collider must operate at cryogenic temperatures. If, on the contrary, the operating temperature of the new, high $T_c$ superconductors can be pushed to $\approx$ 300 °K, then even at 100 TeV the synchrotron radiation load will easily be within the range routinely handled in existing electron storage rings. In that case the ultimate energy of an ELN would not be limited to 100 TeV by synchrotron radiation. Rather a PeV collider could be considered.

## 3. Limitations on collider performance due to physics and technology

With these observations in mind one returns to the central design question, "How should one choose the collider characteristics $\varepsilon_n$, $\beta^*$, $S_B$, and N as a function of the beam energy in order to maximize the luminosity?" What physics and what technologies constrain the choices of these free parameters? The constraints on the accelerator design include the following considerations: 1) Detector limitations – electronics cycling, event resolution, 2) Accelerator physics – tune shifts, beam lifetimes, emittance control, and 3) Accelerator technologies – synchrotron radiation handling, impedance control, radiation damage of components, magnet technologies, fault mode control.

### 3.1. *Detector limitations*

The limitations on collider characteristics imposed by detector constraints derive from setting the bunch population and the bunch spacing as a function of luminosity. The probability of having k events per crossing, $\mathcal{L}_k$, and the most probable number of events per crossing, $\langle n \rangle$, are related to the collider luminosity, the bunch spacing and the machine energy through the total inelastic cross section via

$$\langle n \rangle = \frac{\mathcal{L}\, \sigma_{inel}\, S_B}{c} \tag{7}$$

and

$$\mathcal{L}_k = \mathcal{L} \langle n \rangle^k \frac{\exp(-\langle n \rangle)}{k\,!} \,. \tag{8}$$

Figure 2 displays the probability of one event per crossing and the most probable number of events per crossing versus $S_B$ for SSC, i.e., for an energy of 20 TeV per beam for luminosities of $10^{33}$ and $10^{34}$ cm$^{-2}$s$^{-1}$. The minimum bunch spacing in the collider is that obtained by filling every rf-bucket and by choosing a high frequency for the rf-system. High frequencies have several drawbacks:

1) they imply smaller cavity apertures, implying more difficult control of impedances.

2) the closer bunch spacing will produce more parasitic crossings, raising the long range tune shift and necessitating a larger crossing angle.

3) At the high beam intensities a powerful, wide bandwidth feedback system may be required.

Generally, studies of optimizing the choice of rf-frequency have settled on the range of 400 to 800 MHz as providing the best compromise among all technical considerations.

Regardless of choice of frequency, the designer may not be able to fill every rf-bucket because the detector electronics cannot cycle sufficiently rapidly.

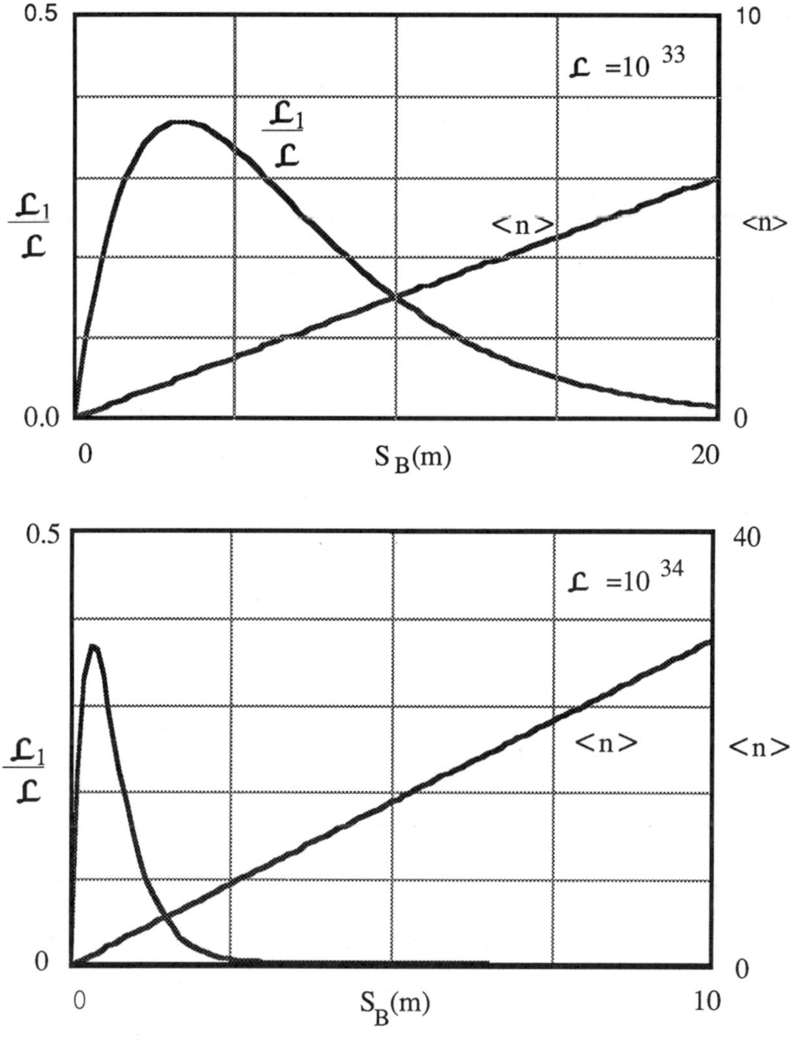


Figure 2. Choosing the bunch spacing for SSC at its design luminosity and for a luminosity ten-fold larger

In the example of Fig. 2 for SSC (or LHC), the minimum cycling time for the electronics is considered to be ≈16 ns between crossings. At a luminosity of $10^{33}$ $cm^{-2}s^{-1}$, this interval also maximizes the probability of obtaining one event per crossing. As the most probable number of events per crossing is also one, the bunch spacing of 5 m would optimize the conditions for event resolution. The lower panel of figure 2 displays the same quantities for $10^{34}$ cm$^{-2}$s$^{-1}$. At a luminosity of $10^{34}$ cm$^{-2}$s$^{-1}$, the bunch spacing of the baseline SSC design leads to a most probable number of events $\langle n \rangle \approx 20$, making event resolution difficult. Scaling the behavior described by Eq. (7) and (8) to an ELN at 100 TeV per beam with a luminosity of $10^{35}$ cm$^{-2}$s$^{-1}$, one finds that the detector designers will be required to live with $\langle n \rangle \approx 200$ events per crossing. Nonetheless, if one distinguishes routine vs. peak luminosity running, such large numbers of events per crossing may be acceptable for specialized event searches. Therefore, the accelerator designer need not be driven to fill every rf-bucket and the electronics wizard need not seek to cycle the detector electronics at rates approaching or exceeding 1 GHz.

3.2 *Radiation effects from the hadronic shower*

Extremely high luminosities will expose certain detector sub-systems close to the beamline, such as the central trackers, to large radiation doses as calculated below. The technical issues of whether these components can be designed to survive the large radiation doses is a further issue which is beyond the scope of this paper. With respect to machine design the radiation can also affect the characteristics of the magnets and the collimators that surround the interaction point. The radiation emanating from interaction point due to the beam collisions comes predominantly from the hadronic shower. Following the approach of Groom[3.1], one can compute the dose in the following manner.

$$\text{Dose} \propto N_{\text{collision}} \times \sigma_{\text{inel}} \times \text{Charged multiplicity/event} \times \frac{dE}{dx} .$$

Thus,

$$\frac{d^2 N_{\text{charged}}}{d\eta \, dp_\perp} \approx H f(p_\perp) \tag{9}$$

where

$$\frac{d^2 N_{\text{charged}}}{d\eta \, dp_\perp} \approx H f(p_\perp) \tag{10}$$

in which $\eta$ is the psuedo-rapidity, H is the height of the plateau in the psuedo-rapidity distribution, and f is the distribution function of the transverse momentum, $p_\perp$. In terms of the scattering angle, $\theta$,

$$\eta = - \ln(\tan \theta/2) . \tag{11}$$

Detailed Monte Carlo studies[3.2] have shown that the dose is insensitive to the functional form of $f(p_\perp)$. Typically in calculations of radiation dose it is sufficient to use a delta function distribution,

$$f(p_\perp) = \delta( p_\perp - \langle p_\perp \rangle ) \tag{12}$$

where for a beam energy E a fit to the Monte Carlo calculations yields

$$\langle p_\perp \rangle = 0.12 \log_{10} 2E + 0.06. \tag{13}$$

The height of rapidity plateau may be taken to be

$$H = 0.78\ s^{0.105} \approx \text{ constant for } \eta < 6 \ \ (\theta > 5 \text{ mrad}) \tag{14}$$

where s is the usual kinematic variable,

$$s = 4\left(\frac{E}{mc^2}\right)^2 \tag{15}$$

For angles smaller than 5 mrad ($\eta > 6$) which are subtended by the collimator and the focusing elements of the interaction region, one can approximate the distribution H as a linear ramp that goes to zero at the kinematic limit of the rapidity, $y_{max}$;

$$y_{max} = \ln[s] \ . \tag{16}$$

With these approximations, the dose that will be deposited in a detector element or in a beamline component loca ted at a distance r from the interaction point is

$$D(E,r) = 26.1\frac{\text{Gy}}{\text{yr}}\left(\frac{\mathcal{L}}{10^{33}}\right)\left(\frac{\sigma_{inel}(s)}{90\text{ mb}}\right)\left(\frac{H(E)}{7.5}\right)\left(\frac{\langle p_\perp\rangle}{0.6\text{ GeV}}\right)^{0.9}\frac{\cosh^{2.9}\eta}{r^2} \ . \tag{17}$$

In Eq. (17) the inelastic cross section may be approximated as

$$\sigma_{inel} = 90\text{ mb} + 90\text{ mb}\left[\log_{10}^2\left(\frac{E}{20\text{ TeV}}\right)\right] . \tag{18}$$

The electromagnetic shower from the decay of $\pi^o$'s also contributes to the radiation dose. Approximately half as many $\pi^o$'s as charged hadrons are produced with

$$H_{\pi^o} \approx \frac{1}{2} H_{charged} \ . \tag{19}$$

In the most general sense the difficulties of dealing with the radiation from the collision point are most simply expressed by the power in charged particle debris (per side); namely,

$$P_{debris} = 350\text{ W}\left(\frac{\mathcal{L}}{10^{33}}\right)\left(\frac{\sigma_{inel}}{90\text{ mb}}\right)\left(\frac{E}{20\text{ TeV}}\right) . \tag{20}$$

The limitations that the hadronic shower places on the design of the interaction region optics are considered in section 4.2 below.

3.3 *Tune shift limitations*

The fundamental beam-beam effect that limits luminosity is the tune shift due to the space charge forces of the colliding beams. The head-on tune shift, $\Delta\nu_{HO}$, seen by a bunch colliding with its corresponding bunch from the second storage ring can be expressed as

$$\Delta\nu_{HO} = \frac{N_B r_p}{4\pi\varepsilon_n} \ . \tag{21}$$

This quantity is precisely the quantity, $\xi$, in Eq. (6).

As the beams leave the interaction point they will experience close encounters with other bunches. In these parasitic collisions, schematically illustrated in Fig. 3, the beams will experience an additional tune shift due to the non-linear space charge forces from the other bunches. The cumulative effect or long range tune shift, $\Delta\nu_{LO}$, from parasitic crossings depends on the number of close encounters, $n_{LO}$, experienced by the beams before they pass through septa into separate beam pipes where they are electromagnetically shielded from each other. The long range tune shift also depends on the separation of the beams (in units of the beam radii) and on the strength of the head on tune shift.

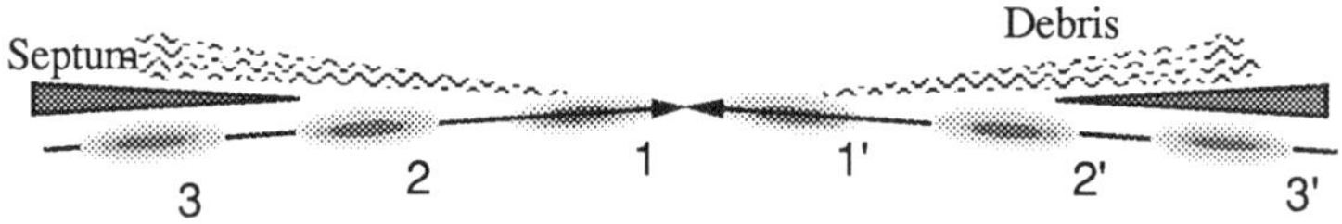


Figure 3. Schematic of parasitic collisions

In terms of the beam radius at the interaction point, one finds that

$$\Delta\nu_{LR} = \Delta\nu_{HO}\, 2n_{LR}\left(\frac{\sigma}{\beta^{*}\alpha}\right)^{2}. \tag{22}$$

As particles at different positions in the bunches experience a wide variation in the magnitude of the space charge forces, the long range shift also generates a tune spread of similar magnitude that cannot be eliminated by changing the tune of the collider. Thus it is the tune spread not the tune shift *per se* that causes the difficulty in locating the operating tune of the collider. Combining Eq. (21) and (22), one computes the total tune shift in a collider with $N_{Hi}$ high luminosity interaction points and $N_{Med}$ interaction points to be

$$\Delta\,\nu_{tot} = (N_{Hi} + N_{Med})\,\Delta\,\nu_{HO} + N_{Hi}\,\Delta\,\nu_{LR}\,. \tag{23}$$

The maximum tune shift with which a collider can operate will generally not be an option that can be chosen by the machine designer. Rather it must be regarded as a "fact of life." Although tune shifts as high as 0.06 have been seen in the PEP electron – positron collider, the experience with hadron beams at the CERN S$\bar{\text{p}}$pS[3.3] and at the Tevatron[3.4] at Fermilab indicates that

$$(N_{Hi} + N_{Med})\,\Delta\,\nu_{HO} \leq 0.024\,. \tag{24}$$

The maximum value has been reached for N = 12 at Tevatron and for N = 6 in S$\bar{\text{p}}$pS, in which case the luminosity was maximized for a fixed tune spread when 3/4 of $\Delta\nu_{tot}$ was allocated to $\Delta\nu_{HO}$ and 1/4 to $\Delta\nu_{LR}$. This observation might suggest that the ultimate luminosity can be reached for $N_{Hi} = 1$ and $N_{Med} = 0$. Unfortunately, the validity of such an extrapolation is unknown. A more conservative assumption which was made in the Snowmass study is that the maximum value of $\Delta\nu_{HO}$ per interaction point is 0.01.

3.4 *Crab–crossing*

One can consider two means of increasing the luminosity as limited by tune shifts. The first way might be to tilt the bunches with respect to the trajectory of the beam at the interaction point. This so-called "crab crossing" scheme, illustrated in Fig. 4, was first introduced by Palmer[3.5] for use in TeV, electron – positron, linear colliders. Crab crossing is now under active investigation[3.6] for use in high luminosity, heavy flavor factories. In the "crab crossing" scheme the bunches collide head-on so that one can employ a large angle between beam trajectories avoiding the significant reduction in luminosity which would otherwise obtain. As the crossing angle is increased, the septum can be moved inward towards the interaction point, thereby reducing the number of parasitic crossings. The net result is that the long range tune spread is reduced and the tune shift limit can be reached from the head-on contribution alone. A further advantage of the crab crossing is that it decouples the transverse and the longitudinal motion of the beams (synchrobetatron coupling), which is usually associated with beams crossing at an angle.

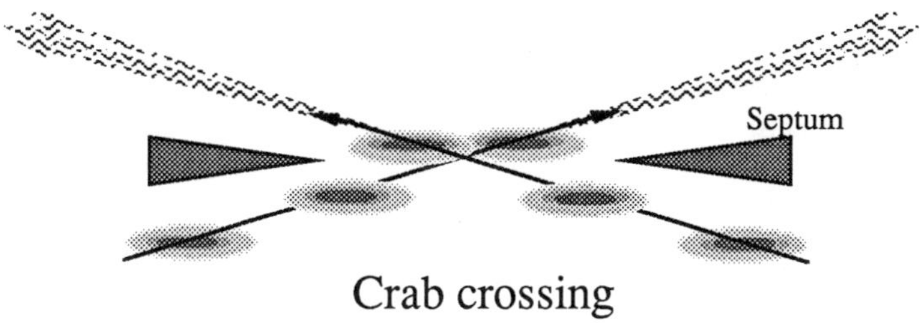


Figure 4. A schematic representation of a "crab crossing"

The tilting of the bunch is produced by radio frequency deflecting cavities located at a betatron phase angle of $(n + 1/2)\pi$ from the interaction point. The voltage in the crab cavity is related to the "crab angle", rf-wavelength of the cavity, and the magnitude of the beta function at the cavity by

$$V_{\perp\, crab} = \frac{\theta_{crab}\, \lambda_{rf\text{-}crab}\, (E/e)}{4\pi \sqrt{\beta^* \beta_{crab}}} \quad . \tag{25}$$

Disadvantages of crab crossing are that the cavities add to ring impedance and that voltage and phase errors can lead to emittance growth. The voltage and phase tolerances to avoid emittance growth depend on the beam radius at the interaction point, $\sigma^*$, via

$$\Delta\phi_{crab} << \frac{4\pi\sigma^*}{\lambda_{rf\, crab}\, \theta_{crab}} \tag{26}$$

and

$$\frac{\Delta V_{crab}}{V_{crab}} << \frac{\sigma^*}{\sqrt{N_{damp\, turns}}\, \sigma_z} \quad . \tag{27}$$

As the number of turns to damp energy fluctuations in a hadron supercollider will be very large, Eq. (27) will impose an especially severe constraint. A further practical difficulty in employing crab crossing is that the hadronic debris from the collisions is "aimed" directly at the high voltage crab cavities.

3.5 *Radiation damping of beam emittance*

The second consideration related to the limitation of luminosity imposed by tune shift limits derives from the observation that the limiting tune shift depends on the normalized emittance of the beam and on the radial distribution of particles in bunch. The emittance damping time is half the magnitude of the amplitude damping time; hence, it is equal to the energy damping time, $\tau_\varepsilon$;

$$\frac{d\varepsilon}{dt} = -\frac{1}{\tau_\varepsilon}\varepsilon \tag{28}$$

where

$$\frac{1}{\tau_\varepsilon} \propto \gamma^3 . \tag{29}$$

The exponential decrease in the emittance of the beam does not continue indefinitely; it is limited by quantum fluctuations, by beam-beam effects, by gas scattering, by intra-beam scattering and by noise in the feedback system. Eventually the beam will assume its natural equilibrium emittance in the storage ring. One consequence of emittance damping is evolution of beam distribution to broad bi-gaussian profile. The form of the profile will determine the degree of non-linearity of the space charge forces and will thereby influence the tune spread associated with the head–on and long–range tune shifts.

In a hadron collider operating at ≥ 20 TeV, radiation damping will become an important determinant of the radial distribution of the beam because the damping time can be much smaller than the luminosity lifetime. Were radiation damping the only effect, the transverse distribution of charge would resemble that seen in electron storage rings. In such a case the maximum tune shift in a hadron collider with a single interaction point might not be limited to 0.01. Instead it might approach the value of 0.03 (or even 0.06) that has been achieved in high energy electron-positron colliders. Unfortunately diffusion of particles from the central region of the bunch proceeds more rapidly than radiation damping even at 100 TeV. Hence there is the possibility that in the range of 20 - 500 TeV the radiation damping during collisions could even be a deleterious effect in that the beam could form a tight, cold core that would limit the tune shift, thereby keeping the "effective tune shift" for the majority of the current less than 0.01. Such an effect could be counteracted by rf-heating of the beam (stochastic heating) to keep the emittance larger than its natural value.

An excellent way to study the limits of stability of a storage ring is to make frequent excursions into the unstable operating regime. Unfortunately the consequent loss of beam would be impermissible in existing storage rings that must also provide high integrated luminosity to users. To develop the requisite experimental database one might consider using a low energy $e^+$–$e^-$ collider operating at ≈ 100 MeV per beam with high currents and with several bunches. Such a storage ring with a flexible lattice would permit both high and low emittance tunes. With the installation of radio frequency deflection cavities ("crab" cavities), the sensitivity of crab-crossing to synchrobetatron resonances could be tested. These features are similar to those of DAΦNE[3.7], the $\phi$ factory now under construction at the Laboratorio Nazionale Frascati (see figure 5).

In an experiment using an $e^+$-$e^-$ store ring as an $e^-$-$e^-$ collider, the lifetime of the beams must be controlled in a manner to simulate the short luminosity lifetime (measured in terms of damping times) of hadron colliders. For "DAΦNE" the lifetime can be set by varying the Touschek scattering lifetime of the beams. The range of lifetimes can be controlled by changing the rf-overvoltage at a given operating energy. A sample calculation for the SMC lattice[3,8] has been calculated with the code ZAP and is displayed in Fig. 6.

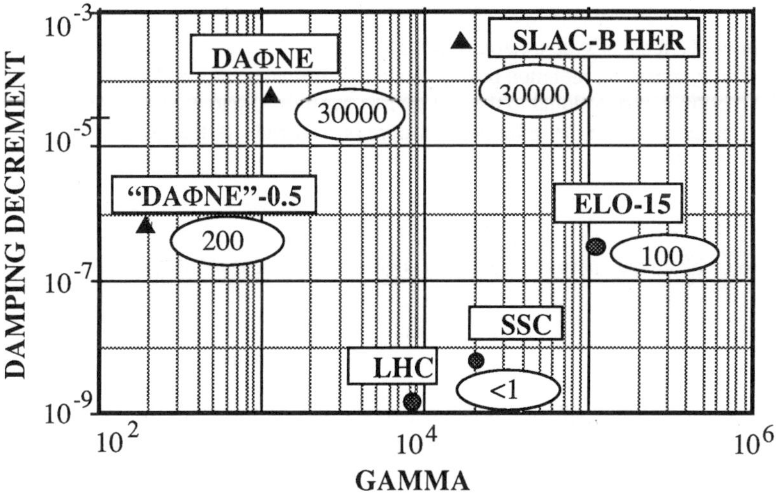


Figure 5. Low energy $e^+$- $e^-$ colliders mimic the behavior of 100 TeV collider for $\Delta\nu$ studies. The circled numbers are the damping time divided by the luminosity lifetime @ $\mathcal{L} = 10^{34}$

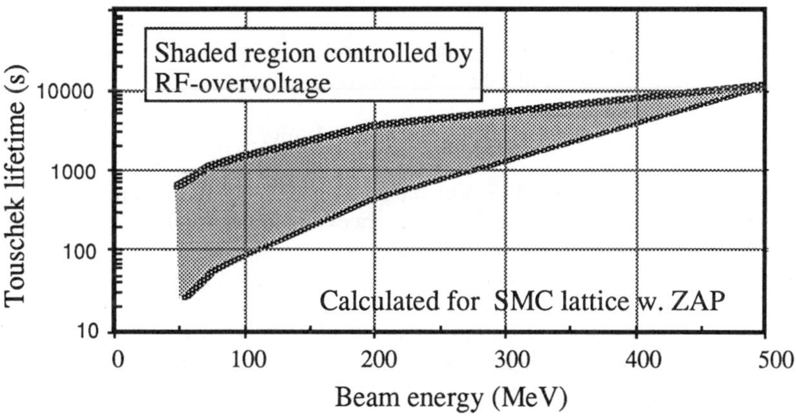


Figure 6. The Touschek lifetime as a function for beam energy in a low energy lattice. The range corresponds to the range of allowable overvoltages in the rf-cavity

3.6 *Effects of synchrotron radiation*

High luminosity supercolliders will have to cope with a phenomenon that has been unimportant in previous generations of proton colliders, but which is a central design issue in high luminosity electron–positron colliders; that phenomenon is the production of intense synchrotron radiation in the vacuum UV to hard X-ray range. The radiation leads to two important effects, heating of the walls of the vacuum chamber and photodesorption of gas from the chamber.

3.6.1 *Thermal effects of synchrotron radiation*

The synchrotron radiation power generated per meter of bend for a beam in uniform circular motion is proportional to the energy lost per turn, $U_0$, to the beam current. I, and is inversely proportional to the radius of curvature of the bends, $\rho$; that is,

$$\mathcal{P}_{synch} = \frac{U_0 I}{2 \pi \rho} . \tag{30}$$

For a proton beam the energy lost per turn is given in practical units by

$$U_o = 6.03 \times 10^{-18}\,\text{GeV/turn} \frac{\gamma^4}{\rho\ (\text{meters})} . \tag{31}$$

In the supercollider the radiation is deposited onto the cold walls of the vacuum chamber, hence the power must be removed with a efficiency which is limited by the Carnot efficiency of the compressors that supply the cryogenic cooling fluid. The power that is consumed during operation of the collider is, thus, related to the radiation power and to the temperature of the wall, $T_{wall}$, by

$$\mathcal{P}_{compress} \approx 5.4 \left( \frac{300\ °\text{K} - T_{wall}}{T_{wall}} \right) \mathcal{P}_{synch} . \tag{32}$$

To limit compressor power to a range that is considered to practical the designers of the SSC and LHC have considered that the acceptable range of radiation power is

$$\mathcal{P}_{synch} = 0.4 \ \text{to} \ 1.0\ \text{W/m} . \tag{33}$$

The present SSC design allows 0.13 W/m (upgradable to 0.4 W/m) incident on the magnet walls at 4.2 °K. The LHC design which incorporates a radiation shield at 20 °K inside the vacuum chamber walls would allow $\approx$1 W/m.

For a supercollider to operate at the highest possible luminosity, the maximum thermal load deposited on the walls must be allowed to exceed 1 W/m. In practice, one must find a way to increase the temperature of the surface on which the radiation is deposited. For example one might consider increasing the temperature of the radiation shield to 70 °K. Unfortunately raising the temperature of the radiation shield is not without serious consequence to collider luminosity due to beam dynamics. That connection comes via the transverse resistive wall instability.

3.6.2 *Transverse resistive wall instability*

Transverse displacements of the beam from the centerline of the beam chamber will grow due to the finite conductivity of the wall, $\sigma_{wall}$. The growth time[1.2] of the instability in a beam pipe of radius, b, is

$$\tau_{RW}^{-1} = \frac{N_B\, M\, r_p\, \beta_{ave}\, \omega_o\, c}{\left(2\,\pi\,\sigma_{wall}\ \omega_o\,\delta\nu\right)^{1/2}}\ \frac{\mathrm{Im}\left[(1+i)\,\zeta\right]}{2\,\pi\,\gamma\,b^3} \tag{34}$$

where M is the number of bunches, $\sigma_{wall}$ is the conductivity of the inner layer of the beam tube, $\omega_o$ is the revolution frequency, $\delta\nu$ is the fractional tune, $\nu - n$, (use 0.1), and $\zeta$ is a correction for the multiple conductivity layers in wall (equal to 2.87 + 2.87i for SSC). As a scaling law, Eq. (34) can be written as

$$\tau_{RW} \approx 1.5\times 10^{16}\left(\frac{83\ \mathrm{km}}{C_{ring}}\right)\left(\frac{S_B(\mathrm{m})}{c\ (\mathrm{m/s})}\right)\left(\frac{b^3(\mathrm{cm})}{8\ N_B}\right)\left(\frac{E}{20\ \mathrm{TeV}}\right)\left(\frac{T_o}{0.277\ \mathrm{ms}}\right)\left(\frac{\sigma_{wall}}{1.8\times 10^9}\right)^{0.5} . \tag{35}$$

In Eq. (35) $T_o$ is the revolution period of the beam in the storage ring. As the resistive wall instability is an absolute instability, its growth cannot be Landau damped by spreading the betatron frequencies in the beam. Controlling the instability, therefore, requires the use of a powerful feedback system. The limits of such a feedback system have not been established. However, many experts consider $\tau_{RW} \approx 6\ T_o$ to represent the limits of presently available feedback systems.

The resistivity of the wall (RRR) for copper plated onto stainless steel varies as with temperature as shown in Fig. 7; where $RRR_{wall}$ is normalized to its value at 300 °K, ($1.59\times 10^{-6}$ ohm cm$^{-1}$); an excellent fit to the curve of Fig. 7 is

$$\mathrm{Ln}\ (RRR_{wall}) = 3.05\left[1 + \frac{41.55\ T^{-0.93} - 0.24\ T^{0.93}}{41.55\ T^{-0.93} + 0.24\ T^{0.93}}\right] + 0.24\ . \tag{36}$$

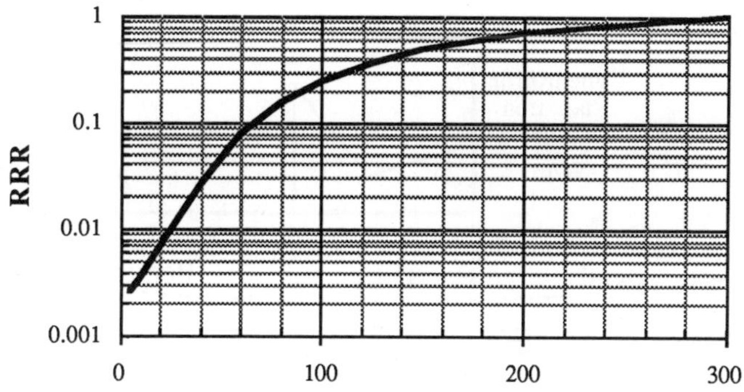


Figure 7. Residual resistivity ratio (RRR) for copper from Ref. 3.9.

In evaluating the design of the vacuum system of a supercollider, it is convenient to specify the maximum permissible power load on the walls of the chamber, $\mathcal{P}_{synch}$. In that case the temperature of the walls is specified by

$$T_{wall}\,(^{\circ}K) = \frac{1.6 \times 10^{3}\, \mathcal{P}_{synch}}{\mathcal{P}_{compress} + 5.4\, \mathcal{P}_{synch}} . \tag{37}$$

At beam energies E >> 10 TeV, for which copious radiation is generated, maximizing the luminosity of the collider while keeping $\tau_{RW} >> T_o$ will place an upper limit on the wall temperature and a lower limit on the vertical aperture of the dipole magnets. Both $T_{wall}$ and b are functions of the choice of $B_{dipole}$. The consequences of the constraints on $T_{wall}$ and the beam pipe radius are both the operating costs of supplying power to the compressors and the capital cost of providing for a large magnetic field volume.

3.6.3 *Longitudinal variation of synchrotron radiation*

Eq. (30) and (31) describe the longitudinal rate at which synchrotron radiation is generated in the collider. The storage rings, however, are not circular, bend magnets of finite length are separated by straight sections. Consequently the rate of synchrotron power generation is not equal to linear flux density of radiation absorbed on the walls. This distinction[3.10, 3.11] has become a key tool in the design of heavy flavor factories. As the peak fields that are practical in superconducting dipoles increase, it may be possible to mimic this configuration in hadron supercolliders. In such a scenario short, very high field magnets might allow most of the radiation to fall on the walls of high temperature drift sections as illustrated schematically in Figure 8.

If a bend-drift configuration is found to be practical, it would have the added advantage that the straight section on which most of the photons fall could be baked *in situ* to 200 °C, considerably improving the average vacuum properties of the beam chamber. As suggested in Fig. 8 radiation masks would intrude into the beam pipe to shield most of the region interior to the dipoles from synchrotron radiation. Such radiation manipulation features have been incorporated into the design of the interaction regions of flavor factories.

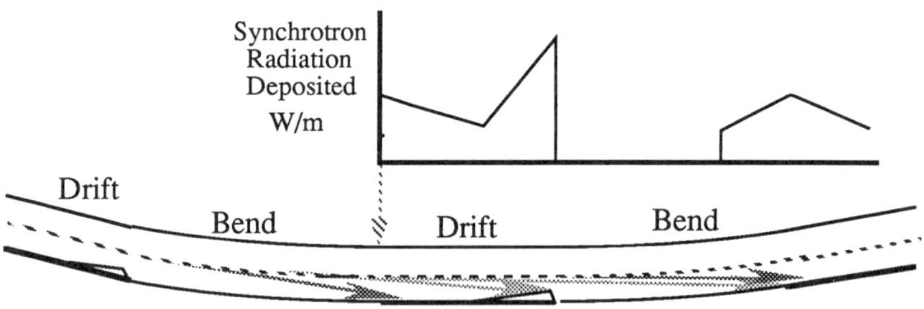


Figure 8. A schematic of a section of the arc in a collider with high field dipoles separated by relatively long straight sections.

3.6.4 *Photo-desorption*

The gas load in the storage rings arises from two processes: thermal outgassing and synchrotron radiation induced photodesorption. Thermal outgassing is common to all vacuum systems and occurs in the absence of synchrotron radiation. While in electron storage rings the thermal outgassing is negligible compared with the dynamic gas load from the photodesorption, the gas load in existing hadron collider is primarily due to thermal out-gassing of the walls. In the next generation of hadron supercolliders, however, the principal gas will be hydrogen which is probably derived from the photo-dissociation of water adsorbed on the walls of the vacuum chamber.

Hydrogen will be desorbed from the walls at a rate in proportion to the photon flux; i.e., the number of molecules, $N_{mol}$, produced per incident photon is

$$N_{mol} \propto \dot{N}_{\gamma} . \tag{38}$$

The average number of photons generated per meter of beamline is directly proportional to the average luminosity;

$$\dot{N}_{\gamma} = \frac{\mathcal{P}}{0.3087\ \varepsilon_{crit}} \propto \gamma I \propto \mathcal{L} . \tag{39}$$

To be more quantitative one can estimate the desorption rate following the approach of Gröbner et al[3.12]. After taking the spectrum of the synchrotron radiation photons into account, one can express the photon flux in the spectral interval (0,x) in the form

$$\dot{N}_{\gamma}(x) = \frac{\sqrt{3}\, r_p}{e\hbar c} F(x)\, E\, I \tag{40}$$

where

$$x = \frac{\varepsilon}{\varepsilon_{crit}} = \frac{\varepsilon}{\left(\frac{3\hbar c}{2}\right)\left(\frac{\gamma^3}{\rho}\right)} \tag{41}$$

with $\varepsilon$ being the energy of the photon and $\varepsilon_{crit}$ being the critical energy of the radiation emitted in the dipoles;

$$\varepsilon_{crit} = \frac{3}{2}\frac{\hbar c}{e}\frac{\gamma^3}{\rho} \tag{42}$$

In Eq. (42) F(x) is the integral over the modified Bessel function

$$F(x) = \int_0^x \int_u^\infty K_{5/3}(y)\, dy\, du \ . \tag{43}$$

For large values of x, $F(x) \rightarrow 5.53$. After rearranging terms and inserting appropriate values for the constants, one obtains the photon production rate in practical units;

$$\dot{N}_\gamma = 4.13 \times 10^{17}\, E{\cdot}I \;\; [\text{photons/s}] \tag{44}$$

where E is in TeV and I is in mA.

Gas molecules are desorbed from the walls in proportion to the photon flux; that is, the number of molecules, $N_{mol}$, produced per incident photon is

$$N_{mol} = \eta\, \dot{N}_\gamma \tag{45}$$

where $\eta$ is the effective photodesorption coefficient that characterizes the vacuum chamber. The Ideal Gas Law relates the number of desorbed molecules to the gas load in Torr-L via a conversion factor of $3 \times 10^{-20}$ (T / 300 °K) Torr-liters/molecule. In practical units, the effective gas load due to photodesorption is found to be

$$Q_{gas} = 3 \times 10^{-20}\, \eta\, \dot{N}_\gamma \left( \frac{T_{wall}}{300\ °K} \right) \frac{\text{Torr - L}}{\text{s - m}}\,. \tag{46}$$

The photodesorption coefficient, $\eta$, is a property of the chamber that depends on several factors:

- the chamber material,
- the material fabrication and preparation,
- the degree of prior exposure of the surface to radiation,
- the angle of incidence of the photon with respect to the surface,
- the photon energy.

In light of these complexities, rather than considering the photodesorption coefficient to be a fundamental material property, one should regard $\eta$ as an effective engineering value that accounts for the differential illumination of the chamber walls by both direct (i.e., beam-produced) photons and diffusely scattered, secondary photons. Using a single value in Eq. (46) yields only a rough estimate of the actual dynamic gas load but an estimate which is nonetheless useful in setting the scale of the engineering task and in choosing the material of the chamber.

As the walls of the vacuum chamber in planned supercolliders (SSC and LHC) are at cryogenic temperatures (generally < 4.2 °K), the walls will cryopump the hydrogen that is formed from the photo-dissociation of water frozen on the walls. As the hydrogen accumulates on the walls, the pressure in the beam tube will then be set by the hydrogen isotherm of the walls. Additional features of this issue are discussed in section 4.3 which concerns limiting features of vacuum technology.

### 3.6.5 *Effects of synchrotron radiation on average luminosity*

The presence of significant synchrotron radiation in a supercollider implies that the luminosity during operation can actually increase because the emittance of the beam will decrease with time. As the number of particles per bunch is depleted by collisions, the luminosity will vary as

$$\frac{1}{\mathcal{L}} \frac{d\mathcal{L}}{dt} = \frac{2}{N_B} \frac{dN_B}{dt} - \frac{1}{\varepsilon} \frac{d\varepsilon}{dt}\,. \tag{47}$$

For SSC emittance damping can lead to a doubling of the luminosity during the first few hours after the beams are brought into collision. As noted above, the cooling of the beam will be limited as the beam-beam interaction approaches the time shift limit. The competition between the radiation cooling of the beam core and the outward diffusion of particles due to collisions under the conditions of a small tune shift limit may require that supercolliders actually employ rf-excitation to heat the beam. In that case, a term to describe the beam heating would have to be added to Eq. (47).

Consumption of the bunches by collisions at the maximum luminosity determines the luminosity lifetime, $t^{-1}{}_{lum}$, via

$$\frac{1}{\mathcal{L}}\frac{d\mathcal{L}}{dt} = \frac{2}{N_B}\frac{dN_B}{dt} - \frac{1}{\varepsilon}\frac{d\varepsilon}{dt} . \tag{48}$$

The time allowable for injection should be short compared to the luminosity lifetime of the collider if the average luminosity is to remain close to the peak luminosity. Therefore Eq. (48) is a principal determinant in setting the minimum current that must be delivered by the injection chain. A more precise evaluation should account for other beam loss mechanisms such as beam–gas scattering, which sets the operating pressure in the collider and Touschek intra-beam scattering. Moreover, other effects that change the emittance should also be taken into account. These effects include small angle elastic scattering during collisions, gas scattering, and multiple intra-beam scattering. Usually to estimate injection requirements one uses the luminosity half–life, which is related to $\tau^{-1}{}_{lum}$ by

$$T_{1/2,\,lum} \approx 0.41\,\tau_{lum}(E) . \tag{49}$$

Then one requires that the time to complete the injection process, $T_{inj}$, is small compared to $T_{1/2,lum}$; e.g.,

$$T_{inj} < \frac{1}{10} T_{1/2,\,lum} . \tag{50}$$

3.7 *Constraints on injection scenario at maximum luminosity*

Short luminosity lifetime at maximum luminosity requires powerful injectors to maintain high average luminosity. The injection chain consists of a linac plus a set of intermediate synchrotrons which are represented schematically in Fig. 9 for a 100 TeV supercollider. The designer of the injection chain must balance several technical and economic issues in furnishing a chain of accelerators that can store several tens of thousands of bunches into the main ring of the supercollider and then bring them to full energy in a time that satisfies Eq. (50). The characteristics of the injection chain that are to be identified include 1) the energy of injection into the main rings, 2) the number of intermediate synchrotrons, 3) the energy of the linac that begins the injection chain, and 4) the emittance that can be delivered from the linac into the collider.

The injection energy into the main ring is constrained by the resistive wall instability and by the limits on ramping the superconducting dipoles. In practice the field in the dipoles should not be varied by more than a factor of twenty. Hence, one may take E/20 as the lowest reasonable injection energy. At this energy, however, the growth of the resistive wall instability may be too rapid for the required current to achieve the design luminosity.

As the wall temperature could, in principle, be lower at injection than at the full energy, the growth time at the injection energy can be written as

$$\tau_{RW,i} \approx 3\times10^{15}\left(\frac{83\,\mathrm{km}}{C_{ring}}\right)\left(\frac{S_B(\mathrm{m})}{c\,(\mathrm{m/s})}\right)\left(\frac{b^3(\mathrm{cm})}{8\,N_B}\right)\left(\frac{E_{inj}}{4\,\mathrm{TeV}}\right)\left(\frac{T_o}{0.277\,\mathrm{ms}}\right)\left(\frac{\sigma_{wall,inj}}{1.8\times10^9}\right)^{0.5}. \quad (51)$$

Generally this time will be shorter than the growth time at the operating energy. If $\tau_{RW,i} < 6\,T_o$, one may either raise $E_{inj}$, enlarge the pipe aperture, or both. Increasing $E_{inj}$ increases the cost of the high energy booster, while increasing b increases the cost of the main ring.

The accelerator energies at the low energy end of the injector chain are governed by assuring that the incoherent space charge (Laslett) tune spread does not exceed 0.02. For a booster accelerating $M_b$ bunches of length $\sigma_z$ spaced by $S_b$ the incoherent tune shift is

$$\Delta\nu_{SC} = \frac{N_B\,M_b\,r_p}{4\pi\left(\beta\gamma^2\right)\varepsilon}\,\frac{S_b}{2\,\pi\,\sigma_z}\,. \quad (52)$$

The Laslett tune shift constraint together with Eq. (50) sets energy of linac and boosters.

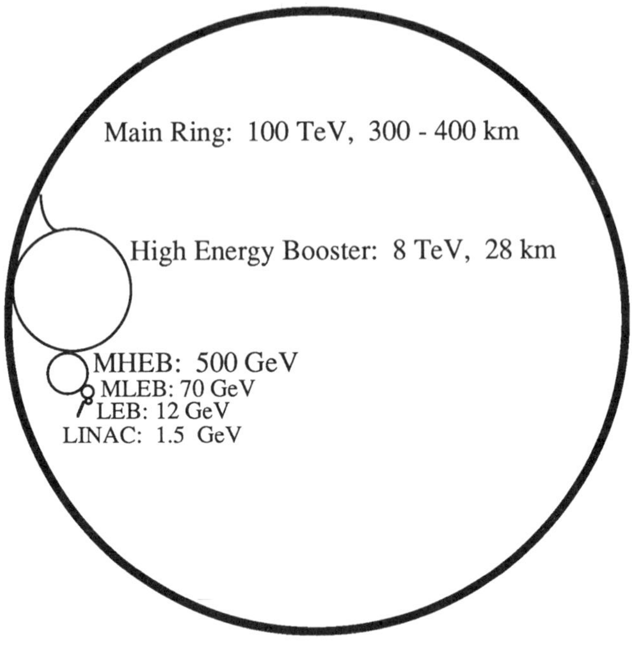


Figure 9. A possible injection chain for ELN with four intermediate synchrotrons

A 100 TeV collider could contain as many as $5 \times 10^5$ bunches. Even with such a large number of bunches a plausible injection sequence could be constructed with the booster chain employing four intermediate synchrotrons that is described in Table 1. In this example the total loading time per main ring is 3000 seconds with 1.5 nC per bunch. For an acceleration time 1000 seconds per main ring, the total fill time of the 100 TeV supercollider would be $T_{inj}$ = 8000 seconds. At a luminosity of $10^{35}$ $cm^{-2}s^{-1}$ the luminosity half-life is $10^5$ s; hence, the requirement that $T_{inj} << T_{lum,1/2}$ is satisfied. One might consider making the highest energy booster with normal conducting magnets and locating it in the same tunnel as the supercollider. This choice probably invites more operational difficulties than the savings in tunnel cost are worth.

The number of intermediatc boosters and the energy of the linac injector could be reduced if the amount of charge that produces the incoherent tune shift could be raised beyond the level implied by Eq. (52). Values of 0.03 may be acceptable with conventional lattices. Recently, Reiser[3.13] has proposed to build the low energy booster with a large dispersion and rather weak focusing. If this approach were adopted, it may be possible to triple the stored charge at injection in the low energy booster (LEB), which is generally the bottleneck in the injector chain.

Table 1. Characteristics of a potential injection chain designed to supply $5 \times 10^5$ bunches to a 100 TeV collider operating at a luminosity of $10^{35}$ $cm^{-2}$ $s^{-1}$

| | Circum (km) | Max E | Min E |
|---|---|---|---|
| **Main Ring** | 270 | 100 TeV | 5 TeV |
| **HEB** | 28 | 5 TeV | 0.5 TeV |
| **MHEB** | 2.9 | 500 GeV | 70 GeV |
| **MLEB** | 0.35 | 70 GeV | 12 GeV |
| **LEB** | 0.1 | 12 GeV | 1.7 GeV |
| **LINAC** | 0.1 | 1.7 GeV | — |

| | Bunches | $\Delta \nu_{SC}$ | Cycle T (s) |
|---|---|---|---|
| **Main Ring** | 500000 | 1.60E-04 | 1000 |
| **HEB** | 50000 | 1.60E-03 | 300 |
| **MHEB** | 5000 | 7.97E-03 | 30 |
| **MLEB** | 200 | 9.61E-03 | 1.2 |
| **LEB** | 10 | 1.23E-02 | 0.06 |
| **LINAC** | 5 | — | 0.03 |

## 4. Technologies limiting collider performance

In addition to the constraints of beam physics, several key technologies can also limit the maximum luminosity and beam energy of multi-TeV supercolliders. This section describes briefly the systems requirements, technical challenges and performance limitations of a number of critical technologies embodied in hadron supercolliders. Where it is appropriate, comparison is made to corresponding requirements in $e^+$–$e^-$ heavy flavor factories.

*4.1 RF and feedback requirements*

The rf-system in the main ring must be specified to perform two primary functions, 1) accelerate the injected beam to full energy in a period small compared with a luminosity–lifetime, and 2) provide enough power to replace the energy lost from the beam due to synchrotron radiation. As the total radiated power increases with the fourth power of the beam energy the second consideration becomes increasingly important as the energy of the supercollider approaches or exceeds 100 TeV. For example in an ELN operating at $10^{35}$ $cm^{-2}\ s^{-1}$ the rf-system must provide ≈20 MW. In comparison the rf-system for the high energy ring (9 GeV) of the PEP-II B–factory[4.1] must provide 5.5 MW at the design luminosity $3 \times 10^{33}$ $cm^{-2}\ s^{-1}$. As in case of the B-factory, the rf-cavities of the supercollider must also be designed with a minimum impedance for the higher order modes that can couple the motion of the thousands of bunches in the storage ring. The impedance reduction must, moreover, be accomplished without excessive diminution of the fundamental accelerating mode.

Options to be considered in the design of the rf-systems include 1) frequency of the fundamental mode, 2) the use of room temperature rf-cavities (more mature technology) versus the use of superconducting cavities (allows for higher voltages per cavity), 3) the number of rf-cavities, and 4) the use of time domain versus frequency domain feedback. Such choices can only be made in the context of a specific accelerator design. It may be instructive, however, to describe the choices made for PEP–II. Minimizing the number of cavities with high gradient implies increasing the power that must be transmitted through each window. For PEP–II and CESR-B, the maximum power per window has been set at 500 kW. If room temperature cavities are used, one must design for thermal loads of ≈ 150 $kW_{therm}$/cavity; hence, considerable care must be taken in thermal engineering. Regardless of the choice of superconducting or room temperature cavities, the shape cavity must be designed to reduce higher order modes. General features of the cavity design include a large bore to minimize trapped modes, waveguides to couple out higher order modes and absorbers for the higher order mode energy. Even such strong passive suppression techniques do not eliminate the growth of the fastest growing modes. Hence, the only way to control the growth of the unwanted modes is via a broad band, feedback system.

4.1a *Choice of frequency for maximum luminosity running*

The choice of rf-frequency for the collider comes from balancing several physical effects. The choice of bunch separation, $S_B$, has been discussed in section 3.1. As the bunches can be no closer together than 1 rf-wavelength, the maximum frequency for the rf-system sets the minimum allowable value of $S_B$. As the beams cross at an angle, $\alpha$, one must allow for the luminosity reduction factor due to the finite crossing angle,

$$R_L = \left(1 + \left(\frac{\alpha\,\sigma_z}{2\,\sigma}\right)^2\right)^{-1/2} . \qquad (53)$$

As the bunch length is proportional to the square root of the time derivative of the rf-voltage, lower frequencies imply increased $\sigma_z$ producing a large reduction factor. The reduction in luminosity can be compensated by increasing the number of particle per bunch. This rise would, however, increase both the synchrotron radiation and the number of events per crossing. Thus important technological considerations both in the accelerator per se and in the detector favor the choice of a high frequency.

The choice of frequency is, however, complicated by several deleterious effects that grow worse with increasing frequency. Parasitic heating of the walls of the collider due to wakefield effects scales as $\sigma_z^{-3/2}$. As the deposited energy must be removed at low temperature, resistive wall heating must not be neglected in the design of the cryogenic system, especially in the energy range from 10 - 20 TeV. As the collider energy increases to 100 TeV this effect becomes a small fraction of the synchrotron radiation power.

The growth rates of the coupled bunch, collective modes scale inversely with the bunch length. Moreover, as the bunch length decreases, fewer modes are Landau damped. Consequently, as the frequency of the rf-system is increased (thereby decreasing $\sigma_z$), one must take ever greater care to reduce the Q of the higher order modes (HOM). The specific technical approach to reduce the Q of the higher order modes varies with the choice of either a superconducting or room temperature rf-system. In the range of frequencies around 500 MHz, groups designing B-factories at SLAC and Cornell have developed new cavity designs with structural components to couple out the unwanted modes. Although such design modifications are necessary, they are not sufficient. Even if the Q of the higher order modes can be reduced to < 100, the growth time, $T_{multi\text{-}bunch}$, can be $\approx$ 1 - 5 ms (less than the synchrotron damping time even in electron rings). The remedy is to provide a powerful, wide bandwidth feedback system to damp the modes. The feedback requirements are briefly described in the next section.

As the bunch becomes shorter, the intra-beam scattering will cause the emittance to increase with a rate that scales as $\sigma_Z^{-3/2}$. In the absence of rapid radiation damping the rate of emittance dilution must be kept small compared with the luminosity lifetime and should be no greater than that from other emittance dilution effects. In an ELN operating at 100 TeV, the stronger radiation damping may allow a slightly higher choice of frequency than has been selected for the range of 10 - 20 TeV.

Balancing all of these effects for a specific storage ring requires a detailed analysis to assure acceptable performance of the supercollider throughout the operating energy range of the accelerator and an appropriate match with the frequencies of the synchrotrons in the injector chain. The general range of optimum frequencies is 200 - 600 MHz. Both the SSC and the LHC designers have opted to choose frequencies near the upper end of this range. For a collider at 100 TeV or more, the choice is unlikely to be very different. As a consequence of this choice, the minimum time interval between bunches cannot be less than $\approx$1.5 ns. To exploit the benefits of a luminosity of $10^{35}$ cm$^{-2}$s$^{-1}$, the detector and its associated electronics will have to operate in a mode consistent with conditions in which the average number of events per crossing is much greater than 1.

4.1.b *High power feedback for instability control*

The technical challenge for the designer of the feedback system is that even with well-damped rf-cavities the growth times of the fastest growing coupled bunch instabilities remain very much shorter than damping times due to natural processes.

4.1.b .1 *Longitudinal feedback*

To suppress the growth of the coupled bunch longitudinal modes the collider must have a powerful feedback system with large bandwidth[4.1],

$$W_f = \frac{1}{2} N_B f_o. \tag{54}$$

Even with the minimum allowed bunch spacing, the feedback system of the 100 TeV ELN would require $W_f \approx 150$ MHz. For comparison one should note that the corresponding values for the SSC and the PEP-II B-factory are ≈ 30 MHz and ≈ 119 MHz respectively. The required power for $\tau_g = 3$ ms is 2 kW; the power of the feedback system scales as

$$P_f \propto (\tau_g)^{-2}. \tag{55}$$

The construction of such feedback systems appears to be well within the range of capability of the new generation of commercially available rf-equipment. The preferred design approach, which has been adopted for PEP–II, is to use fast (≈2 ns rise), bunch-by-bunch, time domain feedback. In this approach one must detect the phase offset of each bunch and feedback kick that bunch after it has executed 90° of synchrotron oscillation. The number of computations can be greatly reduced if the bunch is not kicked on each turn. Rather, a technique of "down-sampling" can be used in which the bunch motion is measured on each turn, but kicks are given every $n^{th}$ turn. Down-sampling factors of 4 to 8 seem reasonable.

The number of channels in such a feedback system is the number of bunches. Hence, keeping the complexity of the system within reasonable bounds argues to keep the spacing between bunches large. For PEP-II the number of channels will be ≈ 2000. SSC will require ≈ 20,000 channels. ELN is likely to require 2 – 3 times as many channels.

A test of such a bunch-by-bunch system with ≈ 200 channels is now planned for the Advanced Light Source at the Lawrence Berkeley Laboratory. A definitive demonstration of the technique would go a long way to establishing the practicality of a bunch-by-bunch feedback system for the ELN, a hadron collider much larger the SSC.

4.1.b.2 *Transverse feedback systems*

The maximum current that can be stored in the collider will be limited by the growth of the transverse resistive wall instability at injection. A fast, narrow band, low frequency feedback system must be supplied to damp the growth of the instability. Conventional approaches with presently available technology may be expected to control growth rates as fast as $6\ T_o$. If even faster damping is needed, a "feed-across the ring" system may be able to exploit the large radius of a supercollider by measuring and kicking the beam in multiple sectors of the collider. The practicality of such a "feed-across" network, which would need fast encoding and decoding electronics, has not been assessed.

A problem unique to hadron colliders without the strong radiation damping typical of electron rings is that vibrations in the magnets, injection errors, or electronic noise in the feedback system itself can lead coherent motions of the beam. These motions must be controlled to limit a) position errors between the beams at the interaction point and b) "decoherence" of the motion that manifests itself as emittance dilution at a rate which is substantially faster than the radiation damping rate. In the Tevatron, which has no feedback systems to control emittance dilution, the normalized emittance is found to grow at a rate ≥ 3 π mm-mrad/hr. Such a growth rate would be completely unacceptable in ELN or SSC where the normalized emittance must be kept to ≤ 1 π mm-mrad. The consequence of emittance dilution would be a loss of luminosity once the beams are brought into collision.

A recent examination[4.2] of the design of feedback systems for proton supercolliders has indicated that care in the design of the feedback electronics system can keep emittance growth to tolerable proportions. The most difficult system for an accelerator beyond SSC appears to be the transverse dampers of injection oscillations. Already for SSC such a system will require ≈ 400 kW. As the energy of the collider is increased the injection energy must also be increased. However, the spacing between the closed orbit and the injection trajectory in the Lambertson magnets is likely to remain constant. The beam size at injection will be smaller because the geometric emittance will be smaller at the higher energy. Hence, the tolerance on the injection error will shrink.

4.2 *Interaction region magnets*

The system of magnets that focus the beams to the collision point have a direct influence on the luminosity of the collider because the luminosity scales as

$$\mathcal{L} \propto \left(\beta^*\right)^{-1}. \tag{56}$$

The minimum value of $\beta^*$ will be set by several practical considerations: the amount of free space required by the detector, the maximum practical field gradient, G, in the first quadrupole, and the radiation hardness of the magnet assuming realistic radiation shielding. One expects that a general purpose, 4π–detector at the high luminosity interaction point of a supercollider with beam energy from 10 – 100 TeV would need 20 to 40 m of free space, $l^*$, from the collision point to the first quadrupole, Q1. A detector designed for a PeV collider would require even more free space.

Brown[4.3] has formulated a set of scaling relations for an optimized focusing triplet as illustrated in Fig. 10. In such a triplet the focal length, *f*, is related to the field strength, B, the radius of curvature of the particles, ρ, and the free space by

$$f = 2\,l^* = \left(\frac{BL}{a^2\,(B\rho)}\right)^{-1/2} \propto \left(\frac{\gamma}{G}\right)^{1/2}. \tag{57}$$

The field gradient in Q1 will depend upon whether the magnet uses normal or superconducting windings (or permanent magnet materials). Presently the maximum gradient that can be obtained is estimated[2.1] to be 250 T/m in superconducting magnets with $Nb_3Sn$ at 4 .2 °K or NbTi at 1.8 °K. Based on this value for G the minimum value of $\beta^*$ follows the scaling relation,

$$\beta^* \geq 0.3\,\mathrm{m}\left(\frac{\gamma}{10^4}\right)^{1/2}. \tag{58}$$

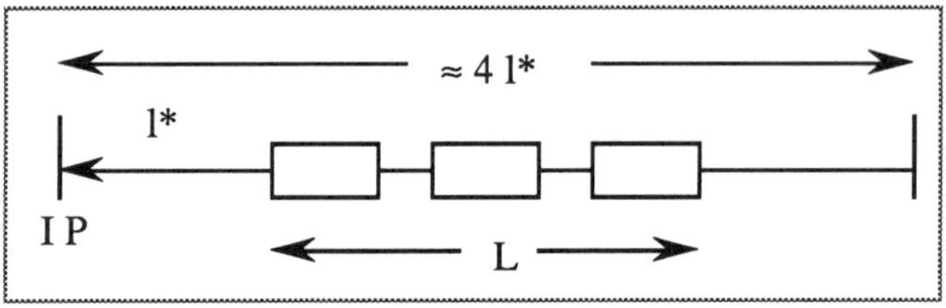


Figure 10. A schematic of an idealized quadrupole triplet

To keep the contributions of the interaction region optics to the chromaticity to manageable levels, the free space, $l^*$ should be chosen to be > 50 $\beta^*$. Hence,

$$l^* = 20\,\mathrm{m}\left(\frac{E}{20\ \mathrm{TeV}}\right)^{1/2}. \tag{59}$$

The beta function at Q1, $\beta_{Q1}$, is related to $\beta^*$ by

$$\beta_{Q1}\,\beta^* = l^{*2}. \tag{60}$$

The value of $\beta_{Q1}$ determines the beam size at Q1 via

$$\sigma_{Q1} = \sqrt{\frac{\varepsilon_n\,\beta_{Q1}}{\gamma}}\ . \tag{61}$$

Thus $\beta_{Q1}$ sets the aperture of Q1. As the aperture of the magnet is not adjustable, once the magnet layout has been selected to minimize $\beta^*$ for the highest machine energy, in computing the achievable luminosity at intermediate energies one should not scale the $\beta^*$ at intermediate energies according to Eq. (58) as the beam would then be too large for the aperture of Q1.

At the maximum luminosity of the collider, superconducting quadrupole magnets may not be practical because of excessive energy deposition from the hadronic debris. The energy deposited not only sets a large heat load for the cryostat to handle, but more importantly it can lead to quenching of the magnet due to deposition in the superconducting windings. The magnitude of the issue can be seen from the following example for ELN.

For an ELN at 100 TeV and at a luminosity of $\approx 10^{35}$ cm$^{-2}$s$^{-1}$, the power in the hadronic debris, $P_{debris}$, will be 180 kW per side. If the aperture of Q1 is 1.5 cm, then in the absence of the collimator the dose in Q1 as computed with Eq. (17) would be $\approx 4 \times 10^8$ Gy/year. At constant luminosity the dose dependence in an unshielded first quadrupole as a function of the operating energy of the supercollider is shown in Fig. 11.

The dose at 100 TeV deposited by the hadronic debris in Q1 corresponds to $\approx$ 45 W/kg. For a superconducting Q1 without a collimator removing this power would require, $\approx$ 20 kW/kg of compressor power. Similarly, one finds that were the first quadrupole a permanent magnet (SmCo), Q1 would have a lifetime of $\approx$ 1 week. A normal conducting, iron Q1 with polyimid insulation of the winding would have a lifetime of ≈ 3 months. Suitable shielding and collimation can attenuate the dose absorbed by the sensitive components by a factor of 10 – 100 depending on the energy of the beam and the details of the design. An additional reduction of the dose can be provided by increasing $\beta^*$ by an amount that depends on the maximum design luminosity; a reasonable suggestion from a recent study[4.4] is to scale $\beta^*$ according to

$$\beta^* \geq 0.3\,\mathrm{m}\left(\frac{\gamma}{10^4}\right)^{1/2} \mathrm{Max}\left[1, \mathrm{Log}_{10}\left(\frac{\mathcal{L}}{10^{33}}\right)\right]. \tag{62}$$

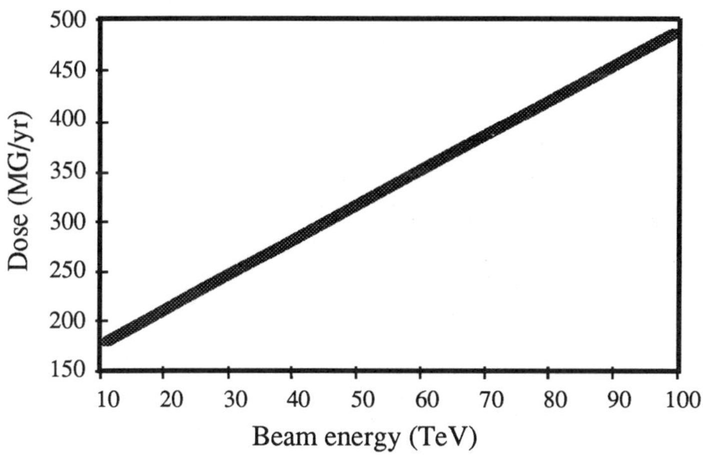


Figure 11. Energy scaling of dose to an unshielded Q1 for a luminosity of $10^{35}$ $cm^{-2}s^{-1}$

While this uncertainty in dose may seem large, one may still conclude that permanent magnet materials will be ruled out and that superconducting quadrupoles are unlikely for use in supercolliders operating at the highest energies and at top luminosities. Moreover, satisfying Eq. (6) implies that the beam current in the collider must be increased to compensate for the increase in $\beta^*$ as specified by Eq. (62). The photon flux and thermal load on the vacuum chamber due to synchrotron radiation will increase proportionally.

4.3. *Vacuum technology constraints*

A well designed beam chamber is the foundation of providing an adequate vacuum environment to permit a long beam-gas lifetime, to minimize emittance dilution and beam halo formation, and to prevent scattered, high energy protons from depositing enough energy in the superconducting magnets to cause quenches. For operation at beam energies exceeding 30 TeV at the highest luminosity, the chamber walls must also be thick enough to serve as the primary shield to absorb the synchrotron radiation from the beam that might otherwise trigger quenches in the superconducting magnets and damage other beamline components. The vacuum chamber also contains several special sections and components such as cross-section transitions, pumping ports, bellows, beam position monitors, all of which must be carefully designed to satisfy the impedance budget for the storage ring.

The surface physics and constituent technologies of the vacuum chamber in dipole arcs of the hadron supercollider will place strong limits on the maximum luminosity and energy of the accelerator. In schematic form, the components of the ultra-high vacuum chamber in the dipoles are illustrated in Figure 12. The primary chamber is a stainless steel tube, which is at the temperature of the liquid He that cools the windings of the superconducting coils. In the LHC (SSC) design, this temperature is 1.9 °K (4.2 °K). As the chamber wall is in intimate contact with the superconducting coils, it appears unlikely that this surface can be subjected to an in situ vacuum bake at 150 – 200 °C. This prohibition will have a strong influence on the photo-desorption properties of this surface.

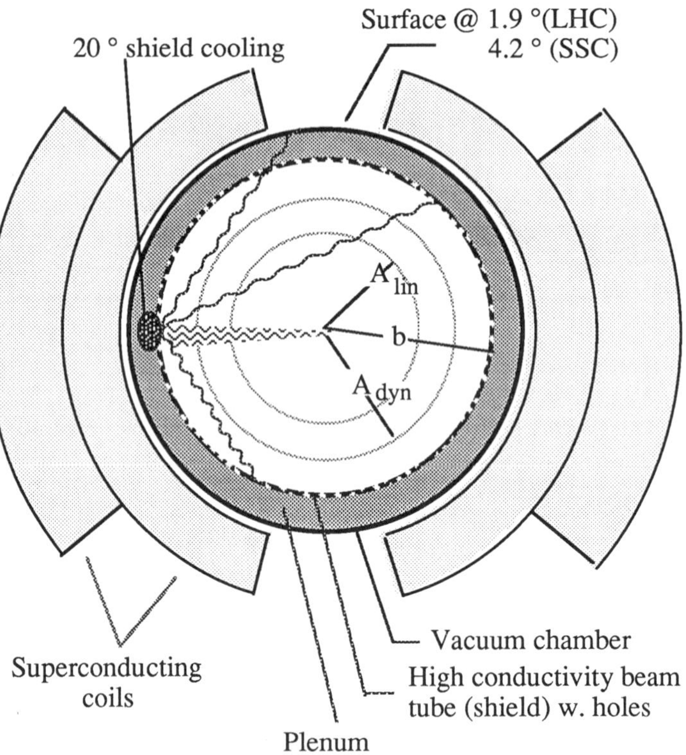


Figure 12. Schematic of the cross section of the vacuum chamber of the dipole arcs in a hadron supercollider with magnets of conventional design

The synchrotron radiation from the beam need not be allowed to fall on the surface of the primary vacuum chamber. Instead one can insert a higher temperature radiation shield (screen) so that the heat can be removed with a greater Carnot efficiency. In the event of a quench of the dipole, the rapidly changing magnetic fields (20 - 30 T/s) will induce currents that could deform the shield. To prevent such deformation the shield must be sufficiently strong. The material of choice for the shield is stainless steel which is strong and has a high resistivity. The shield will form the principal electromagnetic environment seen by the beam. Therefore it must have an inner surface with an electrical conductivity as high as possible to minimize resistive wall effects. Moreover, the thickness of the high conductivity layer must be large enough that the image fields of the beam do not penetrate the layer. For a shield at 10 - 20 °K, 0.1 – 0.2 mm of copper is sufficient. As the design temperature of the shield is raised, the thickness of the copper layer must be increased or the vertical aperture must be enlarged to suppress resistive wall effects. Intrusions for heat removal and the size and arrangement of slots in the screen for pumping must be consistent with maintaining a low electromagnetic impedance[4.5]. If a shield is not used, the copper lining must be placed on the stainless steel vacuum chamber.

4.3.a *Gas load due to photo-desorption*

As described in section 3.6.4, the synchrotron radiation will cause gas molecules to be desorbed from the walls at a rate proportional to the luminosity. For an operating pressure, Π, the required distributed pumping is

$$S = \frac{Q_{gas}}{\Pi} \propto N_\gamma \propto \mathcal{L}. \tag{63}$$

The space between the stainless steel vacuum chamber and the shield forms the plenum for cryopumping the beam tube. The cryopumping at speeds of $>10^3$ L/s/m ($N_2$ equivalent) can be provided by the cold chamber wall.

As the gas is not actually removed from the system a more complex situation develops than in warm walled storage rings. The chamber pressure will be determined by the pumping speed and the desorption coefficient until somewhat more than one monolayer of gas freezes onto the surface. At that point the partial pressure of each gas will be set by the plateau value of its respective isotherm. These plateau pressure levels[4.6] for $H_2$, CO, and $CH_4$ are shown in Fig. 13.

Clearly, if the wall is at a temperature > 3 °K, it will not be possible to maintain the chamber at pressures < 0.1 nTorr for indefinite periods. The use of cryosorption getters such as activated charcoal can increase by orders of magnitude the effective surface area of the pumping surface and the period for the pumping surface to accumulate a monolayer of gas. If the radiation shield comes to a temperature of > 22 °K, the equilibrium vapor pressure of CO will be great enough that the holes in the screen must provide a suitably high conductance for the CO to be pumped by the cold surface. Whether such large holes are compatible with electromagnetic impedance requirements is uncertain. Of course as the plenum becomes larger and more complex the cost of the chamber will rise rapidly.

The rate at which the desorbed $H_2$ accumulates on the wall is determined by the desorption coefficient. As the surface can be vacuum baked in situ the desorption coefficient is large as seen in a sample of the data of Limon and Bintinger[4.7] shown in Fig.

14. For comparison recent measurements of baked copper have shown desorption coefficients lower than $10^{-6}$.

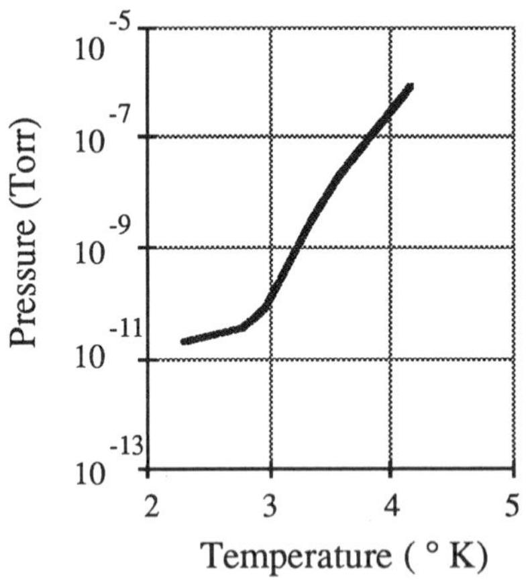


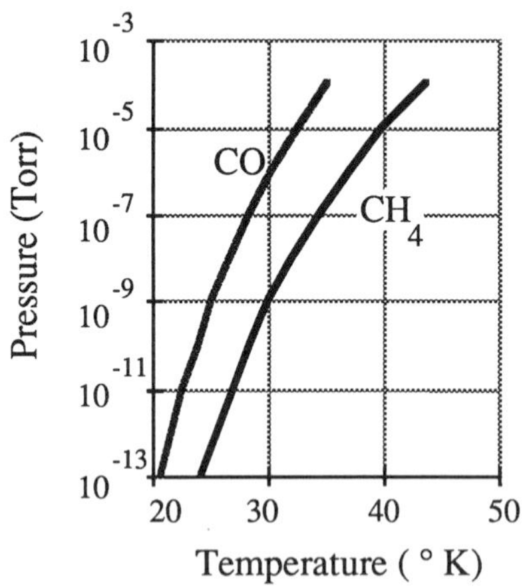


Figure 13. The vapor pressure of hydrogen, carbon monoxide and methane as a function of temperature assuming a wall coverage of ≈ 2 monolayers from Mathewson[4.6]

For the SSC the data of Fig. 14 imply that a monolayer of $H_2$ will accumulate on the chamber walls in a few days during initial operation at a luminosity of $10^{33}$ $cm^{-2}s^{-1}$. Once this point is reached the ring must be warmed and the $H_2$ removed with lumped pumps. The thermal cycling of the chamber results in a reduction of the average luminosity of the collider. For a supercollider meant to achieve the highest possible average luminosity, careful attention must be paid to the chamber design at the outset of the design process, before the magnet design is fixed.

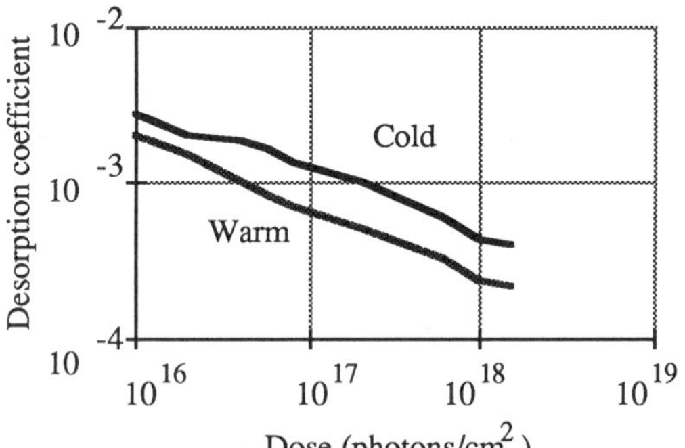


Figure 14. Photodesorption coefficients as a function of photon dose at 270 eV for copper plated stainless steel at room temperature and at 4.2 °K from Ref. 4.7

4.3.b *Operating pressure and beam loss*

The average operating pressure in the collider is set by three principal requirements: 1) the beam-gas lifetime should be long compared to the luminosity lifetime, 2) the loss rate of particles due to beam-gas scattering must be less than the minimum loss that causes magnet quenches, 3) locally the pressure must not exceed the threshold for the "ISR pressure bump" effect . This effect leads to a pressure runaway and a consequent local catastrophic beam loss.

Considering the nuclear scattering to be the dominant loss mechanism, one can compute the beam gas lifetime, $\tau_g$;

$$\frac{1}{\tau_g} = c\, N_g\, \sigma_{nuc} \tag{64}$$

In Eq. (64) c is the speed of light, $N_g$ is the number density of the gas and $\sigma_{nuc}$ is the nuclear scattering cross section for the gas species in question. Assuming a maximum permissible loss rate to satisfy 2) of $10^7$ $s^{-1}$ as in the LHC design[1.1], one can estimate that the background pressure of $H_2$ should be $\approx 1$ nTorr for a luminosity of $10^{34}$ $cm^{-2}s^{-1}$. Eq. (64) implies that as the number of particles in the collider is increased to provide greater luminosity, the collider pressure will have to be decreased proportionally.

Background gas molecules that are ionized by the beam will be accelerated toward the walls by the space charge of the beam and can attain kinetic energies of $\approx$0.5 keV. When the ion strikes the wall, it will sputter off molecules from the surface. As the number of neutrals ejected per incident ion, $\eta_{ion}$, is greater than 1, ionization of the background gas can lead to a local pressure runaway when the product, $\eta_{ion}I$, exceeds a threshold value (of order of 1000 A). $\eta_{ion}$ for clean surfaces can be <10; however, as the number of monolayers of $H_2$ on the beam tube increases, $\eta_{ion}$ can increase to values[4.8] from $10^2$ to $10^4$. The speed at which the monolayers accumulate depends on the photodesorption coefficient, which unfortunately is not well known as a function of photon energy on cold surfaces. Furthermore, the vacuum properties of the surface depand on the surface fabrication and preparation techniques. As the beam tube must have both excellent electrical and vacuum properties, a wide range of experimental studies will be required to establish appropriate materials and fabrication techniques.

This effect, known as the "ISR bump" effect, can produce a localized, catastrophic loss of the beam. As the energy stored in the beam can be several hundred megajoules in a nominal SSC or LHC and 10 – 100 GJ in an ELN at the highest luminosity, catastrophic beam losses can have dramatic, undesirable effects on the structural integrity of the collider.

4.4 *Beam protection systems*

As supercolliders are driven to the highest possible energy and luminosity, the energy stored in the beam could be of the order of 100 GJ. At that level even small fractional beam losses can lead to magnet quenches and damage of sensitive components. The management of the routine losses is accomplished with collimators and beam dumps.

Fault mode protection in a PeV UELN will require meticulous care and a high degree of redundancy. Whether such protection can be provided by completely passive systems is a open question. Although there seems to be no a priori reason why suitable systems cannot be built for supercolliders at the highest energy and luminosity, beam protection systems will require careful study before suitable deigns can be suggested.

## 5. Design code and parameter study

The implications of trying to design supercolliders at the highest energies and luminosities in a self-consistent design can be explored most easily with a computer code based to perform parameter searches. The ELOSCALE code is a spreadsheet–format design code based on the scaling relations described in the previous sections. The formulation is such that the code can easily be expanded via linked spread sheets to contain the details of the design characteristics of each of the lower energy machines that fill the main collider.

Codes such as ELOSCALE in expanded versions can be used during the machine construction phase as "official" parameter lists to assure that changes in the machine characteristics lead to consistent parameter sets that do not violate operational constraints or increase physics risks. The codes can also be used to map design strategies that assure the upgrade potential of the collider to the highest energies and luminosities. Finally during commissioning phase the code can be used to develop turn-on scenarios that bring the machine to full design specifications. Such an expanded spread-sheet based code[5.1] is now being used to guide SSC construction.

The input variables describing the beam are the injection, operating and maximum beam energies, the normalized emittance and the bunch spacing. A critical characteristic of the collider is the maximum permissible tune shift. This value should be treated as a constraint rather than as a parameter in the study. In the absence of experimental evidence for higher values of the maximum head-on tune shift per interaction point, one should take this value to be 0.01 as used in the Snowmass study. The inputs describing the storage rings are as follows: the maximum dipole field, dipole fraction in the ring, and the vertical aperture of the dipoles, the maximum permissible radiation power on the walls and the temperature of the innermost beam tube, the number of interaction points, the crossing angle, the distance from the interaction point to the septum, the scale $\beta^*$ at 20 TeV (0.3 m in Eq. (58)), operating frequency of the collider rf-system. The injection chain is assumed to have a linac and four intermediate boosters. Other characteristics of the collider and the injector chain are computed in the code.

### 5.1 *Parameter searches*

Based on a parameter exploration with ELOSCALE it seems practical to construct a 100 TeV proton supercollider with a luminosity $>10^{34}$ cm$^{-2}$ s$^{-1}$ by using the same technologies that are being realized for the LHC. At energies below its full design value such an ELOISATRON would have an energy vs. luminosity dependence as shown in Figure 15. Reducing the allowable radiation to 2 W/m would reduce the luminosity at 100 TeV by a factor of two. In contrast, raising the value to 20 W/m yields $10^{35}$ cm$^{-2}$s$^{-1}$. Parameters for both of these cases are given in Tables 2 and 3 respectively. In neither case does the use of crab-crossing seem to offer much increase in the design luminosity.

In Fig. 15 the top luminosity at low energies is set by the resistive wall instability which limits the injected current and by the head-on tune shift. At high energies the limitation on luminosity is set by the radiation power on the walls and by the head-on tune shift.

Further design studies indicate that if new materials allow one to increase the radiation load to ≈100 W/m, a 100 TeV supercollider should be able to reach a top luminosity ≈ $10^{36}$ cm$^{-2}$ s$^{-1}$, especially when the collider operates at intermediate energies (see Figure 16). Realizing this latter possibility will depend on the existence of a broadly based experimental research effort prior to the final machine design. Of course it remains to be seen whether even special purpose detectors could tolerate the enormous radiation doses implied by such high luminosity.

Another approach to allowing very large radiation loads is to search for magnet designs that would allow the radiation to be deposited on warm surfaces, for example through the use of "C-magnets" and antechambers. In either case and most importantly, the possibility of achieving top luminosity (> $10^{36}$ cm$^{-2}$ s$^{-1}$) must be incorporated *ab initio* into the design of the supercollider, not considered as an after thought. A similar remark applies to building in the possibility of an energy upgrade.

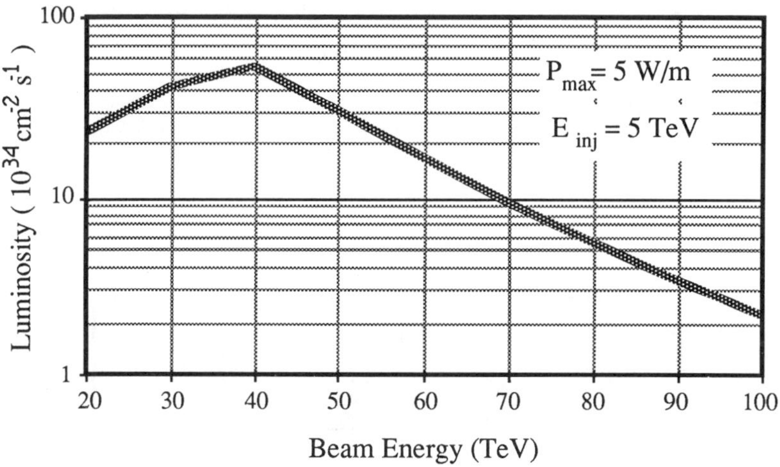


Figure 15. $\mathcal{L}$ versus energy for an ELOISATRON with 5 W/m of radiation allowed on the walls of the vacuum chamber.

The stored energy of the beam which corresponds to the 20 W/m case of Fig. 16 can be over 100 GJ (at 50 TeV) as shown in Fig. 17. In Figure 17 the unit of energy, tonnes of high explosive (4.2 GJ = 1 tonne), has been chosen to emphasize the need for elaborate, fail-safe fault control systems and beam dumps in supercolliders designed to operate at the highest possible luminosities. It also suggests that for normal dumps such as those performed before reloading the collider, the beam should be decelerated before being extracted from the storage ring.

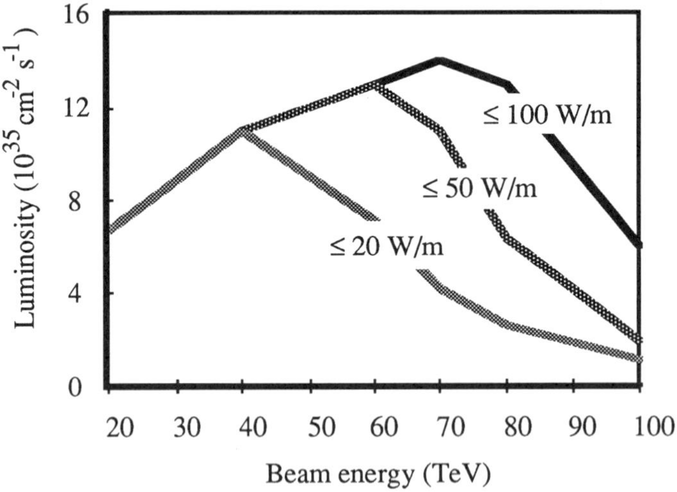


Figure 16. The luminosity of a 100 TeV ELN as a function of beam energy assuming technologies allowing radiation loads from 20 – 100 W/m.

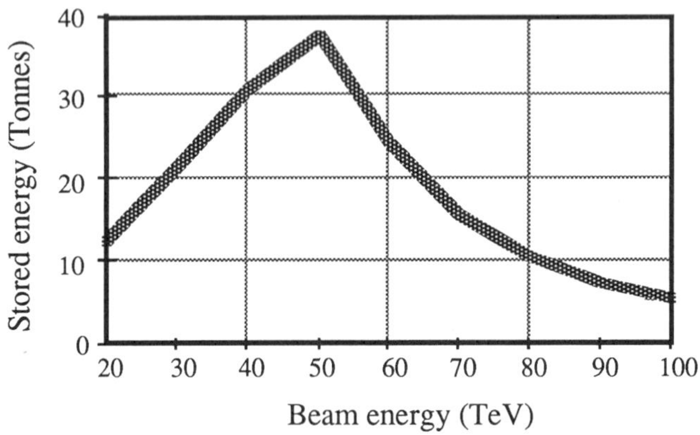


Figure 17. The stored energy in the beam as a function of operating energy for an ELN for the case of 20 W/m shown in Fig. 16

Table 2. Parameter set computed with ELOSCALE for ELN operating at $10^{34}$ cm$^{-2}$s$^{-1}$ using present technology

**Ring charcteristics**

| | | |
|---|---|---|
| * | *Max Energy (TeV)* | *100* |
| * | *B dipole - max (T)* | *7.7* |
| * | *Dipole fraction* | *0.765* |
| | To - rev time (s) | 1.19E-03 |
| | Circum (km) | **355.55** |

**Beam characteristics**

| | | |
|---|---|---|
| * | *Energy (TeV)* | *100* |
| * | *Norm emit@E(π mm-mrad)* | *0.56* |
| * | *Bunch space buckets* | *8* |
| | gamma | 1.1E+05 |
| | Bunch space (m) | 5.82 |
| | Number of bunches | **6.1E+04** |
| * | *NBunch (nC) < 0 for input* | *0.0* |
| | Bunch population | **1.2E+10** |
| | Current (A) | **0.10** |

**Injection / Energetics**

| | | |
|---|---|---|
| * | *Injection Energy (TeV)* | *10* |
| | Fill time (hr) | **4.0** |
| | Stored Energy (GJ) | **11.71** |
| | D(E,Q1) MGy/yr | **21.6** |
| | Hadrons in Q1 (W/kg) | 2.2 |
| | Debris (kW per side) | **18.0** |

**Synchrotron radiation**

| | | |
|---|---|---|
| * | *Max P(W/m) on walls* | *5* |
| | Uo (GeV/turn) | 1.8E-02 |
| | E- Damping time (s) | 6.6E+03 |
| | Power (W) | 1.8E+06 |
| | Power density (W/m) | **5.0** |

**Parasitic heating**

| | |
|---|---|
| Resitive wall (kW/beam) | 0.2 |
| RF-HOMs (kW/beam) | 0.4 |
| P to Compressors (MW) | **125.0** |

**Lattice**

| | |
|---|---|
| <Beta> (m) | 491.9 |
| Dispersion (m) | 5.6 |

*Rev 20.11.92; Σ's = Gaussian widths*

**Interaction points**

| | | |
|---|---|---|
| * | *Number of IPs* | *2* |
| * | *Crossing angle (mr)* | *0.07* |
| * | *L* - long range (m)* | *170* |
| | l* - (m) | 54.8 |
| | Beta* (m) | 1.00 |
| | Beta max (km) | 18.0 |
| | Σ-IP (μm) | 2.3 |
| | 2Σ-Q1 (cm) - flag | 0.31 |
| | Δnu-HO / IP | 2.5E-03 |
| | Δnu-LR | 6.0E-04 |
| | Δnu-tot | **0.006** |
| | Δt crossing (ns) | **19.4** |
| | 95% Lum. region (cm) | 21.9 |
| | Interact / crossing | **27.7** |
| | T-lum/t-damp-perp | 26.1 |
| | Lumin. half-life (hr) | **47.7** |
| | Luminosity / IP | **1.0E+34** |

**Instabilities**

| | | |
|---|---|---|
| * | *R-pipe (cm)* | *4* |
| * | *Operating temp (°K)* | *40* |
| * | *Injection temp (°K)* | *40* |
| | Resistive wall -turns | 555 |
| | RW at injection - turns | **56** |
| | μwave Z/n @ inj (ohms | **109.6** |

**Vacuum**

| | | |
|---|---|---|
| * | *Op. pressure (nTorr)* | *5* |
| * | *Desorb coeff.* | *0.001* |
| | E-crit (eV) | 8.0E+03 |
| | N-gamma (s-1 m-1) | 1.2E+16 |
| | Req. pumping (L/s/m) | **10.0** |

**RF systems/longitudinal**

| | | |
|---|---|---|
| * | *rf -Frequency (Mhz)* | *412* |
| * | V-eln/V-ssc | 8 |
| | Harmonic number | 146500 |
| | Σ-z (cm) | **7.8** |
| | Synch. tune | 1.80E-03 |
| | Mom. compaction | 9.0E-05 |
| | N cavities/ring | 33 |
| | P-max to klystron (MW | **6.1** |

Table 3. Parameter set computed with the code, ELOSCALE for the 100 TeV per beam ELN operating at $10^{35}$ $cm^{-2}s^{-1}$.

| | | |
|---|---|---|
| | **Ring charcteristics** | |
| * | *Max Energy (TeV)* | *100* |
| * | *B dipole - max (T)* | *7.7* |
| * | *Dipole fraction* | *0.765* |
| | To - rev time (s) | 1.19E-03 |
| | Circum (km) | **355.55** |
| | **Beam characteristics** | |
| * | *Energy (TeV)* | *100* |
| * | *Norm emit@E(π mm-mrad)* | *0.37* |
| * | *Bunch space buckets* | *6* |
| | gamma | 1.1E+05 |
| | Bunch space (m) | 4.37 |
| | Number of bunches | **8.1E+04** |
| * | *NBunch (nC) < 0 for input* | *0.0* |
| | Bunch population | **3.5E+10** |
| | Current (A) | **0.40** |
| | **Injection / Energetics** | |
| * | *Injection Energy (TeV)* | *10* |
| | Fill time (hr) | **4.0** |
| | Stored Energy (GJ) | **46.84** |
| | D(E,Q1) MGy/yr | **146.5** |
| | Hadrons in Q1 (W/kg) | 14.6 |
| | Debris (kW per side) | **161.8** |
| | **Synchrotron radiation** | |
| * | *Max P(W/m) on walls* | *20* |
| | Uo (GeV/turn) | 1.8E-02 |
| | E- Damping time (s) | 6.6E+03 |
| | Power (W) | 7.1E+06 |
| | Power density (W/m) | **20.0** |
| | **Parasitic heating** | |
| | Resitive wall (kW/beam) | 3.8 |
| | RF-HOMs (kW/beam) | 5.0 |
| | P to Compressors (MW) | **307.9** |
| | **Lattice** | |
| | <Beta> (m) | 491.9 |
| | Dispersion (m) | 5.6 |

| | | |
|---|---|---|
| | **Interaction points** | |
| * | *Number of IPs* | *2* |
| * | *Crossing angle (mr)* | *0.12* |
| * | *L* - long range (m)* | *170* |
| | l* - (m) | 109.6 |
| | Beta* (m) | 2.00 |
| | Beta max (km) | 36.0 |
| | Σ-IP (μm) | 2.6 |
| | 2Σ-Q1 (cm) - flag | 0.35 |
| | Δnu-HO / IP | 1.1E-02 |
| | Δnu-LR | 4.1E-04 |
| | Δnu-tot | **0.023** |
| | Δt crossing (ns) | **14.6** |
| | 95% Lum. region (cm) | 21.9 |
| | Interact / crossing | **187.2** |
| | T-lum/t-damp-perp | 11.6 |
| | Lumin. half-life (hr) | **21.2** |
| | Luminosity / IP | **9.2E+34** |
| | **Instabilities** | |
| * | *R-pipe (cm)* | *4* |
| * | *Operating temp (°K)* | *60* |
| * | *Injection temp (°K)* | *40* |
| | Resistive wall -turns | 77 |
| | RW at injection - turns | **14** |
| | μwave Z/n @ inj (ohms | **36.5** |
| | **Vacuum** | |
| * | *Op. pressure (nTorr)* | *5* |
| * | *Desorb coeff.* | *0.001* |
| | E-crit (eV) | 8.0E+03 |
| | N-gamma (s-1 m-1) | 4.9E+16 |
| | Req. pumping (L/s/m) | **60.1** |
| | **RF systems/longitudinal** | |
| * | *rf -Frequency (Mhz)* | *412* |
| * | V-eln/V-ssc | 8 |
| | Harmonic number | 146500 |
| | Σ-z (cm) | **7.8** |
| | Synch. tune | 1.80E-03 |
| | Mom. compaction | 9.0E-05 |
| | N cavities/ring | 33 |
| | P-max to klystron (MW | **24.4** |

*Rev 20.11.92; Σ's = Gaussian widths*

*5.2 Optimization studies*

The ELOSCALE design code also allows the machine designer to search for optimum values of the machine characteristics. An example with major cost implications is the choice of the strength of the dipole field. A representative examination of the variation with $B_{dipole}$ of several characteristics of an ELN operating at a luminosity of $10^{35}$ $cm^{-2}$ $s^{-1}$ is shown in Figure 18. It is difficult to draw any final conclusions about the choice of $B_{dipole}$ without an adequate model of the variation in magnet cost with dipole strength, dipole length, and vertical aperture.

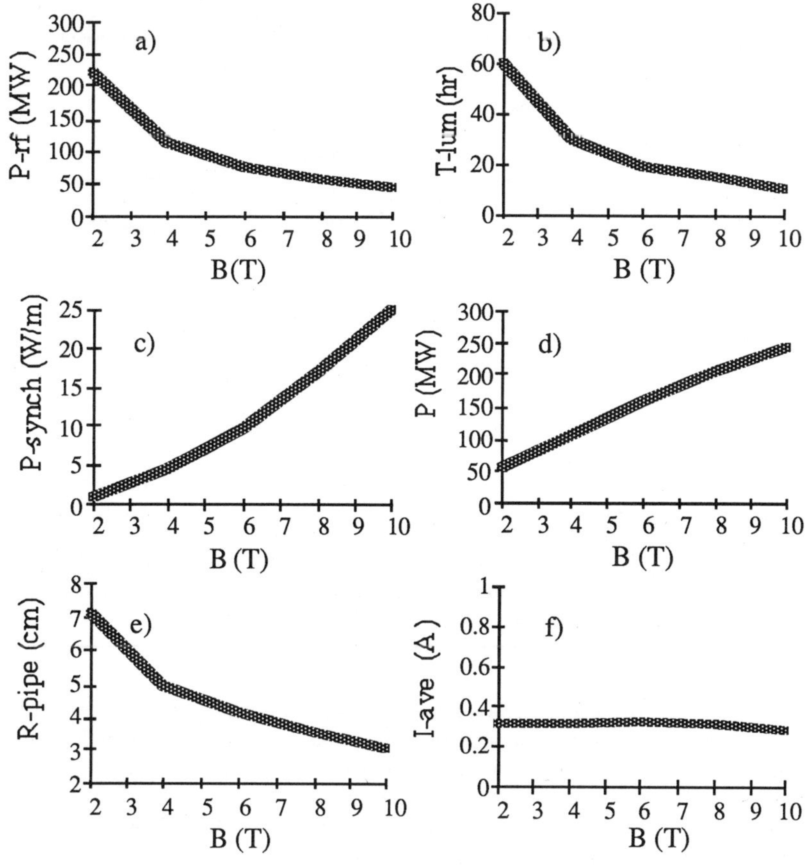


Figure 18. Variation of collider characteristics with dipole field for ELN at 100 TeV per beam operating at $10^{35}$ $cm^{-2}$ $s^{-1}$. a) RF-power v. $B_{dipole}$, b) Luminosity lifetime v. $B_{dipole}$, c) Thermal load on vacuum chamber v. $B_{dipole}$, d) Total operating power from mains v. $B_{dipole}$, e) Vertical aperture to control growth of resistive wall modes v. $B_{dipole}$, and f) Average current v. $B_{dipole}$

A quantity that is not plotted in Fig. 18 is the energy stored in the beam. As the average current in the collider is nearly independent of $B_{dipole}$, the number of particles in the ring and thus the stored energy will be proportional to the circumference as displayed in Fig. 19. At 2 T, the circumference would be roughly 1500 km and the stored energy would be ≈ 200 GJ. Even such a cursory look suggests that a 100 TeV collider should not be based on a low field dipole. However, there is also little reason to base the design on the very highest fields that have ever been achieved in accelerator-type magnets.

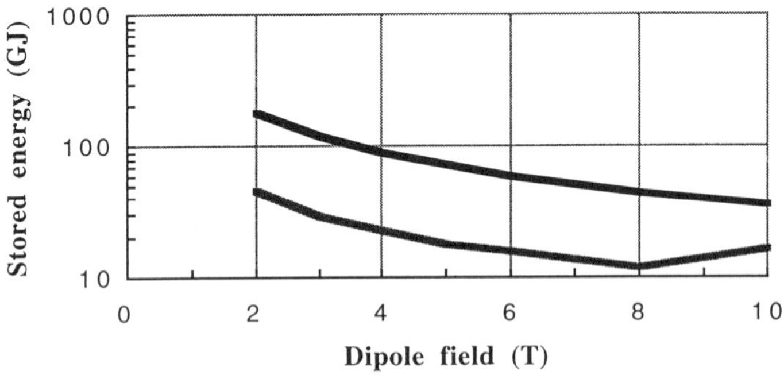


Figure 19. The energy stored in the beam of the ELOISATRON operating at luminosities of $10^{34}$ $cm^{-2}s^{-1}$ (grey) and $10^{35}$ $cm^{-2}s^{-1}$ (black). For reference, note the stored energy for the SSC operating at $10^{33}$ $cm^{-2}s^{-1}$ is ≈ 500 MJ.

As the dipole magnets will constitute the most expensive technical sub-system of the hadron supercollider, the optimization of $B_{dipole}$ should be carried out under several scenarios, each of which will require back-up hardware studies to assure the practicality of the choice of field strength. In particular, the optimization for the case of a design with C-magnets is likely to yield a much different result than that for closed dipoles once all the ramifications of field choice are taken into consideration.

### *5.3 Ultimate supercollider*

With respect to the longest term future, an examination of the systematics of collider design suggests the ultimate potential of conventional storage ring technology in the exploration of the high energy frontier of elementary particle physics. If the vacuum chamber of the proton storage ring can be designed to be operated at or slightly above room temperature as is typical in electron-positron colliders, then it should possible to construct a hadron collider with a center of mass energy of 1 PeV and a luminosity exceeding $10^{36}$ $cm^{-2}$ $s^{-1}$ at its peak energy. This machine, with a circumference twenty times larger than the SSC and requiring ≈ 2 GW of mains power might well be the ultimate supercollider (UELN).

As the survival of detector components is questionable at such a high luminosity, a more probable scenario for UELN is to keep the luminosity at $10^{34}$ $cm^{-2}$ $s^{-1}$. In that case all of the technical sub-systems are much closer to the present state of technology. In particular the vacuum sub-system should be fairly close in character to that of the ELN. The walls could run at 150 °K to allow the power to the compressors to be kept at 500 MW. A comparison of the high and low luminosity options is presented in Table 4.

Table 4. Two possible sets of characteristics of the Ultimate ELOISATRON (UELN)

| | | |
|---|---|---|
| Center of mass energy | 1 PeV | |
| Circumference | 1500 km | 1015 km |
| $B_{dipole}$ | 8 T | 13.5 T |
| Beam energy | 500 TeV | |
| Beam current | 800 mA | 10 mA |
| Mains power | 2 GW | 500 MW |
| $\langle \mathcal{P}_{synch} \rangle$ | 1400 kW/m | 55 W/m |
| Interaction regions (IR) | 2 | |
| Limiting technology | IR survival | Project management |
| Tune shift per IR | 0.01 | 0.006 |
| Luminosity ( $cm^{-2}\,s^{-1}$) | $\approx 10^{36}$ | $10^{34}$ |

The physics reach and discovery potential of an Ultimate ELN, which is based on credible extrapolations of conventional proton synchrotron technology, would be very far beyond that of a 10 TeV electron–positron collider for which no believable design scenario exists. With the walls operating at cryogenic temperatures the power requirements of UELN are within a factor of 2 of what has been considered for 1 - 2 TeV linear colliders.

This exercise of exploring the limits of proton synchrotron technology also raises the question of the energy regime for which the linear collider approach might become applicable to proton colliders. If the length of each linac of a linear UELN is to be no longer than the radius implied by Table 4, then the linac must be able to operate with accelerating fields > 1 GeV/m. Such high gradients imply a frequency of the linac structure in excess of 100 GHz. The luminosity is directly related to the average power in the beams by

$$\mathcal{L}(10^{33}\,\mathrm{cm}^{-2}\,\mathrm{s}^{-1}) = \frac{D\,H_D}{30}\left(\frac{1\,\mathrm{mm}}{\sigma_z}\right)\left(\frac{P_{beam}}{1\,\mathrm{MW}}\right) \tag{65}$$

where $H_D$ is the luminosity degradation due to the pinch effect and where D is the disruption parameter that measures the pinch;

$$D = \frac{r_p\,N_B\,\sigma_z}{\gamma\,\sigma_{x,y}^2} = r_p\,N_B\left(\frac{\sigma_z}{\beta^*\,\varepsilon_n}\right). \tag{66}$$

The disruption is related to the tune shift in a storage ring by

$$D(\mathrm{ring}) = 4\pi\,\Delta\nu_{HO}\left(\frac{\sigma_z}{\beta^*}\right). \tag{67}$$

For D < 2, the value of $H_D \approx 1$. At 500 TeV $\beta^*$ is limited to be $\approx$ 2m. The normalized emittance $\varepsilon_n$ will be difficult to make smaller than $10^{-7}$ m-rad. For an accelerating field at a frequency of 100 GHz the bunch length, $\sigma_z \leq 10^{-6}$ m. Hence the quantity in parentheses in Eq. (66) is of order 1 $m^{-1}$. Even were it possible 1) to generate bunches of 100 nC (more than 10 times greater than in present high quality, proton sources) with such low emittances and 2) to preserve the emittance in the presence of extremely large wake fields in the linear accelerator, it would be difficult for $r_pN_B$ to exceed $10^{-6}$ m. One concludes that the disruption in a proton linear collider will be exceedingly small. From Eq. (65) one sees that to achieve a luminosity of $10^{33}$ $cm^{-2}$ $s^{-1}$ the average power must be $\approx$30 GW per beam. Even assuming 100% efficiency of transformation of mains power to beam power, this requirement is 30 times greater than the power requirement for the synchrotron. Actual power conversion efficiencies are unlikely to be much greater than 10%. Hence, a PeV linear proton supercollider would be 300 times as "power hungry" as a synchrotron at the same luminosity. As D decreases with increasing energy, the practicality of employing a linear proton collider actually diminishes at even higher energies. Hence, the ultimate supercollider, UELN, should be a synchrotron.

## 6. Conclusions

On the basis of a systematic parameter search with the computer code ELOSCALE it appears possible to build hadron supercolliders with an energy and luminosity much higher than those of the SSC based on conventional approaches. In particular, an ELOISATRON (ELN) operating at 100 TeV per beam with a luminosity exceeding $10^{34}$ $cm^{-2}s^{-1}$ could be constructed by using presently available technologies such as those which are being incorporated in the designs of SSC and LHC. Assuming moderate advances in the state of accelerator technology during its design cycle, one could expect to operate an ELN supercollider at luminosities $\approx 10^{35}$ $cm^{-2}s^{-1}$ at 100 TeV/beam. Even higher luminosities would obtain at lower energies. Such a hadron supercollider based on conventional technology would have the physics reach and discovery potential at least as great as a 10 TeV $e^+$–$e^-$ linear collider, for which no reasonable design concept now exists. With further advances in a few key technologies a PeV collider, UELN, based upon conventional proton synchrotron approaches appears to be technologically possible.

While a 100 TeV collider with a luminosity of $10^{34}$ $cm^{-2}$ $s^{-1}$ may be quite conventional, one must emphasize that achieving the highest possible luminosity (>$10^{36}$ $cm^{-2}$ $s^{-1}$) will push both the beam characteristics and the constituent accelerator technologies considerably beyond present practice. To reach the highest possible luminosities, existing technologies will have to be pushed into new regimes (for example, by finding practical, high $T_c$ superconductors suitable for magnet windings). In that case the design strategies listed above and their limitations must receive detailed *experimental* exploration for one to arrive at a level of confidence commensurate with the cost of a 100 TeV collider with a luminosity in the range of $10^{36}$ $cm^{-2}$ $s^{-1}$. Adopting the design strategies described in this report would yield a collider with a very large number of bunches with associated beam crossing rates of the order of 0.1 – 1 GHz with several tens of collisions per crossing. Determining whether detectors and data acquisition and processing systems can be designed to accommodate (or even survive) the incredible rates implied by extremely high luminosities needs a serious evaluation by a group of detector specialists working with accelerator physicists familiar with both hadron and electron–positron colliders.

## Acknowledgments

The author wishes to thank Prof. Antonino Zichichi for his encouragement in undertaking this study and for his kind hospitality at the Ettore Majorana Centre for Scientific Culture. Conversations with Robert Siemann (SLAC), Robert Palmer (BNL/SLAC), and Alex Chao (SSCL) have also been extremely helpful in framing the analytical approach. Alaister Mathewson and Oswald Gröbner of CERN have provided many valuable insights concerning vacuum issues. Glen Lambertson has been a major source of information about feedback systems. The author's participation in Snowmass 1990 and the ELOISATRON Workshop, Maximizing the Luminosity of Hadron Colliders at 10 – 100 TeV, also provided a broad source of the physics embodied in the ELOSCALE design program. This work was partially performed under the auspices of the Lawrence Livermore Laboratory for the U. S. Dept. of Energy under contract W-7405-eng-48.

## References:

1.1 Design Study of the Large Hadron Collider, CERN 91-03
1.2 SSC Site-specific Conceptual Design Report, 1990
1.3 Zichichi, A., The ELOISATRON Project (August, 1988)
2.1 *Proceedings of the Snowmass Summer Study of High Energy Physics*, 1990
2.2 Keil, E. "Hadron Colliders between 1 and 100 TeV Beam Energy", CERN report LEP-TH/83-39, Aug. 1983
2.3 Peterson, J. "Parameter Choices for the SSC", *Physics of Particle Accelerators*, AIP Conference Proceedings 184, 1988
2.4 Chao, A. "Approaching Ultimate Luminosity at the SSC", *Proceedings of the APS Division of Particles and Fields*, 1990
3.1 Groom, D., "Radiation Levels in SSC Detectors", *Proceedings of the Snowmass Summer Study on High Energy Physics in the 1990's* (1988)
3.2 Report of the SSC Task Force on Radiation Levels in the SSC Interaction Regions, SSC Central Design Group Report, SSC-SR-1033 (June, 1988)
3.3 Schmidt, R., and Harrison, M., Proceedings of the European Particle Accelerator Conference, Nice, June, 1990
3.4 Evans, L., and Gareyete, J., CERN report CERN/SPS/82-8 (1982)
3.5 Palmer, R., SLAC-PUB-4707
3.6 Bartalucci, S., et al. "A Crab Crossing-Option", LNF internal report
3.7 DAΦNE Conceptual Design Report, Laboratorio Nazionale Frascati, 1990
3.8 Pellegrini, C. et al., "A Superconducting Mini-collider for the Copious Production of Phi Mesons", *Proceedings of IEEE Particle Accelerator Conference* (San Francisco, 1991)

3.9 Gareyete, J., These proceedings
3.10 Barletta,W., "Modeling Photo-desorption in High Current Storage Rings", *Proceedings of IEEE Particle Accelerator Conference*, San Francisco (May, 1991)
3.11 Barletta, W. A., Calderon, M., Jenkins, T., Wong, R., "Design of the Vacuum System for the High Energy Ring of for an Asymmetric B Factory based on PEP", *Proceedings of IEEE Particle Accelerator Conference*, San Francisco (May, 1991)
3.12 Gröbner, O., Mathewson, A., Störi, H., Strubin, P., Souchet, P., *Vacuum*, **33**, 397 (1982)
3.13 Reiser, M., "Use of Dispersion to Increase the Space Charge Limit in Circular Accelerators, Un. of Maryland CPB Technical Report #92-003
4.1 An Asymmetric B-Factory Based on PEP: Conceptual Design Report, SLAC-372 (1991)
4.2 Lambertson, G., These proceedings, 1992
4.3 Brown, K., Private communication as reported in Palmer, R., "Interdependence of Parameters for TeV Linear Colliders", *Proceedings of Workshop on New Developments in Particle Accelerator Techniques*, Orsay (June, 1987)
4.4 Report of IR Working Group, *Proceedings of the 19th ELOISATRON Workshop – Maximizing the Luminosity of Hadron Colliders at 10 – 100 TeV*, Erice, Sicily (Nov.,1991)
4.5 Gluckstern, R., "Longitudinal Coupling Impedance of a Liner with Holes inside a Beam Pipe", *Proceedings of the 19th ELOISATRON Workshop – Maximizing the Luminosity of Hadron Colliders at 10 – 100 TeV*, Erice, Sicily (Nov.,1991)
4.6 Mathewson, A., Private communication, 1991
4.7 Bintinger, D., Limon, P., and Rosenberg, R., **A7**, J. Vac. Tech. p.59 (1989)
4.8 Hillert, N. and Calder, R., "Ion Desorption of Condensed Gases", *Proc. 7th Inter. Vac. Congress*, Vienna, 1977
5.1 Chou, W. and Peterson, J., Private communication, 1992

# PERFORMANCE LIMITATIONS BEYOND LHC/SSC

J. Gareyte,
*CERN, SL Division*
*Geneva, Switzerland*

ABSTRACT

The detailed designs of both the LHC and the SSC have shown that luminosities in the $10^{33}$ cm$^{-2}$ s$^{-1}$ - $10^{34}$ cm$^{-2}$ s$^{-1}$ range can be obtained. Extrapolating present knowledge and technology to higher energy reveals the existence of luminosity limitations. The beam-beam effect allows luminosities of up to a few $10^{35}$ cm$^{-2}$ s$^{-1}$ at 100 TeV. However, other limitations, some technological, others of a more fundamental nature, take precedence over beam-beam at these extreme energies. They interplay is discussed.

## 1. Introduction

Design studies for the LHC[1] and SSC[2] have shown that luminosities ranging from a few $10^{33}$ to a few $10^{34}$ cm$^{-2}$s$^{-1}$ can be obtained. This already puts a severe demand on physics detectors, since at this level each bunch collision generates several events, and also because the collisions themselves are a powerful source of damaging radiation. The possible detector limitations will be ignored in the following, to concentrate the discussion on accelerator issues related to very high luminosity in hadron colliders. The case of the 100 TeV x 100 TeV Eloïsatron proton collider will be considered for illustration purposes.

The luminosity is given by

$$L = \frac{N^2\gamma}{4\pi\varepsilon_n\beta^*\Delta t} \tag{1}$$

where N is the number of particles per bunch, $\gamma$ the relativistic factor, $\varepsilon_n = \gamma\sigma^2 / \beta$ the normalized emittance, $\beta^*$ the betatron function at the interaction point and $\Delta t$ the bunch spacing in seconds.

The most fundamental limitation of a collider, the beam-beam effect, will be examined first. Problems related to the large beam currents which are necessary to reach high luminosity, essentially synchrotron radiation and collective instabilities will then be discussed.

## 2. Previous experience on the beam-beam effect

The effect of the beam-beam interaction in hadron colliders has been extensively studied in the CERN $\mathrm{Sp\bar{p}S}$[3] and in the Fermilab Tevatron[4], two machines which operate with a few dense bunches colliding head-on in 3 to 12 crossing points.

The most significant effect of this interaction is a slow diffusion of large amplitude particles into the tails of the transverse distributions. This eventually leads to particle losses which decrease the beam lifetime and create background in the experimental detectors. In the future colliders the occurrence of such losses in the superconducting part of the machine could lead to magnet quenches and jeopardize the operation. This diffusion is created by isolated high order betatron resonances excited by the strongly non linear beam-beam force. The resonance condition is given by

$$n_x Q_x + n_z Q_z = p \qquad (2)$$

where $n_x$, $n_z$ and p are integers and $Q_x$ and $Q_z$ are the horizontal and vertical tunes. The excitation strength of the resonances is proportional to the beam-beam parameter, which is for a round beam

$$\xi = N r_p / 4\pi\varepsilon_n \qquad (3)$$

where $r_p$ is the classical proton radius. In addition, the resonance strength depends on the particle amplitude expressed in r.m.s beam size of the opposing beam, as shown in Fig. 1.

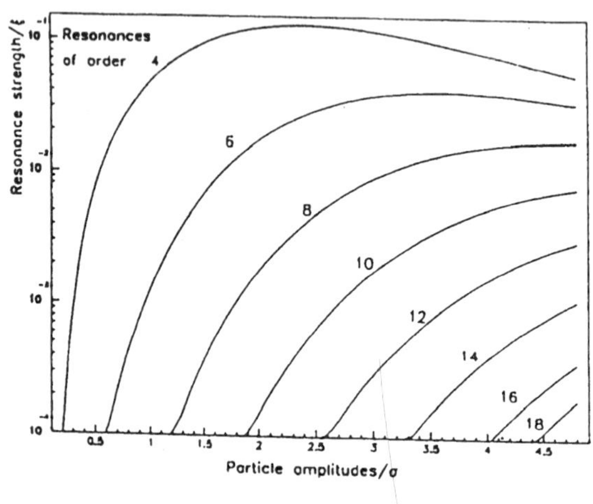


Fig. 1 Strength of high order beam-beam induced resonances as a function of the particle amplitude normalized to the r.m.s. radius of the opposing beam.

Resonances of high order ($n = |n_x| + |n_z|$) are only dangerous for large amplitude particles, and for a given amplitude the strength increases rapidly when resonances of lower and lower order are considered. This explains why the parameter which is of primordial importance in the beam-beam interaction is the resulting total tune spread, because it determines the lowest order of the resonances which cannot be avoided in the tune diagram.

In the $Sp\bar{p}S$ operating at values of $\xi$ between 0.003 and 0.006 and with 3 crossing points resonances of order 10 or less produce a strong diffusion which reduces the beam lifetime to unacceptably low values, while the influence of resonances of order 13 and 16, which cannot be avoided, is clearly noticed. In the Tevatron, resonances of order 12 seem tolerable, but at a smaller value of $\xi$ than in the $Sp\bar{p}S$. Considering this experience, it seems prudent to base the design of the future colliders on a total tune spread in the beam smaller than 0.02; this is about the largest value which allows operation outside resonances of order 12 or less, without going extremely close to integer tune values. The future colliders will operate with a large number of closely spaced bunches and in order to prevent them from colliding in many places all along that portion of the trajectories which is common to both beams around each interaction region, the beams cross at a small angle $\phi$. This slightly reduces the luminosity by a factor $(1 + (\phi.\sigma_s/2\sigma)^2)^{1/2}$ where $\sigma_s$ is the r.m.s. bunch length and $\sigma$ the r.m.s beam radius at the collision point. It provides a separation of the order of 6 to 7 $\sigma$ of the beam trajectories, sufficient to prevent particle collisions but not to suppress the electromagnetic interactions between the bunches, the so-called long range interactions.

The beam-beam resonances are excited mainly by the quasi head-on collisions and in this respect the situation in the future colliders will be very similar to that of the $Sp\bar{p}S$, since both the beam-beam parameter and the number of interaction points are similar in the two cases. However there is a difference as far as the total tune spread is concerned. In the $Sp\bar{p}S$ the tune spread is roughly $\xi$ times the number of collision points, whereas in the future colliders an additional contribution comes from the long range interactions. This is illustrated in Fig. 2 for the case of the LHC.

The combination of the long range and the head-on interactions produces an average tune shift which can be compensated by adjusting the machine tune, and a tune spread which has to be accommodated in between the dangerous resonances. The situation is complicated by the existence of gaps in the beam azimuthal distribution which are required by the finite risetimes of the injection and abort kickers. A few bunches at the beginning and at the end of the bunch trains suffer only half the long range interactions; these edge bunches are therefore displaced in the tune diagram, and as a consequence the total area occupied by the particles is increased. This effect will be neglected in the following to simplify the analysis, but it should be included in a detailed design.

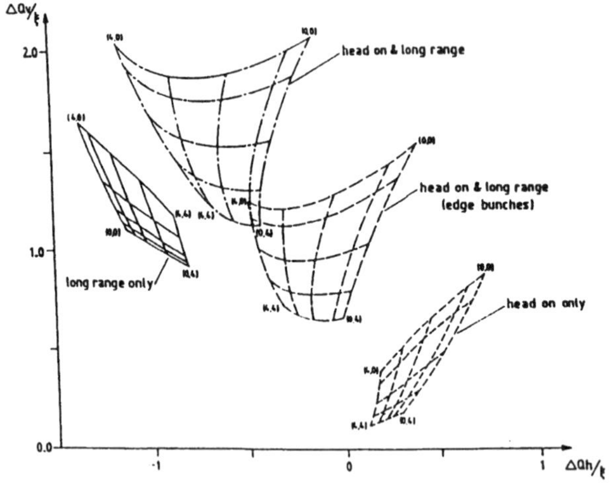


Fig. 2 Beam-beam induced tune shifts per interaction normalized to ξ for the LHC.

## 3. Beam-beam limitations in future colliders

The total allowed tune spread has to be shared by the various interaction points. As a consequence the luminosity attainable in a hadron collider depends on the number of experiments simultaneously active, and is maximum in the case of a single experiment. It is proposed to concentrate the following discussion on the potential of hadron colliders in the case of three interaction regions.

We will be conservative and limit the maximum tune shift parameter to 0.01 and the total tune spread to 0.015. This leaves 0.005 for the tune spread arising from the long range interactions, which is given by

$$\Delta Q_{LR} = \frac{3 r_p n^2 \varepsilon_n N D}{2\pi c \gamma^2 \beta^{*2} \phi^4 \Delta t} \tag{4}$$

where n is the particle amplitude at which ΔQ is evaluated divided by the r.m.s. beam radius, D the effective distance over which the two beams interact and c the velocity of light.

Using relations (3) to express $N/\varepsilon_n$ and (4) to express $N/\Delta t$ the luminosity can be written

$$L = \frac{2}{3}\frac{\pi c}{r_p^2}\,\xi\,\Delta Q_{LR}\,\frac{\gamma^3 \beta^* \phi^4}{n^2 \varepsilon_n D} \tag{5}$$

The values of D and $\beta^*$ depend both on the maximum gradient achievable in the quadrupoles of the final focusing triplet, and on the length of the experimental set up. With the maximum quadrupole gradient of 250 T $m^{-1}$ allowed by present technology and a free length for experiments of 20 m the following approximate scaling laws apply in the range of the LHC and SSC energies :

$$D \approx 100\ [\gamma/10^4]^{1/2}\ \mathrm{m} \tag{6}$$

$$\beta^* \approx 0.3\ [\gamma/10^4]^{1/2}\ \mathrm{m} \tag{7}$$

It will be assumed that these relations hold up to 100 TeV.

The crossing angle $\phi$ is limited by geometrical considerations. Assuming an inner coil diameter of about 40 mm in the final focusing quadrupoles the separation between the two beam centres at a distance D/2 from the interaction point should not exceed 10 mm, which gives a maximum possible value

$$\phi = \frac{0.02}{D} \tag{8}$$

Using relations (5) to (8), $\xi = 0.01/3$ and $\Delta Q_{LR} = 0.005/3$ for n = 4 gives

$$L = 4.5\ 10^{24}\ \frac{\gamma}{\varepsilon_n}\ \mathrm{cm}^{-2}\ \mathrm{s}^{-1}\ . \tag{9}$$

Assuming $\varepsilon_n = 1\ 10^{-6}$ m, which is the value considered in the SSC design, this formula gives a maximum beam-beam limited luminosity of 4.8 $10^{35}$ $\mathrm{cm}^{-2}\ \mathrm{s}^{-1}$ for a 100 TeV machine. As will be shown later, such a high value of the beam-beam limited luminosity is difficult to attain due to other limitations.

Operating the machine at a lower energy requires the value of $\beta^*$ to be increased proportionally to $\gamma^{-1}$ to keep the divergence of the beam in the interaction region constant. This divergence is given by $n[\varepsilon_n / \gamma\beta^*]^{1/2}$ for a particle at $n\sigma$ and is limited by the maximum possible crossing angle. As a consequence the luminosity decreases like $\gamma^2$ at lower energies, as indicated in Fig. 3.

## 4. Synchrotron radiation

The total power radiated by the two beams is

$$P_s = \frac{2}{3}\frac{Z_o e^2 c N \gamma^4}{\rho\,\Delta t} \tag{10}$$

where $Z_0$ is the impedance of free space, e the elementary charge, c the velocity of light and $\rho$ the bending radius. Since the maximum magnetic field B is fixed by the technology available the following formulation is also of interest

$$P_s = \frac{2}{3} Z_o \frac{e^3}{m}\frac{N}{\Delta t} B\gamma^3 \tag{11}$$

where m is the proton mass.

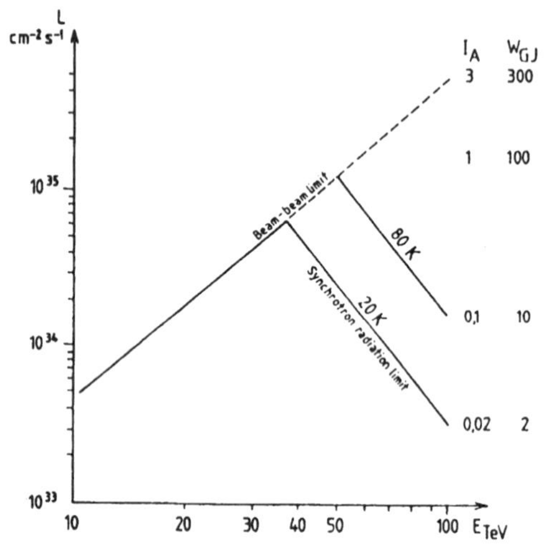


Fig. 3 Luminosity limits in a 100 TeV x 100 TeV proton collider using 10 T magnets and $\varepsilon_n = 1\ 10^{-6}$ m. The beam-beam limit is for three interaction regions and the synchrotron radiation limit corresponds to a maximum cooling power of 60 MW for the two beams and different temperatures of the radiation shield. The beam current I and stored energy W corresponding to the luminosity at 100 TeV are indicated on the right hand side.

For several reasons, some of which will be explained later, this power will probably have to be absorbed at cryogenic temperatures. In this case the total cooling power necessary is $P_s/\eta$ if $\eta$ is the efficiency of the refrigeration system. Typical values for $1/\eta$ are 250 if the synchrotron radiation is absorbed at 5 K, 60 if it is absorbed at 20 K and 12 if it is absorbed at 80 K. It is therefore interesting to provide a radiation shield at a temperature higher than that of the superconducting coil, as is done in the LHC where the coil is at 2K. A higher limit of the temperature of the shield is given by considerations of heat transfer to the coils, pumping problems and the necessity explained below the provide a beam pipe with a low resistivity.

Financial and social considerations will limit the total power consumed by the collider. Let us assume for illustration purposes that the fraction of power used for the absorption of synchrotron radiation is limited to 60 MW. Then the beam current is limited according to formula (10), so that at 100 TeV the attainable luminosity is reduced to the values indicated in Fig. 3 for various temperatures of the radiation shield. At lower values of the energy the current is allowed to increase like $\gamma^4$, and the luminosity as $\gamma^3$.

A lower value of the magnetic field is often advocated to reduce the synchrotron radiation power (formula (11)). This has the additional advantage of a cheaper technology, but increases the machine size, and for a given luminosity the stored beam energy.

## 5. Collective instabilities

Attaining the beam-beam limited luminosity indicated by equation (9) requires very high beam intensities. Assuming that such beam intensities can be furnished by the injectors at reasonable cost, a number of potential instabilities have to be mastered in the collider itself.

These instabilities arise from the electromagnetic interaction of the beam with the surrounding structure. The main contributions to the beam-environment coupling impedance come from the finite resistivity of the beam pipe, from the bellows, monitors, kickers, and from the RF cavities. The surface resistivity of the beam pipe can be lowered by copper-coating and maintaining the pipe walls at low temperature, while the bellows and kickers can be partially shielded from the beam electromagnetic fields. The coupling impedance of the RF structures can be considerably lowered by using superconducting cavities with a large bore. All these measures are being applied to the LHC, with the aim of reducing the coupling impedance to about 0.7 $\Omega$ for the resistive wall and 0.1 $\Omega$ for the broad-band component. In this case the thresholds of the single bunch instabilities, which are determined by the broad-band component, are of the order of a few $10^{11}$ protons per bunch and these effects do not limit the performance. The growth rates of the coupled bunch modes which are excited by the beam pipe resistivity depend on the total beam current. These modes are unstable for large beam currents and must be damped by feedback systems. Whereas this is relatively easy in the LHC where the resistive-wall growth time is about 300 turns, it may pose a problem in larger machines. Let us examine how this phenomenon scales with the machine radius.

For a sufficiently large conductivity of the beam enclosure the wall current $I_w$ induced by the beam is equal to the beam current $I_b$. The power dissipated in the wall is therefore

$$Pw = \frac{2\pi R}{2\pi b} \Re\text{surf.}\ I_b^2 \qquad (12)$$

with the surface resistance $\Re\text{surf.} = \rho/\delta$ , $\rho$ being the resistivity and $\delta = (2\rho\ /\ \mu o\omega)^{1/2}$ the skin depth at the frequency $\omega$. This power must be extracted by the cooling system, and adds to the cryogenic load.

The above expression can also be used to calculate the longitudinal coupling impedance defined by $Z_L = P/\ I_b^2 = \Re\text{surf.}\ R/b$. In addition, if the beam is not centered, differential return currents are created on both sides of the pipe, and this can in turn enhance the original transverse motion, thus creating an instability. Normalized to the revolution period T, the growth time of this transverse resistive-wall instability is [5]

$$\frac{\tau}{T} = \frac{2E/e}{\beta Z_T I}\ , \qquad (13)$$

where E/e is the beam energy in electronvolts, and $Z_T$ is the transverse coupling impedance related at frequency $\omega$ to the longitudinal coupling impedance by

$$Z_T = \frac{2c}{b^2}\frac{Z_L}{\omega}\ . \qquad (14)$$

The fastest growth is achieved for the lowest frequency $\omega = cq/R$ where q is the distance of the machine tune to the next integer. Everything else being equal, the transverse impedance scales like $\rho^{1/2}\ R^{1.5}\ /\ b^3$. For a fixed bending field the energy scales like R and the average beta function is likely to grow approximately like $R^{1/2}$. As a consequence the normalized growth time scales like $b^3/R\rho^{1/2}$. In large, high energy machines, the resistive-wall instability becomes a serious problem, since in addition there is a strong incentive to reduce b in these machines for economical reasons. When the instability grows significantly in one revolution period, it becomes difficult to damp it efficiently with a feedback system. Therefore the resistivity of the beam pipe has to be as low as possible, and this can be achieved by using a pure metal like copper and cooling it below 20K, as can be seen on Fig.4. Then one has the problem of absorbing synchrotron radiation at low temperature.

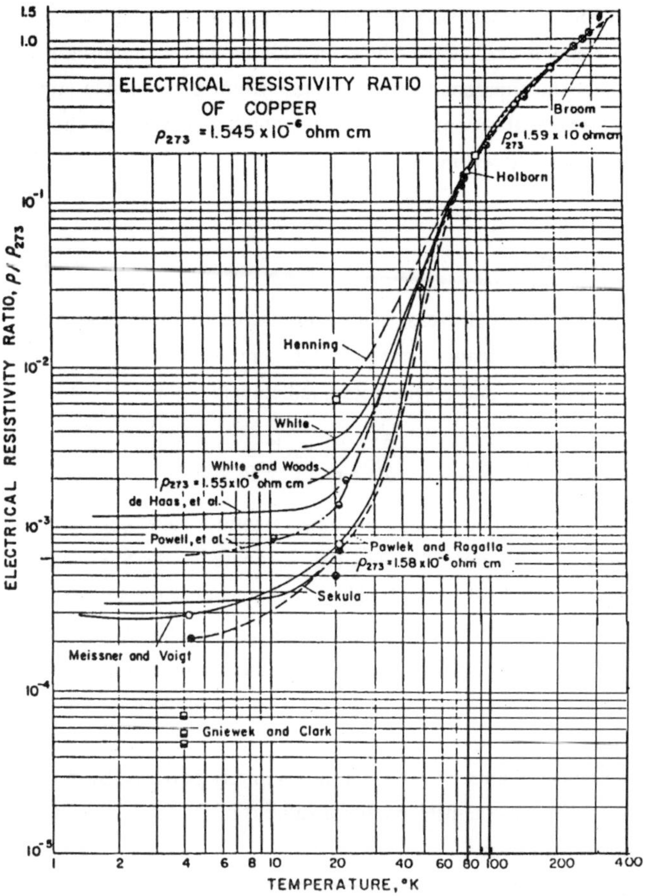


Fig. 4 Resistivity of Copper at low temperature for samples of different levels of purity. There is a strong incentive to cool the beam screen down to 20 K to profit from the low value of the resistivity of copper at this temperature.

## 6. Beam stored energy

The energy stored in each beam is

$$W = \frac{2\pi R\, E I}{c}. \tag{15}$$

For the LHC and SSC W reaches about 500 MJ, and in these machines the design of the beam dumping systems is already not trivial. In addition an elaborate set of collimators is necessary to protect the superconducting magnets from unavoidable continuous beam losses. These problems will become much worse at higher energies and currents, and will certainly limit the achievable performance.

## 7. Luminosity lifetime

The luminosity may decrease with time due to the growth of beam transverse dimensions and the loss of beam currents. In high energy hadron colliders the damping times of the beam emittances due to synchrotron radiation is

$$\tau_\varepsilon = \frac{3}{2}\,\frac{m}{e\,r_p}\,\frac{R}{J_x\,\gamma^2 B} \tag{16}$$

where $J_x$ is the damping partition number and other quantities have been defined before.

This amounts to 18 hours for the LHC, 12 hours for the SSC and 1 hour for our 100 TeV example using 10 T magnets. It is judicious to choose machine parameters so that the growth times of the emittances due to various phenomena like intrabeam scattering stay well above the radiation damping times. In this case, there is no emittance growth, and very long lifetimes can be expected at small enough values of the luminosity. The SSC operates in such a regime.

However, at high luminosity the lifetime is entirely governed by the loss of particles in the beam-beam collisions themselves. For a value $\sigma$ of the total cross-section the initial decay time of the beam intensity is

$$\tau = N_o \,/\, L\sigma\, k_x \tag{17}$$

where $N_o$ is the initial total number of particles, and $k_x$ the number of interaction regions. The luminosity half life is

$$t_{1/2} = \left(\sqrt{2} - 1\right)\tau. \tag{18}$$

In the case of the LHC, with $L = 1.7\ 10^{34}\ cm^{-2}\ s^{-1}$, $\sigma = 10^{-25}\ cm^2$ and $k_x = 3$ one finds $t_{1/2} = 11$ hours, which is still acceptable. However, one sees that any proposal to increase the luminosity without increasing the current, if at all possible, would result in a too low lifetime. This consideration may prove relevant in certain design strategies for high luminosity high energy colliders : at a given value of the luminosity a minimum beam intensity is required to assure an acceptable lifetime.

## 8. Conclusion

The design of Future Hadron Colliders rests firmly on solid foundations, due to the experience and the knowledge acquired in a long chain of precursor machines. With the LHC and the SSC one enters a new domain of complexity and performance, and the challenge for machine builders and operators is exciting. Looking farther in the future one can contemplate even higher energies and higher luminosities using technologies similar to those developed for the and SSC. However, several phenomena tend to limit the performance of very high energy colliders below that authorized by the beam-beam effect alone. These limitations become very serious for a 100 TeV collider at luminosities around $10^{35}$ cm$^{-2}$ s$^{-1}$.

## REFERENCES

[1] CERN 91-03, Design Study of the Large Hadron Collider, 2 May 1991.

[2] SSC-SR-2020. The Superconducting Supercollider Conceptual Design, March 1986. A publication of the SSC Central Design Group, LBL, Berkeley, CA 94720 (U.S.A.).

[3] L. Evans, J. Gareyte, M. Meddahi, R. Schmidt, Beam-beam effects in the strong-strong regime at the CERN SPS. Proc. 1989 IEEE Particle Accelerator Conference, March 20-23, 1989 Chicago, Il. p.1403.

[4] M. Harrison, R. Schmidt, The performance of proton antiproton colliders. Second EPAC, June 1990, Nice, France.

[5] F. Sacherer, Theoretical Aspects of the Behaviour of Beams in Accelerators and Storage Rings. CERN 77-13, pp 198-265.

# STRATEGIES FOR THE ELOISATRON BASED ON BEAM-BEAM AND SYNCHROTRON RADIATION CONSIDERATIONS

J. Gareyte,
*CERN, SL Division*
*Geneva, Switzerland*

ABSTRACT

By fixing the value of a few parameters which are severely constrained anyway, one can simplify considerably the discussion on the main Eloïsatron parameters. A few features linked to the choice of the magnetic field are discussed.

## 1. Introduction

From the discussions at this workshop, it appears that the incertitude about some of the main Eloïsatron parameters is still very large, while other parameters can already be fixed with a good precision. Among the parameters which suffer large variations are the value of the main magnetic field and the amount of synchrotron radiation power which can be absorbed, these two numbers being closely related. Among the parameters which cannot vary much are the bunch spacing $\Delta t$, the betatron function at the collision point $\beta^*$, and the length D of the trajectory around the interaction points common to both beams. By fixing these parameters to $\Delta t = 15$ ns, $B^* = 1$ m, D = 316 m, one can considerably simplify the search for an optimum in the other parameters. Of course a value which is almost fixed is the maximum luminosity, which should approach $10^{35}$ cm$^{-2}$ s$^{-1}$.

Here we will discuss luminosity versus magnetic field B and synchrotron radiation power $P_s$ in the spirit of ref. 1. We will consider two extreme values of the field, B = 2T and B = 10T, and three values of the synchrotron radiation power $P_s$ = 1 MW, 5 MW and 20 MW.

## 2. Luminosity limitations

### *2.1 Beam-beam effects*

The total beam-beam tune spread is limited to 0.015. We will show below that the contribution of the long range effects can always be made smaller then 0.005 in the Eloïsatron, and therefore we can safely ascribe $\Delta Q = 0.01$ to the head-on collisions. We consider the case of 3 collision points and therefore

$$\xi = \frac{r_p}{4\pi}\frac{N}{\varepsilon_n} = 0.0033 \qquad (1)$$

where $r_p$ is the classical proton radius, N the number of particles per bunch and $\varepsilon_n$ the normalized emittance. This relation links the emittance to the beam current $I = eN/\Delta t$ :

$$\varepsilon_n = 3.4\ 10^{-6}\ I\ \ (m) \qquad (2)$$

with $\Delta t = 15$ ns.

The luminosity is given by

$$L = \frac{\xi}{er_p} \frac{\gamma}{\beta^*} I \tag{3}$$

which gives for $\beta^* = 1$ :

$$L = 1.45\ 10^{35}\ I\ \left(cm^{-2}s^{-1}\right) \tag{4}$$

The long range beam-beam tune spread is

$$\Delta Q_{LR} = \frac{3r_p}{2\pi ce} \frac{n^2}{\gamma^2} \frac{D}{\beta^*} \frac{\varepsilon_n}{\varnothing^4} I \tag{5}$$

where c is the speed of light, n the particle amplitude divided by the R.M.S. beam width (n = 4) and $\varnothing$ the crossing angle.

with D = 316 m one has :

$$\Delta Q_{LR} = 6.8\ 10^{-15} \frac{\varepsilon_n I}{\varnothing^4} \tag{6}$$

and the minimum crossing angle which gives $\Delta Q_{LR} \leq 0.005/3$ is using (2) and (4),

$$\varnothing_{min} = 51 \left(\frac{L}{10^{35}}\right)^{1/2} \mu rad. \tag{7}$$

Now if we allow a maximum beam separation of 1 cm due to the crossing angle the maximum tolerable angle is

$$\varnothing_{max} = \left(\frac{0.02}{316}\right) = 63\ \mu rad. \tag{8}$$

Therefore up to a luminosity of $10^{35}$ cm$^{-2}$ s$^{-1}$ the tune spread due to the long range beam-beam effect stays below 0.005 for 3 crossings. This justifies the procedure used above, which considerably simplifies the scaling.

*2.2 Synchrotron radiation power limitations*

From ref.1 :

$$P_s = \frac{2}{3} Z_o \frac{e^2}{m} \gamma^3 I B \tag{9}$$

where $Z_0$ is the impedance of free space, and e and m are the charge and mass of the proton.

In our case :

$$P_s = 4.7\ 10^6\ I\ B \tag{10}$$

and using (4),

$$L = 3.1\ 10^{34}\ \frac{P_s(MW)}{B} \tag{11}$$

Using the above formulae one can build table I to illustrate the parameters of the Eloïsatron operating at the beam-beam limit at 100 teV.

Table I

| B | Ps (MW) | 1 | 5 | 20 |
|---|---|---|---|---|
| 10T | L ($cm^{-2}\ s^{-1}$) | $3.1\ 10^{33}$ | $1.5\ 10^{34}$ | $6.2\ 10^{34}$ |
| | $\varepsilon_n$ ($10^{-6}$m) | .07 | .35 | 1.4 |
| 2T | L ($cm^{-2}\ s^{-1}$) | $1.5 10^{34}$ | $7.7\ 10^{34}$ | $3.1\ 10^{35}$ |
| | $\varepsilon_n$ ($10^{-6}$m) | .35 | 1.8 | 7.3 |

In the case of a 10T field, the emittance required is very small and probably cannot be obtained from the injectors. However, the synchrotron damping time is of the order of 1 hour and the beam-beam limit will be reached naturally after a while. After that one must provide a mechanism to prevent a further decrease of the transverse emittance to avoid a catastrophic evaporation of the beams (see below).
In case of a 2T field the emittances can be obtained from the injectors.

### 3. Beam-beam and synchrotron radiation damping

If the Eloïsatron uses a large value of the magnetic field, both transverse and longitudinal emittances will decrease fast with respect to the beam lifetime, due to synchrotron radiation damping (emittance damping time is one hour for a field of 10T). This can be used to obtain small emittances or small bunch lengths, but may also have dramatic consequences.

In the longitudinal direction, a too large line density may increase intra-beam scattering, provoke longitudinal instabilities (turbulence) or transverse instabilities (transverse mode coupling). While the longitudinal effects are usually self-stabilizing, the transverse instabilities may destroy the beam beyond a certain density threshold.

In the transverse planes, the synchrotron damping is far from sufficient to counteract the fast diffusion in the tails of the distributions, at, say, 3 to $5\sigma$, which arises from the beam-beam interaction. In the CERN SPS collider one observed diffusion rates of the order of one R.M.S. beam size per few seconds or per few minutes depending on the conditions and on the starting amplitude. There is every reason to believe that a similar phenomenon will affect the Eloïsatron operating close to the beam-beam limit. Therefore the danger is that due to synchrotron damping the core of the beams shrink (there is no fast diffusion due to beam-beam forces in the core) and the beam-beam limit is exceeded, increasing dramatically the diffusion rate in the tails. The beams would in this case "evaporate" completely. Some mechanism has to be found to control the beam size in presence of synchrotron damping.

## 4. References

1. J. Gareyte, Performance limitations beyond LHC/SSC, these proceedings.

# SCALING LAWS AND COST OPTIMIZATION FOR HADRON COLLIDERS AROUND 100 TeV

EBERHARD KEIL
*CERN, Geneva, Switzerland*

ABSTRACT

The scaling laws and performance of hadron colliders at an energy of about 100 TeV are derived in terms of only seven physics and engineering parameters. A cost optimization which balances the cost of the cryogenics plant needed to cope with the synchrotron radiation power and the cost of the collider tunnel and arcs is used to determine the optimum field in the dipoles in the arcs and the total cost of the collider. It is found that the optimum dipole field may be much smaller than the maximum field which can be achieved in superconducting dipoles.

## 1. Introduction

This paper discusses the scaling laws and cost optimization for hadron colliders in an energy range up to about 100 TeV. Using a few simplifying assumptions which should be reasonable approximations, the equations for the parameters and performance of such colliders are expressed in terms of only seven physics and engineering parameters in Chapter 2. The scaling laws for these parameters and several derived quantities with energy are obtained in Chapter 3. The scaling laws for the dipole field in the collider arcs is obtained by cost optimization arguments in Chapter 4. Specific parameters for the Eloisatron project[1] at 100 TeV are given in Chapter 5.

## 2. Machine Parameters and Performance

This chapter contains basic relations between the engineering parameters of a circular pp collider and its performance which were already given earlier.[2] To simplify the presentation, the following assumptions are made:

- The two counter-rotating beams are bunched and collide head-on.
- The dispersion vanishes at the crossing points in both planes and the rms beam radius $\sigma$ there is only determined by the emittance.
- The horizontal and vertical emittances are the same.
- Equal horizontal and vertical amplitude functions $\beta_t$ at the crossing points.
- The bunchlength is small compared to $\beta_t$.

These assumptions are a reasonable approximation to reality. The results arrived at below should not change much if more realistic assumptions were made.

### *2.1. Physics and Engineering Parameters*

The equations derived below will be manipulated such that they are expressed in terms of seven "physics" and "engineering" parameters. Typical values used in the examples are shown below.

- energy expressed $E = 100$ TeV in terms of $\gamma$
- luminosity $L = 10^{35}$ cm$^{-2}$s$^{-1}$
- bunch spacing $s$
- beam-beam tune shift parameter $\zeta = 0.003$
- amplitude function $\beta_t = 1$ m at the interaction points
- dipole field $B = 2 \ldots 10$ T
- dipole filling factor $2\pi\rho/C = 0.765$ as in the SSC

These parameters are not necessarily independent. Indeed, it will turn out later that they are not. In the following, $\xi$ and $2\pi\rho/C$ will be assumed independent of energy. The scaling of $L$, $s$, $\beta_t$, and $B$ will be determined.

### *2.2. Luminosity and Beam-Beam Tune Shifts*

With the assumptions made above, the beams are round at the collision point. Hence, the beam-beam tune shift parameters $\xi$ are the same in both planes:

$$\xi = \frac{N r_p \beta_t}{4\pi\gamma\sigma^2} \tag{1}$$

Here $N$ is the bunch population, $r_p$ is the classical proton radius, $\gamma$ is the usual relativistic factor, and $\sigma$ is the rms beam radius at the interaction points. The luminosity $L$ is given by:

$$L = \frac{N^2 k f}{4\pi\sigma^2} \tag{2}$$

Here $f$ is the revolution frequency, and $k$ is the number of bunches in one beam. With respect to[2] the meaning of $N$ has changed in Eqs. (1) and (2). By eliminating one power of $N$ from (2) by using (1) one arrives at the standard equations:

$$L = \frac{N k f \xi\gamma}{r_p\beta_t} = \frac{I_b k \xi\gamma}{e r_p \beta_t} \tag{3}$$

where $I_b$ is the bunch current, and $e$ in the proton charge. By solving these equations for the bunch population $N$ and the total current in one beam $I_b k$, respectively, and replacing the bunch collision frequency $fk$ by the obvious expression $fk = c/s$ where $s$ is the bunch spacing, and $c$ is the velocity of light, the following relations are obtained:

$$N = \frac{r_p}{c}\frac{L\beta_t s}{\gamma\xi} \qquad\qquad I_b k = e r_p \frac{L\beta_t}{\xi\gamma} \tag{4}$$

Both expressions are independent of any parameter related to the size of the collider.

*2.3. Emittance*

Reaching the beam-beam limit $\xi$ at the bunch population $N$ implies that the normalized emittance, defined as $\epsilon = \pi\gamma\sigma^2/\beta_t$, takes a value given by (1), which can be written as follows:

$$\epsilon = \frac{r_p^2}{4c}\frac{Ls\beta_t}{\gamma\xi^2} \tag{5}$$

With respect to[2] the numerical factor in the definition of $\epsilon$ has changed.

*2.4. Synchrotron Radiation*

The synchrotron radiation loss per turn $U_s$ is given by:

$$U_s = (4\pi/3)r_p c\gamma^3 B \tag{6}$$

The synchrotron radiation power for one beam $P_s = U_s I_b k$ can be brought into the following form:

$$P_s = \frac{4\pi r_p^2 ce}{3}\frac{\gamma^2 BL\beta_t}{\xi} \tag{7}$$

The synchrotron radiation power becomes $P_s = 1.795B[\mathrm{T}]$ MW with the typical parameters.

## 3. SCALING LAWS

*3.1. Insertion Scaling*

A scaling law for the amplitude function $\beta_t$ at the crossing point[3] was generalized later[2]: An existing low-$\beta$ insertion operating up to an energy $E_1$ may be scaled to a higher energy $E_2$ by increasing the amplitude functions through the insertion and the lengths of the quadrupoles and drift spaces in the ratio $\sqrt{E_2/E_1}$, and keeping constant all quadrupole gradients $G$ and apertures $a$. The physical acceptance $\pi a^2/\hat{\beta}_t$ decreases in the ratio $\sqrt{E_1/E_2}$, while the normalized physical acceptance $\alpha = \pi\gamma a^2/\hat{\beta}_t$ increases in the ratio $\sqrt{E_2/E_1}$, where $\hat{\beta}_t$ is the maximum value of the amplitude function $\beta$ in the insertion. The chromaticities of both insertions are the same since the energy dependencies in the chromaticity integrals cancel.

*3.2. Lattice Scaling*

The same scaling law may also be applied to the FODO cells in the arcs of the collider. Looking at the focusing first, we find that a FODO cell operating up to an energy $E_1$ can be scaled to a higher energy $E_2$ at constant quadrupole gradient $G$ and constant quadrupole aperture $a$ by increasing all lengths in the ratio $\sqrt{E_1/E_2}$. This applies in particular to the cell length $L_p$, the quadrupole length $l_Q$, the $\beta$-functions through the cell, and the normalized physical acceptance $\alpha$. The standard equations for a thin-lens FODO cell[4] can be manipulated[2] to arrive at the following expression for the cell length $L_p$:

$$L_p^2 = \frac{8E_p}{ce} \frac{\gamma \sin \mu/2}{GC_Q(1-\rho/R)} \tag{8}$$

Here, $E_p/e$ is the rest voltage of the proton, $R$ is the average radius of the arc cells, and $C_Q(1-\rho/R)$ with $0 < C_Q < 1$ is the fraction of the cell length which is occupied by quadrupoles. Eq. (8) confirms that at constant $C_Q$, $G$, $\mu$, and $\rho/R$, the cell length $L_p$ varies like $\sqrt{\gamma}$.

Similar manipulations of the equations for a FODO cell with thin dipoles and quadrupoles yield an expression for the extreme values of the dispersion $D$:

$$D = \frac{B}{GC_Q(R/\rho - 1)} \frac{2 \pm \sin \mu/2}{\sin \mu/2} \tag{9}$$

The dispersion $D$ depends on the dipole field $B$ and quadrupole gradient $G$, but is independent of the energy. Hence, if the cell length $L_p$ is scaled like $\sqrt{E}$, and if the dipole field $B$ and quadrupole gradient $G$ are kept constant, the dispersion will be the same in the two collider arcs. In this case the "physical momentum aperture" $\delta = a/D$ will also be independent of the energy. If the dipole field $B$ decreases, then $\delta$ will increase with energy. The scaling law for the circumference $C$ is:

$$C = \frac{2\pi E_p}{ce} \frac{\gamma}{B} \frac{C}{2\pi\rho} \tag{10}$$

The ratio $2\pi\rho/C < 1$ is the fraction of the circumference $C$ occupied by dipoles. With the typical parameters, the circumference becomes $C = 2738/B[\mathrm{T}]$ km.

### *3.3. Scaling of Emittance and Bunch Spacing*

The argument has been made that the luminosity of a pp collider must increase like the square of the energy because the cross section for the interesting physics drops like $1/E^2$. As discussed in Section 3.1, the amplitude function $\beta_t$ scales like $\sqrt{E}$. With these assumptions it follows from Eq. (5) that the normalized emittance scales with energy like $\epsilon \propto E^{3/2}s$. Unless the bunch spacing $s$ is reduced in proportion to $1/E$, the normalized emittance $\epsilon$ grows more rapidly with energy than the normalized physical acceptance of both the low-$\beta$ insertions and the arcs, discussed in Sections 3.1 and 3.2. In order to avoid this, we assume in the following that the bunch spacing $s$ varies like $1/E$. A lower limit on $s$ is given by the fact that it must be a multiple of the RF wavelength. A further consideration entering into the choice of $s$ is the parasitic beam-beam collisions which occur at all distances from the interaction points which are a multiple of $\pm s/2$. If one imposes the condition that the lateral separation between the bunches at the parasitic collision points is a fixed number of rms beam radii, and assumes that $\beta_t \ll s/2$, one finds that the crossing angle $\phi$ is proportional to $(\epsilon/\beta_t)^{1/2}$.

Further discussions of this aspect can be found elsewhere in these proceedings.[5] With the scaling laws already obtained for $\epsilon$ and $\beta_t$, this implies $\phi \propto 1/\sqrt{\gamma}$.

With the above assumptions on the variations of $s$, $L$, and $\beta_t$ with $E$, it follows from Eq. (4) that the bunch population is proportional to $\sqrt{E}$. The beam brightness $N/\epsilon$ is independent of the energy. This implies that the injectors for pp colliders of increasing energy need not provide bunches of ever increasing brightness, but they do have to provide bunches of increasing population.

### *3.4. Summary of Scaling laws*

The scaling laws arrived at so far are summarized in Tab. 1. The flow of arguments is from top to bottom, and from left to right. The left column contains assumptions and requirements. The central column contains dimensional parameters which all have the same scaling law. The right column shows other derived quantities. It should be noted that the scaling law for the dipole field $B$ is not yet included.

Table 1: Scaling Laws with Energy $E$

| | | |
|---|---|---|
| $\xi \propto E^0$ | $\beta_t \propto \sqrt{E}$ | $s \propto E^{-1}$ |
| $2\pi\rho/C \propto E^0$ | $\alpha \propto \sqrt{E}$ | $N \propto \sqrt{E}$ |
| $L \propto E^2$ | $\epsilon \propto \sqrt{E}$ | $N/\epsilon \propto E^0$ |
| $a \propto E^0$ | $L_p \propto \sqrt{E}$ | $\phi \propto 1/\sqrt{E}$ |
| $G \propto E^0$ | | |

## 4. Cost Optimization

The synchrotron radiation power $P_s$ is proportional to the product $E^2BL\beta_t$. Using the scaling laws shown in Tab. 1, one finds that the synchrotron radiation power scales like the 9/2-th power of the energy, at constant dipole field $B$, reaching values in the range of several MW for typical parameters at 100 TeV. This variation is much steeper than that of the collider circumference $C$ which only scales in proportion to the energy $E$ at constant dipole field $B$.

In going from a collider at some energy, in which the synchrotron radiation power is already a non-negligible contribution to the cryogenic load, and in which the value of the dipole field $B$ is optimized to achieve a balance between the cost of the cryogenic system and the cost of the fully equipped tunnel, to another collider at a higher energy, it is inevitable that the cryogenic load – and presumably also its cost – rise much faster than the length of the tunnel equipped with magnets etc. – and its cost – when both colliders are designed to operate at the same dipole field $B$. Therefore the cost of the cryogenic system will be out of balance with the cost of the equipped tunnel. This balance can be re-established by reducing the dipole field $B$.

### 4.1. *Scaling Laws from Cost Optimization*

The aim of this section is to obtain the scaling laws of the parameters of a high-energy pp collider with the energy $E$, and in particular the dipole field $B$ which minimizes the total cost $K$, which is written in the following form:

$$K(E) = K_0(E) + K_1(E)B^{k_1} + K_2(E)B^{-k_2} \tag{11}$$

The constants $K_0$, $K_1$ and $K_2$, and the total cost $K$ are all functions of $E$. The cost contributions rising and falling with the largest power of $B$ have been isolated and written in the terms involving indices 1 and 2, respectively, while all cost contributions varying with intermediate powers of $B$, in particular cost contributions which are independent of $B$, are grouped into the $K_0$ term. They will have little effect on the optimum value of $B$. Very similar arguments were used by Richter[6] for the cost optimization of $e^+e^-$ storage rings. Finding the optimum value of $B$ is straightforward, by requiring that the derivative of $K$ with respect to $B$ vanishes. The result is:

$$B(E) = \left(\frac{k_2 K_2(E)}{k_1 K_1(E)}\right)^{1/(k_1+k_2)} \tag{12}$$

To make further progress, the cost contributions which increase and decrease most steeply with the dipole field $B$ must be identified.

### 4.2. *Cost Variation with Synchrotron Radiation Power*

The prime candidate for a cost contribution which increases most steeply with the dipole field $B$ is the synchrotron radiation power $P_s$ (7). In the following, it is assumed that this power must be absorbed in material at cryogenic temperatures. This assumption is justified by the observation that most of the wakefields must terminate on a cold surface with low enough resistivity.[7,8] Otherwise, the transverse resistive wall instability cannot be suppressed by a feedback system,[9] since its growth time would be much less than the revolution time. It appears very difficult to install "synchrotron radiation catchers" in the vacuum chamber and to connect them to heat sinks at much higher than cryogenic temperature without simultaneously increasing the transverse and longitudinal impedance beyond tolerable values. Controversial discussions of this issue can be found elsewhere in these proceedings.

Hence, the compressor power in the cryogenic plant is larger than $P_s$ by a large factor related to the Carnot cycle. Recent cryogenic plants reach about 30% of that efficiency.[10] Over the range from 1 to 20 kW at 4.2 K, the cost $C$ of recent cryogenic plants in Msfr is well approximated by the following scaling laws[11,12] with the power $P$ in kW:

$$C = 0.5 + 0.5P + \sqrt{P} \tag{13}$$

$$C = 2P^{0.7} \tag{14}$$

In the range quoted these scaling laws are indistinguishable. However, the two laws yield completely different results when the power of the cryogenic plant is extrapolated to higher powers over several orders of magnitude.

### 4.3. *Cost Variation with Circumference*

Looking for cost contributions which decrease most steeply with the dipole field $B$, we observe that the length of the arcs of the collider is $2\pi R$ where $R$ is their average radius with $\rho < R < C/2\pi$. It is approximately proportional to the reciprocal of $B$. Therefore, the cost of the fully equipped tunnel, including everything except the components of the collider arcs, will be taken to be proportional to $1/B$.

The cost of the components in the collider arcs is dominated by the cost of the dipoles. Consider the following three models for the cost of dipoles:

1. In the high field model, the dipole cost per running metre is independent of the dipole field $B$. In this case the dipole cost behaves like the cost of the fully equipped tunnel, and unit prices for the tunnel including the collider components can be used.

2. In the low field model, the dipole cost per running metre is proportional to the dipole field $B$. In this case, the total cost of the dipoles is independent of $B$, becomes a contribution to the $K_0$ term in Eq.(11), and does not enter into the cost optimization.

3. In the third model, the dipole cost per running metre rises more steeply than linearly with the dipole field $B$. In this case, the total cost of the dipoles increases with $B$, and belongs to the other part of the cost optimization. The dipole field $B$ in the SSC was optimized by balancing the cost of the dipoles against the cost of the tunnel. The fact that this was possible indicates that for the SSC the dipole cost rises more steeply than linearly with the dipole field $B$. However, the shallowness of the cost minimum indicates that the exponent of the dipole field was not much larger than unity. This model will not be considered further.

### 4.4. *Cost Optimization of Dipole Field*

The total cost $K$ is now expressed in the following form:

$$K(E) = K_0(E) + K_s\,(2P_s)^{k_s} + K_c C \tag{15}$$

Here $K_s$ and $K_c$ are the unit prices for synchrotron radiation power and circumference respectively. A factor two has been included in front of $P_s$ since there are two beams. The exponent $k_s$ allows to play with the cost variation of the cryogenic plant. Substituting from (7) and (10), we find:

$$K(E) = K_0(E) + K_s'\left(\frac{\gamma^2 BL\beta_t}{\xi}\right)^{k_s} + K_c'\frac{C}{2\pi\rho}\frac{\gamma}{B} \tag{16}$$

Here, the natural constants are included in $K_s'$ and $K_c'$. Optimizing the cost $K(E)$ leads to the following scaling law for $B$:

$$B \propto \left(\frac{K_c'}{K_s' k_s}\frac{C}{2\pi\rho}\right)^{1/(k_s+1)}\left(\frac{\gamma^{1/k_s-2}\xi}{L\beta_t}\right)^{k_s/(k_s+1)} \tag{17}$$

Specifically, the dipole field $B$ scales as follows with the physical parameters:

$$\begin{aligned} &B \propto \gamma^{-1/2}L^{-1/2}\beta_t^{-1/2} && \text{for } k_s = 1 \\ &B \propto \gamma^{-0.25}L^{-0.41}\beta_t^{-0.41} && \text{for } k_s = 0.7 \end{aligned} \tag{18}$$

With the scaling laws for $L$ and $\beta_t$ shown in Tab. 1, the dipole field $B$ scales like $E^{-7/4}$, and the cost like $E^{11/4}$ in the case $k_s = 1$, while $B$ scales like $E^{-1.27}$ and the cost like $E^{2.27}$ in the case $k_s = 0.7$. As one would expect, the cost contributions from synchrotron radiation and from the circumference scale in the same way with the energy. The reason for these steep variations with $E$ is the assumed variation of the luminosity proportional to $E^2$. If the luminosity is assumed to be constant, then $B$ scales like $E^{-3/4}$, and cost and circumference like $E^{7/4}$ in the case $k_s = 1$, while $B$ scales like $E^{-0.45}$ and circumference and cost like $E^{1.45}$ in the case $k_s = 0.7$. These scaling laws may be compared to those known from room-temperature $e^+e^-$ storage rings.[6] There, the dipole field scales approximately like $1/E$, and the cost and the circumference scale approximately like $E^2$.

## 5. Cost Optimization of a 100 TeV Collider

Only the first of the three dipole cost models discussed in Section 4.3 – in which the dipole cost per running metre is independent of the dipole field $B$ – satisfies the condition that the total cost of the dipoles is proportional to a negative power of the dipole field $B$. In the following, it will be called the "high-field model". The second model – in which the total cost of the dipoles is independent of $B$ and does not enter into the cost optimization – will be called "low-field model" because a lower value of the dipole field $B$ will be obtained when balancing the cost of the synchrotron radiation power against the cost of only the fully equipped tunnel.

The following unit prices are assumed for the cost optimization of a 100 TeV collider:

1. Complete collider arcs in high field model: 100 Msfr/km.

2. Fully equipped tunnel in the low field model: 7 Msfr/km.

The results of the cost optimization are displayed in Tab. 2. The cost coefficients $K_c$ and $K_s$, and the exponent $k_s$ are defined in Eq. 15. The contributions of the cryogenics plant and of the arcs are listed separately. The total cost does not include the constant term $K_0$. Since the cost of the dipoles is included in this term in the low-field model, it does not appear in the table.

The cost optimization leads to dipole fields within the currently feasible range for superferric and/or superconducting magnets in the case of the low field model. The two assumptions about the scaling of cryogenic systems lead to dipole fields which differ by almost a factor of three. However, in the case of the high field model, the cost optimization leads to dipole fields just beyond and much beyond the currently feasible range, depending on the scaling law for the cost of the cryogenic system. Therefore, it is inevitable in this case to limit the dipole field to some technologically feasible level, say 10 T, and to abandon the idea of cost optimization. The two columns labelled "with $B = 10$ T" show the machine parameters for this case.

Barletta[13] has studied the variation of the parameters of a 100 TeV collider with the dipole field in much detail, for $2\text{ T} \leq B \leq 10\text{ T}$.

## 6. Conclusions

Starting with the requirement that the luminosity scales like the square of the energy, the scaling laws for the parameters of hadron colliders beyond 10 TeV were derived. The

Table 2: Optimum Parameters for a 100 TeV Collider

| | High-Field Model Cost Optimized | | High-Field Model With $B = 10$ T | | Low-Field Model Cost Optimized | | |
|---|---|---|---|---|---|---|---|
| $K_s$ | 0.5 | 2.0 | 0.5 | 2.0 | 0.5 | 2.0 | Msfr@1kW |
| $k_s$ | 1.0 | 0.7 | 1.0 | 0.7 | 1.0 | 0.7 | |
| $K_c$ | 100 | 100 | 7 | 7 | 7 | 7 | Msfr/km |
| $B$ | 12.4 | 44.5 | 10.0 | 10.0 | 3.3 | 9.3 | T |
| $C$ | 222 | 61 | 274 | 274 | 838 | 294 | km |
| $P_s$ | 22.2 | 79.9 | 18.0 | 18.0 | 5.9 | 16.7 | MW |
| Cryogenics | 22.2 | 8.8 | 18.0 | 3.1 | 5.9 | 2.9 | Gsfr |
| Arcs | 22.2 | 6.1 | 27.4 | 27.4 | 5.9 | 2.1 | Gsfr |
| $K - K_0$ | 44.3 | 14.9 | 45.3 | 30.5 | 11.7 | 5.0 | Gsfr |

results are displayed in Table 1. Since there are no accelerator physics arguments which determine the scaling law of the dipole field $B$, it was derived by using cost optimization. Two models each were used to determine the scaling of the cost of the collider arcs and those of the cryogenic system. The results are displayed in Table 2. For the low field model, the dipole fields arrived at are within the range of superferric and/or superconducting magnets. For the high field model, they are beyond that range. In this case the parameters are also given for a dipole field limited at $B = 10$ T.

It would have been nice to include in this study the scaling laws for the dynamic aperture which is of major concern in machines like the LHC and the SSC.

## Acknowledgements

I should like to thank the participants at the workshop on "Maximizing the Luminosity of Hadron Colliders at 100 TeV" for stimulating discussions.

## References

1. K. Johnsen, in *New Techniques for Future Accelerators*, ed. G.Torelli (Plenum, New York, 1990) 1.
2. E. Keil, Proc. 12-th Intern. Conf. on High Energy Accel. (Fermilab 1983) 98.
3. T. Collins, in Report of the 20 TeV Hadron Collider Technical Workshop (Cornell University 1983) Appendix II-2.
4. E. Keil, CERN 77-13 (1977) 22.
5. J. Gareyte, these proceedings.
6. B. Richter, Nucl. Instrum. Methods **136** (1976) 47.
7. The LHC Study Group, CERN 91-03 (1991).
8. L. Resegotti, in *New Techniques for Future Accelerators*, ed. G.Torelli (Plenum, New York, 1990) 115.
9. G. Lambertson et al., these proceedings.
10. P. Lebrun, private communication.
11. J. Schmid, private communication.
12. P. Limon, private communication.
13. W. Barletta, these proceedings.

# DESIGN CRITERIA FOR THE ELOISATRON DIPOLE MAGNET

KARLHEINZ KOEPKE
*Fermi National Accelerator Laboratory**
*P.O Box 500, Batavia, Illinois 60510, USA*

ABSTRACT

This paper reviews exisiting accelerator dipole designs for possible use in the Eloisatron. The synchroton light power expected at a luminosity of $10^{35}$ $cm^{-2}sec^{-1}$ is also calculated along with cryogenic and room temperature methods for extracting the light power from the magnets.

## 1. Introduction

The peak operating energy (100 TeV) and luminosity ($10^{35}$ $cm^{-2}sec^{-1}$) of the proposed Eloisatron proton-proton collider may push the designs of superconducting lattice magnets and their associated cryogenic systems into an operating range unexplored by existing accelerators. This paper reviews currently operating superconducting accelerator dipoles for possible use in the Eloisatron collider and on the basis of current technology, suggests a maximum operating field and a minimum aperture. This is not to suggest that the maximum achievable dipole field is the proper choice for this machine. Such a determination can only be made after a cost optimization that includes not only the cost of the magnets but also the tunnel and all the systems associated with the accelerator.

The Eloisatron is the first proton accelerator whose design energy and/or luminosity are potentially limited by synchrotron light power. Unlike electron machines with conventional magnets that permit the extraction of light power at ambient temperature, modern proton accelerators like the Eloisatron use superconducting magnets. Superconducting accelerators, like conventional accelerators, need sufficient rf power to cancel the energy lost due to synchrotron radiation, but unlike conventional machines, must extract the synchrotron light power at cryogenic temperature at a fraction of Carnot efficiency unless a warm extraction method can be devised. This paper will

*Operated by Universities Research Association, Inc., under contract with the U.S. Department of Energy

explore the possibility of extracting the light power with a cryogenic liner within the bore tube. A possible room temperature light power extraction method is also suggested.

## 2. Superconducting Lattice Dipoles

Insofar as the cost of the collider is proportional to its size, it is desirable to minimize its circumference.This is accomplished, once the peak operating energy of the collider is determined, by increasing the peak operating field of the bend dipoles and by increasing the density of the dipoles within the magnet lattice. The dipole density is limited by the need for magnet interconnection spaces, by the need to introduce quadrupoles into the lattice to focus the beam, and by the need to introduce straight sections that contain beam injection, extraction, rf, abort, and beam collision regions. The peak field in the dipoles is limited primarily by the average achievable current density within the superconducting coils and by the high stresses within the magnet structure due to the Lorentz force.

### *2.1. Magnet Bore Field*

Although other geometries are possible, superconducting accelerator magnets to date have been cylindrical in cross section. Idealized cylindrical magnets are obtained with current sheets whose current density is proportional to cos(mθ) where θ is the azimuthal angle relative to the horizontal line of symmetry, and m is a positive integer (a dipole if m=1, a quadrupole if m=2, a sextupole if m=3, etc). Practical magnets approximate a cos(mθ) current distribution with current shells. The simplest form of a cylindrical dipole composed of a current shell is shown in Figure 1. Higher fields are obtained with two or more concentric shells. The field quality within the bore of the magnet can be improved by adding spacers to the coil

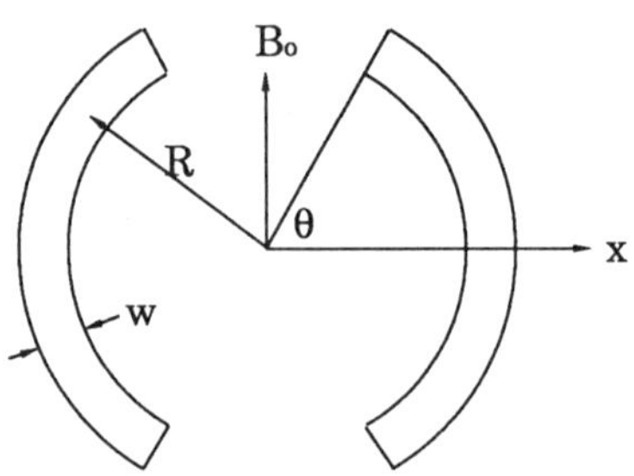


Figure 1. Cylindrical dipole with single shell coil.

Table 1. Accelerator Dipole Coil Comparison

| | Inner Coils | | | | Outer Coils | | | |
|---|---|---|---|---|---|---|---|---|
| | J | w | Fe | B | J | w | Fe | B |
| Tevatron[1] | 369 | 0.78 | 1.21 | 2.66 | 369 | 0.78 | 1.30 | 1.78 |
| HERA[2] | 374 | 1.0 | 1.22 | 3.53 | 355 | 1.0 | 1.34 | 2.45 |
| SSC[3] | 253 | 1.21 | 1.2 | 2.84 | 389 | 1.17 | 1.4 | 3.9 |
| LHC[4] | 266 | 1.7 | 1.1 | 3.86 | 492 | 1.7 | 1.25 | 6.31 |

J Coil average current density (A/sqmm)
w Coil radial width (cm)
Fe Effective current density increase due to iron
B Cenral field due to one coil (T)

to more closely approximate the ideal cos(mθ) distribution and/or by adjusting the coil shell angles.

The central magnetic field obtained from a single current shell of the type depicted in Figure 1 is equal to

$$B_o = \frac{2}{\pi}\,\mu_o\, J\, w\,\left(1+\frac{R^2}{R^2_{Fe}}\right)\sin\theta \qquad (1)$$

where $\mu_o$ is the permeability of free space, J is the current density (averaged over the whole coil including the stabilizing copper, insulation, spacers and voids), w is the radial width of the coil, R is the average radius of the coil, $R_{Fe}$ is the inner radius of the magnetic iron, and θ is the coil shell angle. This equation is strictly valid only for coils with uniform current density but good agreement, a few percent, is obtained when applied to actual coils with coil spacers. Note that except for the presence of magnetic iron, that the achievable bore field is proportional to the product of the average current density and the radial coil width. The coil radius is a free parameter. Its choice is a compromise between competing requirements; increasing the radius improves the obtainable field accuracy while decreasing the radius reduces the magnet cost and coil stress.

The average current density, coil width, iron field enhancement factor and bore field have been compiled in Table 1 for the inner and outer coils of several operating or proposed accelerator magnets. The current density at 5 T and 4.2 K in the NbTi of superconducting strands has almost doubled since the design of the Tevatron dipole, but the average current density in the inner coils of recent high field designs has gone down relative to the current density of the Tevatron. The increase in field in post Tevatron magnets has been achieved by increasing the coil widths; the

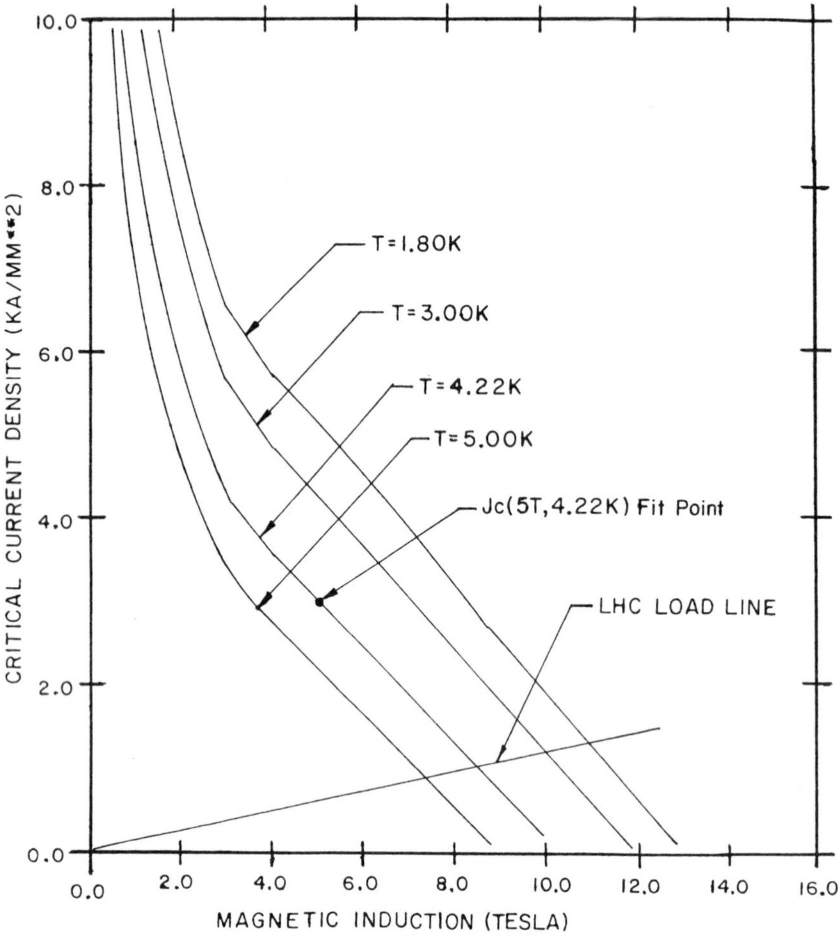


Figure 2. Critical current density[5] for commercially available NbTi superconductor and the LHC dipole magnet[4] load line.

coil width of the LHC magnet is twice the coil width of the Tevatron magnet. Widening of the magnet coil widths even further requires cable with a higher cross sectional aspect ratio and/or higher magnet currents. Eventually, it is more efficient to increase the number of coil shells in a magnet. The inner shells shield the outer shells from the inner fields and allow the outer coil shells to operate at a higher current density. This effect is evident in the outer coil current densities of the SSC and LHC magnets.

It is doubtful that an accelerator magnet operating at a field significantly higher than the 10 T design field of the LHC will be available at an affordable cost in the immediate future. This is evident if one superimposes the LHC load line onto the NbTi critical current density[5] values presently available in cables (Figure 2). The LHC magnet geometry, which has been optimized for high field, is limited to a maximum field of approximately 11 T. Increasing the coil volume through wider cables and more coil layers could in principle raise the field closer to the 14.5 T critical field of NbTi. However, such inefficient use of superconductor makes this approach, because of high cost, unsuitable for accelerator magnets.

Another approach to higher fields[6] is to utilize $Nb_3Sn$, a superconductor, whose critical field is 22.5 T, and therefore capable of conducting currents at fields higher than NbTi. Unfortunately, $Nb_3Sn$ cable is significantly more brittle than NbTi cable and therefore more susceptible to fracture during fabrication. It also remains to be shown that this material will reliably survive the high stresses present in superconducting coils during fabrication and operation. Nonetheless, $Nb_3Sn$ has the potential of higher fields in the future and a maximum field of 15 T is assumed in the synchrotron light power calculations.

### 2.2. *Magnet Stress*

During fabrication, the conductors of a superconducting coil are formed under pressure and heat within a precision mold, and then accurately positioned and clamped within a precise and rigid coil collar. The coils must be compressed sufficiently during the collaring operation at room temperature to retain sufficient azimuthal compression (preload) to counter the azimuthal component of the magnetic force when the magnet is powered at cryogenic temperature. Otherwise the coils, as they are compressed towards the horizontal line of symmetry by the magnetic force, will separate from their azimuthal collar support surfaces (keys), thereby reducing field quality and possibly causing a quench. The minimum azimuthal force per unit length F and the related minimum azimuthal pressure P (preload), that barely keeps the coils from separating from their keys at cryogenic temperature, are approximated by the equations:

Table 2. Inner Coil Preload Comparison

$$\text{Pressure} \sim \frac{R\ (\text{cm})}{w\ (\text{cm})} B^2(\text{T})$$

| | Cold | Warm |
|---|---|---|
| Tevatron | $\frac{3.75}{.78}$ $4.4^2 = 93$ | x 1.5 = 140 |
| SSC (4cm) | $\frac{2.0}{.93}$ $6.6^2 = 93$ | x 1.5 = 140 |
| SSC (5cm) | $\frac{2.5}{1.23}$ $6.6^2 = 89$ | x 1.5 = 134 |
| HERA | $\frac{3.75}{1.0}$ $6.0^2 = 135$ | x 1.0 = 135 |
| LHC | $\frac{2.5}{1.7}$ $10.^2 = 147$ | x 1.0 = 147 |

$$F \cong N\,I\,B/3 \sim R\,B^2$$

$$P \sim R\,B^2/w \tag{2}$$

where N is the number of cables per coil quadrant, I is the magnet current, and B is the magnetic field at the conductor next to the key. These equations require a minimum preload of 2.48 $10^3$ N/m, 32 MPa (1420 lbs/in, 4625 PSI) for the inner coil of the Tevatron dipole during 4.4T operation. The actual azimuthal pressure applied to this coil at room temperature during collaring varies between 69 MPa (10 KPSI) and 138 MPa (20 KPSI). Nondestructive yielding occurs during the collaring process as the onset of coil yielding is approximately 69 MPa. The variation in collaring pressure results from the need to force molded coils with a size tolerance into the same collar and still assure proper preload for the smaller coils. The minimum 69 MPa collaring pressure assures adequate remaining preload as the coil is cooled to 4.6 K. The Tevatron collar is made with stainless steel. With stainless steel collars, approximately one third of the preload is lost during cooldown as the coil's coefficient of thermal expansion is larger than the coefficient of thermal expansion for stainless steel[7].

Table 2 compares the minimum required cold azimuthal preload pressure for the inner coils of several accelerator magnets and the corresponding warm pressure required during collaring at room temperature. The HERA and LHC magnets utilize aluminum coil collars which in principle should increase the preload during cooldown. The table makes the assumption that the preload with aluminum collars stays constant during cooldown[8]. Despite the large variation in field, aperture and cable width, the applied pressure during collaring is the same for these magnets to avoid damaging the coils. Utilizing a superconductor capable of higher currents and fields in these magnets requires higher warm collaring pressures and increases the risk of coil failure through shorts and/or broken strands. In principle, however, acceptable stresses should be achievable in a future 15 T dipole by distributing the Lorentz force over even wider cables or additional coil shells, or by adding additional collars between coil shells with metal spacers that segment the coils azimuthally and bear some of the azimuthal pressure.

2.3. *Aperture*

The systematic and random field errors of the magnets in the accelerator lattice determine the available beam aperture if physical obstructions are absent. If we assume that the lower order systematic field errors can be compensated with correction coils, then the aperture is limited by the random field errors of the magnets. An analysis[9] of random errors in exisiting accelerator magnets indicates that the random errors can be scaled as a power of the inner coil radius $d_c$. In addition to the beam aperture, the coil aperture must house the beam pipe, a helium channel to cool the coils, and perhaps for the Eloisatron coil, a cryogenic channel to intercept synchrotron light.

The relationship between available beam aperture and coil aperture can be obtained by scaling the measurements obtained from operating accelerators or as in the case of the SSC, by computer simulation of the design lattice with statistically anticipated field errors. The empirical result of extensive SSC tracking calculations, which relate the linear beam aperture radius $A_{linear}$ to the inner coil radius[10], is

$$A_{linear}(\mathrm{mm}) = (13 \pm 3)\,[\,L\,(\mathrm{m})\,]^{-0.76}\,[d_c\,(\mathrm{cm})\,]^{1.89} \qquad (3)$$

where L (m) is the lattice half-cell length.. This equation can be applied directly to the Eloisatron magnets if we assume a similar lattice, the same injection energy to peak energy ratio, and similar field errors as a function of coil radius as in the SSC tracking calculations.

The assumed Eloisatron injection parameters are listed in Table 3. A stronger than SSC quadrupole strength is assumed to retain the SSC maximum $\beta$ value (300 m) and half cell length (100 m), thereby reducing the required coil aperture. The assumed emittance is a pessimistically large value used to calculate synchrotron light power in the next section. The 5 TeV injection energy was chosen in order that the magnet field has a dynamic range of 20 as in the SSC tracking studies. Except for the beamsize, the parameters summed in Table 3 to obtain the total required linear radius

Table 3. Dipole Aperture Requirements Using the Method of the SSC Conceptual Design Report 1986

| | |
|---|---|
| Assumptions | |
| bmax | 300 m |
| Emittance ($\varepsilon$) | $5\ 10^{-6}$ rad m ($\sigma^2 = \varepsilon\beta/\gamma$) |
| Injection energy | 5 TeV |
| Needed Aperture | |
| Maximum beam size $\sqrt{6}\ \sigma$ | 1.39 mm (95% particles) |
| Max. injection position error | 1.5 mm |
| Maximum orbit distortion | 1.25 mm |
| Miscellaneous | 0.5 mm |
| Total required linear rad. | 4.64 mm |
| From SSC Tracking Studies (without sorting) | |
| Required coil diameter | 3.7 cm (50% confidence level) |
| | 4.3 cm (95% condidence level) |
| Approx. dynamic radius * | 7.7 cm |

* Obtained by scaling the linear aperture with the SSC dynamic to linear aperture ratio.

are taken directly from the SSC Design Report and are reproduced here without additional justification. The above equation then yields the necessary coil diameter, 3.7 cm, that will result in the desired aperture with a 50% confidence level. A 95% confidence level increases the required coil aperture to approximately 4.3 cm.

## 3. Synchrotron Light

The protons, as they are guided by the magnetic field of the lattice dipoles, radiate light in a direction tangent to the circular orbit. The total synchrotron light power radiated per ring $P_{ring}$ and the watts radiated per meter of bend dipole $P_m$ are given by the equations:

$$P_{ring} = N_a \frac{Z_o}{6\pi} e^2 c^2 \frac{\gamma^4}{\rho^2} = N_a\ 4.6\ 10^{-20} \frac{\gamma^4}{\rho^2}$$

$$P_m = \frac{P_{ring}}{2\pi\rho} = N_a\ 7.3\ 10^{-21} \frac{\gamma^4}{\rho^3} \qquad (4)$$

where $N_a$ are the total number of protons in the arcs, $Z_o$ is the impedance of free space (377 $\Omega$), e is the unit electric charge, c the speed of light, $\gamma$ the ratio of proton energy to proton rest energy, and $\rho$ (m) is the bend radius of the lattice dipoles. $P_{ring}$ must be supplied by the rf system to maintain the proton orbit. It is also the power that the refrigeration system has to

Table 4. Synchrotron Light Radiated at a Luminosity of $10^{35}$ $cm^{-2}sec^{-1}$

| Field (T) | Power (W/m) | | Power/Ring (MW) | |
|---|---|---|---|---|
| | Beam Crossings | | Beam Crossings | |
| | One | Six | One | Six |
| 6.6 | 5.8 | 35 | 1.9 | 11 |
| 10.0 | 13.3 | 80 | 2.8 | 17 |
| 15.0 | 30.0 | 180 | 4.2 | 25 |

remove at cryogenic temperature unless a room temperature means can be found to remove the light power. $P_m$ is a measure of how difficult it is to remove the light power from the magnets. The larger the value of $P_m$, the more aperture is required within the magnet for cryogenic pipes to transport the light energy to the heat exchangers that remove the power and the shorter the distance between heat exchangers.

The number of protons in the arcs, and therefore the synchrotron light power, can be related to the allowed beam-beam tune shift and desired luminosity per interaction point. The equations for the beam-beam tune shift $\Delta\nu$, the luminosity L, the beam size $\sigma$, and the number of protons in the arcs $N_a$ are:

$$\Delta\nu_{6\ \text{crossings}} = \frac{6 N_b r_p}{4\pi\varepsilon} < .01 \qquad \sigma^2 = \frac{\varepsilon\,\beta^*}{\gamma}$$

$$L = \frac{N_b^2 c}{4\pi\sigma^2 S_b} = 10^{35}\ \text{cm}^{-2}\text{sec}^{-1} \qquad N_a = \frac{N_b\, 2\pi\rho}{S_b} \tag{5}$$

where $N_b$, $r_p$, $\varepsilon$, $S_b$, and $\beta^*$ are the number of protons per bunch, classical radius of the proton, normalized transverse emittance, bunch spacing and $\beta$ value at the interaction point respectively. We require that the beam-beam tune shift be less than .01 and that the luminosity per operating interaction region be $10^{35}$ $cm^{-2}sec^{-1}$. Assuming a value for $\beta^*$ (.5 m) and $S_b$ (4.7 m), we can simultaneously solve for the emittance and the number of protons per bunch. If 6 interaction regions are in operation, each one operating at a luminosity of $10^{35}$ $cm^{-2}sec^{-1}$, the required emittance and protons per bunch are 5 $10^{-6}$ mrad and 7 $10^{10}$ protons per bunch respectively. If only one interaction region needs to operate at a luminosity of $10^{35}$ $cm^{-2}sec^{-1}$, these values reduce to 0.14 $10^{-6}$ mrad and 1.18 $10^{10}$ protons per bunch.

The radiated synchrotron power, assuming one or six interaction regions, is given in Table 4. The listed wattages are extremely high, especially if the power has to be removed at cryogenic temperature. If we assume a peak bend field of 10 T with only one interaction region operating at a time, the total cryogenic heat load for two rings is 5.6 MW which, if it is removed at LN temperature, equals approximately 75 MW of wall power for the refrigerator; high but still acceptable. The removal of this power level at 4 K is prohibitively expensive because of the much lower heat extraction efficiency at 4 K.

The value of $P_m$ is also significantly higher than in previous accelerators. For example, the Tevatron has a heat load of 2 W/m (heat leak and cyclic loss); the more efficient cryostat of the SSC magnets has a heat load of 0.4 W/m (heat leak plus synchrotron light).

### 3.1. *Cryogenic Heat Extraction*

The removal of heat from the bore of a superconducting magnet can be accomplished with single phase cryogens whose temperatures rise as they absorb heat or with two phase cryogens whose liquids vaporize as they absorb heat. In both cases, the heat needs to be removed at intervals along the accelerator in order to limit the temperature rise or to replenish the liquid. Because of the high heat load of the Eloisatron, it is assumed that the heat is extracted at every half cell of the lattice, approximately 100 m. A longer heat exchange interval will require larger cryogenic pipes within the bore to accommodate the additional mass flow rate; a shorter heat transfer interval, for example, after every dipole, will complicate the structure of the cryostat.

Except for a two phase helium heat intercept operating at the same temperature as the magnet coils, the cryogen that removes the synchrotron heat must be insulated from the coil cryogens. Figure 3 shows a simple

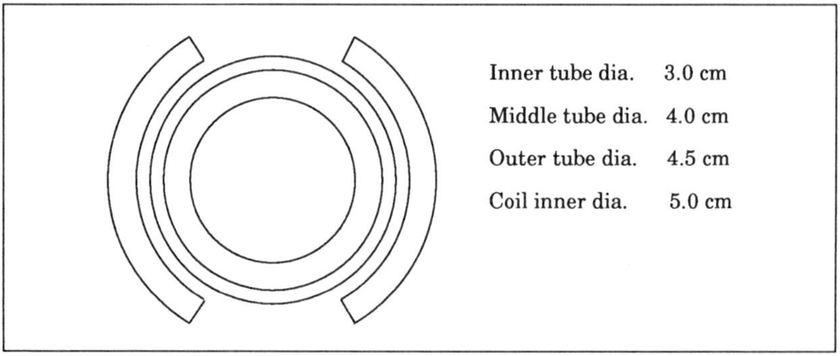


Figure 3. Example of 3-pipe cryogenic system to extract the synchrotron light power.

Table 5. Pressure Drop Per 100 Meters For the Annular Heat Intercept Depicted in Figure 3.

| | Pm (W/m) | ΔT (K) | $\dot{M}$ (g/s) | ΔP/100m (MPa) |
|---|---|---|---|---|
| 1φ LN 75 K - 85 K | 5.8 | 10. | 29. | .0007 |
| | 180. | 10. | 900. | .30 |
| 2φ LN 78 K* | 5.8 | 0.0 | 2.9 | .001 |
| | 180. | 0.0 | 91. | .50 |
| 1φ LHe 4.2 K - 5. K | 5.8 | 0.8 | 121. | .028 |
| | 180. | 0.8 | 3750. | 21.0 |
| 1φ He 15 K - 25 K | 5.8 | 10. | 11.6 | .009 |
| | 180. | 10. | 360. | 5.2 |
| 2φ He 4.2 K* | 5.8 | 0.0 | 27.6 | .009 |
| | 180. | 0.0 | 857. | 6.50 |

*Pressure calculation assumed 100% gas.

3-pipe system that is compatible with a coil temperature range of 2 K to 4 K and a heat intercept range of 4 K to 78 K. The inner pipe is the evacuated beam pipe. The cryogen within the annular space bounded by the inner and middle pipes intercepts the synchrotron light heat. The space between the middle and outer pipes is again evacuated to prevent thermal contact with the coil. The pipe dimensions were constrained primarily to fit within an arbitrary 5 cm inner coil diameter.

The cryogenic suitability of the cooling channel depicted in Figure 3, at least the cross sectional area of the cooling channel, can be determined by calculating the pressure rise for a half cell length with a cryogen flow rate sufficient to remove the listed heat load. The pressure drop per unit length is given by the equation

$$\frac{\Delta P}{L}\left(\frac{\text{MPa}}{\text{m}}\right) = 2.0\ 10^{-5}\,\frac{f\,(\dot{M}\,/\,A\,)^2}{n\,D_h} \tag{6}$$

where f is a dimensionless friction factor, $\dot{M}$ is the mass flow rate (g/s), A is the annular area ($cm^2$), n is the cryogen density (g/cc) and $D_h$ is the difference between the inner and outer annular diameter (cm).

The calculated pressure drops for the annular heat intercept of Figure 3 with single phase and two phase helium and nitrogen as the intercept cryogens are listed in Table 5. The two phase pressure calculations assumed 100% gas along the whole length and therefore should be reduced by approximately a factor of 2. The pressures indicate that helium, given the intercept area of Figure 3, can handle heat loads up to approximately 5 W/m

Table 6. Maximum Allowed Dipole Length That Allows Geometric Interception of the Synchrotron Light Versus Magnet Field, Bore Radius, And Absorber Radius.

| | Bore Radius (mm) = | 15. | 20. | 25. |
|---|---|---|---|---|
| Field (T) = 6.6 | | | | |
| Absorber Radius (mm) = 5.0 | Magnet Length (m) = | 16.4 | 22.4 | 27.6 |
| 10. | | 7.1 | 13.1 | 18.4 |
| 15. | | 0.0 | 6.0 | 11.3 |
| Field (T) = 10.0 | | | | |
| Absorber Radius (mm) = 5.0 | Magnet Length (m) = | 13.3 | 18.2 | 22.5 |
| 10. | | 5.8 | 10.6 | 14.9 |
| 15. | | 0.0 | 4.9 | 9.2 |
| Field ( T ) = 15.0 | | | | |
| Absorber Radius (mm) = 5.0 | Magnet Length (m) = | 10.9 | 14.8 | 18.3 |
| 10. | | 4.7 | 8.7 | 12.2 |
| 15. | | 0.0 | 4.0 | 7.5 |

if the periodic heat extraction is done every 100 m. However, the cost of heat extraction at this temperature is prohibitive. Nitrogen, given the same area and heat extraction interval, can remove heat loads up to approximately 200 W/m. If nitrogen is used, the annular pipe must contain periodic holes to allow the 2 K to 4 K coil temperature to cryo-pump the bore vacuum.

### 3.2. *Room Temperature Heat Extraction*

Cryogenically cooled liners have the potential of removing the synchrotron light power, but because of the inefficiency of heat extraction at cryogenic temperature, are costly to operate. Operating a heat intercept similar to that shown in Figure 3 at room temperature is difficult because cryo-pumping of the bore vacuum is lost. An alternate possibility of extracting the light power at room temperature may be the addition of light absorbers in the dipole interconnect regions.

Synchrotron light is emitted within a cone angle of $\gamma^{-1}$ and at the operating energy of the Eloisatron, can be approximated as a line-like ray. The ray is emitted tangentially to the proton orbit. It therefore separates transversely from the orbit by an amount x

$$x = \frac{s^2}{2\rho} \qquad (7)$$

where s is the distance along the closed orbit and $\rho$ is the local radius of the closed orbit. The fraction of the light that preferentially hits the light target instead of the bore tube is a function of the dipole length, the dipole bend radius, the bore tube radius and the amount that the light targets project into the bore tube. Total geometric interception by the light targets is obtained, if the light ray, that just misses the target in front of the magnet, just misses the bore tube at the target downstream of the magnet. Table 6 relates a range of magnetic fields, bore tube radii and absorber radii to maximum magnet lengths that allow geometric interception of all light rays. The table assumes that the beam is centered in the bore tube. As one would expect, as the absorber radius (distance of absorber tip to center of the bore tube) decreases, the distance between absorbers and therefore the length of magnet, can increase.

The proposed light targets physically project into the bore tube and reduce the physical aperture available to the beam. They also periodically disturb the image currents that flow in the wall of the bore tube. The physical aperture reduction is not a sever problem as the dynamic aperture of a cylindrical magnet is typically a small fraction of its coil diameter. The image current discontinuity and the resulting potential beam instability needs further study. The instability may be reduced by adding a liner with a continuous longitudinal slot that allows a path for the synchrotron light to reach the light targets while shielding the targets from the beam.

## 4. References

1. *A Report on the Design of the Fermi National Accelerator Laboratory Superconducting Accelerator* ( May 1979 ) p. 29
2. R. Meinke, *IEEE Trans. Mag.* **27**:1728 (1991)
3. C. Goodzeit, R. Gupta, J. Jayakumar, S. Kahn, J. Kerby, K. Leung, G. Morgan, F. Nobrega, D. Orrell, P. Sanger, R. Schermer, G. Snitchler, G.Spigo, J. Strait, J. Turner, and E. Willen, *IEEE Trans. Mag.* **27**:2016 (1991)
4. R. Perin, *IEEE Trans. Mag.* **24**:734 ( 1988)
5. M.A. Green, SSC Internal Report SSC-N- 502
6. H. ten Kate, A. den Ouden, D. ter Avest, S. Wessel, R. Dubbeldam, W. van Emden, C. Daum, M. Bona, and R. Perin, *IEEE Trans. Mag.* **27**:1996 (1991)
7. J. Strait, R. Bossert, J. Carson, K.J. Coulter, S. Delchamps, T.S. Jaffery, W. Kinney, W. Koska, S. Gourlay, M.J. Lamm, M. Wake, R. Sims, and M. Winters, SSC Internal Report SSCL - 542
8. M.D. Anerella and R. Jackimowicz in *Proceedings of CERN Accelerator School, "Superconductiviy in Particle Accelerators"* ( June 1988) p.110
9. H.E. Fisk, *SSC Central Design Group Report* SSC-7 ( 1985 )
10. *SSC Conceptual Design* ( March 1986 ) p. 133

# INJECTOR FOR A 100 TeV COLLIDER ("ELOISATRON")

K.SCHINDL
*CERN, CH-1211 Geneva, Switzerland*

ABSTRACT

Next Generation Hadron Colliders - under construction or proposed - require injector chains capable of producing high-intensity, small-emittance proton beams. Only beams of unprecedented brightness can be accomodated in the very small dynamic aperture of such colliders and yet provide high luminosity at collision. Inspired by recent work done on the injector issues, in particular for the SSC and LHC projects, challenges, limitations, technologies, other interesting features are discussed. Various possible avenues to produce small emittance beams ($H^-$ sources and charge-exchange injection, beam cooling techniques) as well as phenomena tending to spoil them along the accelerator chain (space-charge effects, transverse instabilities, transfer between machines) are addressed. These consideration are applied to the ELOISATRON, a project for a two-ring 100 TeV proton collider, for which a luminosity of $10^{35}$ cm$^{-2}$ sec$^{-1}$ is envisaged. The basic performance parameters are: about 43000 bunches per ring with a population of $5.10^{10}$ protons, bunch spacing 8 m, circulating current 0.3 A, transverse emittances of $\varepsilon_n$ = 0.8 µm in either plane. An injector chain capable of producing such beams and loading the ELOISATRON within about 2 hours is sketched; although no fundamental difficulties emerge, several issues will require further R&D effort.

## 1. Injector Chain: Tasks, Limitations

Given the performance parameters of the collider, is it possible to sketch an injector chain satisfying them? If so, are there degrees of freedom enabling the injector to help maximizing the luminosity? This quantity,

$$L = \frac{N_b^2\, c\, \gamma}{4\, \pi\, \varepsilon_n\, \beta^*\, s_b} \tag{1}$$

where $s_b$ is the bunch spacing in m, $\varepsilon_n = \sigma^{*2}/\beta^*(1/\beta\gamma)$ the normalised r.m.s. emittance (either plane), $\beta^*$ the betatron amplitude function and $\sigma^*$ the r.m.s. beam size at the collision point, has to be optimized.

The bunch population and bunch spacing are determined by respectively the acceptable synchrotron radiation power, and the maximum event rate the detector can manage. Other parameters relevant for the injector are:

(i) The *dynamic aperture* at collider injection: superconducting magnets tend to suffer from strong field non-linearities, the effect of which is worst at lowest energy, in particular with large emittance beams. Thus the transverse emittance (in general equal in both phase planes) delivered by the injector has to be small so as to avoid beam diffusion and transverse blow-up due to these field imperfections. Clearly this limitation favours a high injection energy of the collider.

(ii) The *beam-beam tune shift at collision*. This quantity,

$$\xi = \frac{N_b \; r_p}{4 \, \pi \, \varepsilon_n} \qquad (2)$$

where $r_p$ is the classical proton radius, is a measure for the tune spread induced by the electro-magnetic fields of the colliding bunches. Experience from operational colliders sets it to a limit of 0.02 (sum of all collision points). Thus in the collider the transverse beam density $N_b/\varepsilon_n$ is limited by the permissible beam-beam tune shift at collision.

(iii) *Resistive wall instability* at collider injection: the ratio of instability growth time $\tau$ to collider revolution time T scales as[1]

$$\frac{\tau}{T} \sim \frac{(E/e)\, b^3}{R^{1.5}\, \rho^{0.5}} \qquad (3)$$

where E is the beam energy at collider injection, b the beam pipe radius, R the machine radius, $\rho$ the specific resistivity of the beam pipe inner surface. This number of machine turns for one e-folding rate should not be lower than, say, 5, so as to enable a conventional transverse feedback system to keep the resistive wall instabilities under control without excessive power requirements[2]. The scaling law suggests that a higher injection energy as well as a larger dimension of the beam pipe (liner, or screen) lead to more favourable conditions.

(iv) The *incoherent space charge tune shift* in the low energy synchrotrons of the injector chain. This figure is again a measure for the total tune spread one has to accommodate in the betatron tune diagram without straddling lower order stop-bands (orders 2, 3, possibly 4 to be avoided). The tune shift is

$$\Delta Q_{sc} = \frac{N_b \, M \, r_p}{4\pi \, \beta\gamma^2 \, \varepsilon_n \, B_f} \; (G) \qquad (4)$$

where M is the number of bunches in the synchrotron, $B_f$ the mean/peak line density ("bunching factor"), (G) a geometric form factor indicating the distribution in the real x-y plane (1 for bi-gaussian, 0.5 for uniform). In the machine with strongest space charge, the Low Energy Booster (LEB), $\Delta Q_{sc}$ should not exceed 0.25 to 0.4 for the nominal beam transverse density $N_b/\varepsilon_n$. The situation in the LEB can be alleviated by setting its injection energy to higher $\beta\gamma^2$

(requiring a higher injection linac) as well as by faster cycling so as to reduce the intensity per pulse, $N_b.M$ . Moreover, keeping in mind that injectors are cheap compared to the collider cost, the possibility of producing the nominal bunch population $N_b$ in an emittance even smaller than nominal should be explored: this further reduction in beam at collider injection eases the dynamic aperture problem and potentally gives rise to some cost savings of the collider magnet. Clearly the beam would then be too small to comply with the beam-beam tune shift (Eq. 2), thus it would have to undergo controlled transverse blow-up during the collider ramp (how to do this without beam loss is an open question).

(V) *Filling time.* Although not a hard limit, the loading time should be kept short with respect to the time for which beams collide so as to achieve high average luminosities. Even more importantly, dynamic aperture considerations at collider injection flat bottom suggest short dwell times as well.

In summary, the injector chain has to load the collider with many thousands of bunches within very small transverse normalized emittances, where bunch population and emittances are entirely determined by the collider characteristics. A high collider injection energy and a short filling time are favoured.

## 2. Injector chains for SSC and LHC

The study of 100 TeV colliders obviously relies on the huge amount of work being done for the near-future colliders SSC and LHC, and this also applies for the injector chain. Table 1 summarizes some of the more significant parameters and properties of these injectors[3,4,5,6], and displays a few interesting features maybe relevant for the ELOISATRON injector. Clearly the proposed SSC injector system, being built on green field, is a far better model for the ELOISATRON than the LHC one, but some features of the LHC chain (existing, thus not optimized) are relevant, in particular because of hands-on experience with actual beams.

Apart from the considerable difference in design emittances (see Chap. 3), the most striking features are:

(i) Both in the Medium Energy Booster (MEB) of the SSC and the PS of LHC, the beams have to pass through transition where the bunches become very short with large energy spread and thus large horizontal beam dimensions in the dispersive sectors of the machine. In the CERN PS, there are indications of transverse emittance growth on transition, possibly due to particles spilled beyond the dynamic aperture.

(ii) The LHC injectors have to put up with existing RF systems (which will be kept for cost saving), thus a delicate debunching-rebunching process in the PS is required so as to produce the final bunch distance of 4.5 m (15 ns). By contrast, in the SSC injector chain, the final bunch distance (5m) is generated right at the adiabatic RF capture in the LEB, which indeed is the best place to do this.

Table 1: SSC and LHC Injector Chains

| **SSC ($L=10^{33}$ $cm^{-2}$ $sec^{-1}$)** | | | | **LHC ($L=1.65\ 10^{34}$ $cm^{-2}$ $sec^{-1}$)** | | | |
|---|---|---|---|---|---|---|---|
| **Acceler.** | **Energy**<br>Intensity<br>*ε- norm.* | **cycling**<br>RF<br>*harmonic* | Special<br>features | **Acceler.** | **Energy**<br>Intensity<br>*ε – norm.* | **cycling**<br>RF<br>*harmonic* | Special<br>features |
| RFQ | **2.5 MeV**<br>30 mA $H^-$<br>*<0.2μm* | **0.1 sec**<br>427 MHz | | RFQ | **0.75 MeV**<br>200 mA p<br>*0.5 μm* | **1.2 sec**<br>200 MHz | |
| LINAC | **600 MeV**<br>25 mA $H^-$<br>*0.4 μm* | **0.1 sec**<br>427 MHz<br>1281 MHz | <br>DTL<70<br>CCL>70 | LINAC2 | **50 MeV**<br>200 mA p<br>*1.2 μm* | **1.2 sec**<br>200 MHz | DTL |
| LEB<br><br>6 pulses | **12 GeV/c**<br>9.1E9 p/b<br>1E12 p<br>*0.6 μm* | **0.1 sec**<br>~ 60 MHz<br>*h=108* | $H^-$ Inject.<br>5 m bunch<br>spacing | PSB<br><br>2 pulses | **1.4 GeV**<br>7.2E12<br>1.8E12 p/r<br>*2.5 μm* | **1.2 sec**<br>0.6-1.8 MHz<br>*h=1/h=2* | 4 rings<br>3-turn inj. |
| MEB<br><br><br>3 pulses | **200 GeV/c**<br>8.6E9 p/b<br>6.8E12<br>*0.7 μm* | **3 sec**<br>~60 MHz<br>*h=792* | **transition**<br>**crossing** | PS<br><br><br>3 pulses | **26 GeV/c**<br>1E11 p/b<br>1.4E13<br>*3.0 μm* | **3.6 sec**<br>66 MHz<br>*h=8=>140*<br>*140 b* | **transition**<br>*harmonic*<br>*change*<br>4.5 m bunch<br>spacing |
| HEB<br><br>2 * 8<br>pulses | **2 TeV**<br>8.5E9 p/b<br>1.84 E13<br>*0.8 μm* | **120 sec**<br>60 MHz<br>*h=2160* | **s.c.**<br>**Polarity**<br>**change** | SPS<br><br>2 * 12<br>pulses | **450 GeV**<br>1E11 p/b<br>4E13<br>*3.5 μm* | **16.8 sec**<br>200 MHz<br>*h=4620*<br>*1540 b* | **Two s.c.**<br>**lines to**<br>**LHC**<br>**(3 km)** |
| SSC | **20 TeV**<br>8.4E9 p/b<br>1.3E14/ring<br>*1 μm* | **113 min**<br>360 MHz<br>*h=104544*<br>*15500 b/r* | **load**<br><br>Dyn. aperture<br>**~10 σ** | LHC | **7.7 TeV**<br>1E11 p/b<br>4.7E14/rin<br>*3.75 μm* | **10 min**<br>400 MHz<br>*h=36540*<br>*4700 b/r* | **load**<br><br>Dyn. aperture<br>**~6 σ** |

(iii) For loading both rings of the collider, one either needs to change - once per fill - the polarity of the (superconducting) High Energy Booster (HEB) (SSC solution), or one has to build two long transfer lines (LHC) with a total length of about 3 km and equipped with superconducting magnets. The SSC solution is cheaper and at first sight preferable; it remains to be seen whether the beam delivered to the SSC will be up to the specifications in terms of stability and reproducibility despite of the frequent polarity changes of the superconducting HEB.

(iv) Nominal load times quoted: 113 minutes for the SSC, 10 minutes for the LHC. A factor 3 is explained by the circumference ratio; a further factor 2 is due to the cycle time of the HEB (120 sec) which is much more than the figure quoted for the (normal-conducting) CERN SPS (16.8 sec). Note that the fairly short LHC load time will be badly needed because the large ratio between top and injection field levels (about 17) will lead to more detrimental effects of the stronger persistent currents: the available dynamic aperture is a mere 6 $\sigma$ of the injected beam ($\sigma$ is the r.m.s. beam size).

### 3. How to produce small transverse emittances

A considerable amount of experience from other machines has gone into the design of the SSC (LHC) injectors. Probably their most difficult task is to provide beams with the high intensity and small emittance required while keeping the collider loading time reasonable. In Table 2, a few key parameters relevant for the emittance issue are collected (for convenience, corresponding figures of the ELOISATRON injectors are included; they are discussed in Chap. 4).

A first ingredient towards small emittances is the use of *$H^-$ sources*; state-of-the-art devices generate some 50 mA with $\varepsilon_n$ ~ 0.2 - 0.3 µm, and there are indications that further progress will be made. Note that for all injector chains under scrutiny here, $\varepsilon_n$ is determined by the space-charge effects in the LEB, and the source emittance is in general substantially smaller. However, source emittances of $\varepsilon_n < 0.1$ µm are desirable so as to enable beam painting at the charge-exchange injection into the LEB, thus alleviating space charge by optimizing the transverse beam distribution.

The *incoherent tune spread*(Eq. 4) is largest in the LEB and represents a major obstacle towards high brilliance beams (= large ratio $N_b/\varepsilon_n$). Ways out are

- a higher injection energy, at the expense of a longer linac;
- a faster cycling of the LEB, thus fewer particles per pulse to fill the MEB, at the expense of a much higher accelerating RF voltage. However, fast cycling machines are more tolerant to betatron stop-bands, thus one may contemplate to accomodate a beam with a tune shift of up to $\Delta Q_{sc}$ ~ 0.5 in the tune diagram (has proven to work even on slow-cycling machines like the PS Booster[7], the proposed LEB of the LHC).

Another figure decisive for the performance of the injectors is the transverse *"emittance budget"* available: per synchrotron (including beam transfer) the permissible increase of normalized emittance $\Delta\varepsilon_n$ is ~0.4 in the LHC chain, and merely ~0.1 for the SSC (Table 2). This calls for great efforts on accelerator components such as fast kicker magnets, septa, power supplies, on the precision of beam size measurement devices, on the quality of optical and dispersion matching between machines. Steering errors at injection can largely be eliminated by powerful (so as to damp within the decoherence time), wide-band (up to half the bunch repetition frequency, typically ~30 MHz) transverse feedback systems. If the error to be corrected is kept smaller than the r.m.s beam size, these devices are fairly conventional and their specifications rather similar

Table 2: Emittance and Brightness in Hadron Collider Injectors

| | SSC | LHC | ELOISATRON |
|---|---|---|---|
| Luminosity ($cm^{-2} sec^{-1}$) | $10^{33}$ | 1.65 $10^{34}$ | $10^{35}$ |
| $\varepsilon_n$ (r.m.s.) at collision | 1 μm | 3.75 μm | 0.8 μm |
| bunch spacing (m) | 5 | 4.5 | 7.9 |
| $N_b$ (p/bunch) | 0.84 $10^{10}$ | $10^{11}$ | 4.9 $10^{10}$ |
| **$N_b/\varepsilon_n$ ($10^{10}$/μm) coll.** | **0.84** | **2.7** | **6.1** |
| Space-charge LEB | 0.38 | 0.35 (actual distrib.) | 0.56 (bi-Gaussian) |
| $\Delta Q_{sc}$ MEB | 0.08 | 0.20 (actual distrib.) | 0.04 (bi-Gaussian) |
| Permissible blow-up | | | |
| $\Delta\varepsilon_n$ per machine | ~0.1 | ~0.4 | ~0.1 |
| Dyn. aperture collider | ~11 σ | ~6 σ | >10 σ |

to dampers for transverse multi-bunch instabilities[2]. Thus these dampers will have to be assisted by a system which corrects slowly varying injection conditions in order to avoid saturation of power amplifiers.

Is there a way to alleviate the emittance specifications - or produce even smaller beam sizes - by *beam cooling techniques*? The very high beam intensities involved imply *stochastic cooling times* of the order of an hour[8], incompatible with useful collider filling times. *Electron Cooling* appears better suited for this job, because it more easily deals with already cold beams and tolerates large beam currents. However, as the LEB is in general space-charge limited and fast-cycling, it is not the right place for electron cooling, thus the earliest opportunity in the chain is the MEB injection front porch. This possibility is indeed under scrutiny for the SSC MEB (12 GeV/c), where a ~5 MeV, 2A electron beam could halve the emittance within one minute, while raising the total SSC filling time by less than a factor 2[9].

## 4. A Possible Injector Chain

During this Workshop, several consistent parameter sets for 100 TeV colliders were proposed[10], all of them seemingly complying with many stringent and conflicting requirements. The proposed injector chain is based on one of these sets, emerging as most plausible. The corresponding ELOISATRON performance parameters are summarized in table 2, and ensuing characteristics of a possible injector chain are collected in table 3. Fig. 1 shows a sketch of the injector system.

The parameter choice is based on the following considerations:

(i) The ELOISATRON transverse beam brightness is significantly higher than that anticipated for SSC and LHC, thus the even more severe space charge effects in LEB have to be alleviated by increasing its energy and making it fast-cycling. The 1.27 GeV, 30 mA $H^-$ Linac (10 Hz, duty factor $10^{-3}$) would be the *highest energy hadron linac* ever built. Also its emittance is very small, requiring in turn an $H^-$ source of 50 mA within $\varepsilon_n$ = 0.1 μm.

Table 3: A possible ELOISATRON Injector, key parameters

| Machine | cp (GeV) (top) | B (T) (top) | C (km) | #bunches (1) | fill time (sec) | p/ring | $\varepsilon_n$ (μm) | cycle (sec) (2) |
|---|---|---|---|---|---|---|---|---|
| **Linac** | 2 (T=1.3) | | 0.4 long | | | 30 mA $H^-$ | < 0.2 | |
| **LEB** *16 pulses* | 25 | 1.6 | 0.43 | 55 | 15 μs | $2.75\ 10^{12}$ | 0.4 | 0.1 |
| **MEB** *5 pulses* | 400 | 1.6 | 6.9 | 860 | 1.5 | $4.3\ 10^{13}$ | 0.5 | 10.5 |
| **HEB** *2*10 pulses* | 10 000 | 8 | 34.2 | 4 300 | 42 | $2.5\ 10^{14}$ | 0.6 | 350 (?) |
| **Collider** | 100 000 | 8 | 342 | 43 000 | ~2h both | $2.1\ 10^{15}$ | 0.8 | |

(1) neglecting gaps for kickers (2) fill time included

(ii) The calculated space charge detuning of the LEB is above 0.56 for a bi-gaussian distribution in the (real) x-y plane, which may be reduced to (acceptable) 0.4 by *transverse painting* - possible if $\varepsilon_n$(LEB) > 2 $\varepsilon_n$(Linac).

(iii) An *HEB energy of 10 TeV* is preferred to, say, 5 TeV, mainly so as to keep the beam more safely in the dynamic aperture at collider injection. At 10 TeV, the non-linear effects of persistent currents in the collider are weaker, and $\sigma$ will be down to ~0.2 mm, ensuring a larger safety margin. A further argument in favour of 10 TeV is the longer growth time of the transverse resistive wall instability in the collider, which is 4.4 machine turns at 10 TeV, and down to 2.2 turns at 5 TeV (Eq. 3), barely manageable by a damper.

(iv) *HEB superconducting or normal?* A 10 TeV normal conducting machine would be very large, but there are arguments in favour[11]: one is the possibility of housing the HEB in the collider tunnel (requiring 0.8 T magnets), the other is that one could switch off the HEB magnets in the long idling periods between collider fillings for energy saving, in contrast to superconducting systems which have to be kept cold all the time. Arguments which favour a superconducting HEB are: this 10 TeV machine will be extremely attractive for fixed-target physics and is more than likely to have a physics programme on its own, and thus should be accessible also during collider running. Moreover, a normal conducting HEB, of the same size as the collider, would suffer from a dramatically large resistive wall impedance, with growth times smaller than a machine revolution period. Thus a superconducting HEB is favoured.

(v) The *filling time* is mainly determined by the HEB cycle time which depends on the rate of current change dI/dt the superconducting filaments may tolerate. The figure of 300 sec cycle time (+ 50 sec filling) of the HEB given in Table 3 corresponds to today's technology; by the time the machine might be built, faster cycling - reducing the load time of the collider - may be feasible.

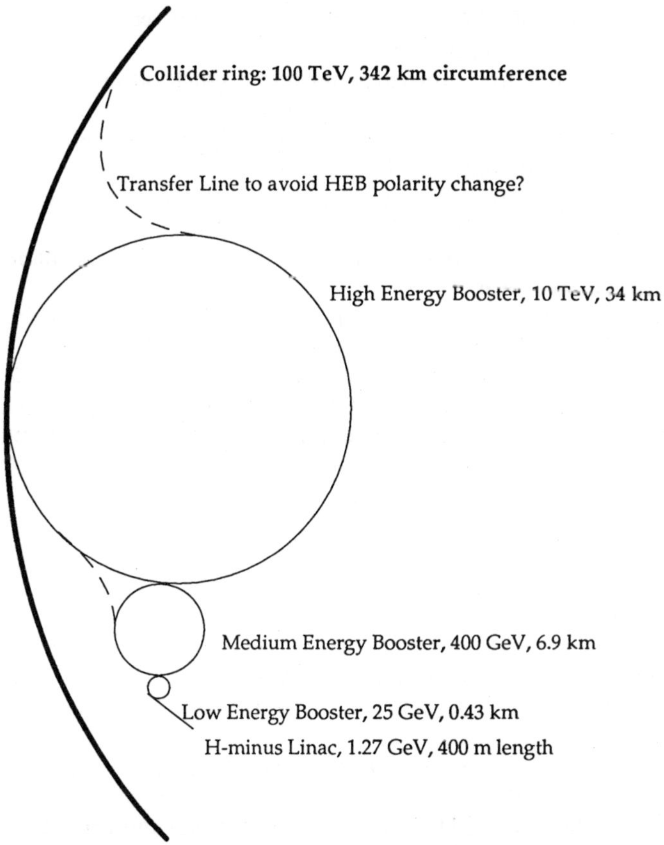


Fig.1: An injector chain for a 100 TeV collider, sketch

(vi) The permitted *transverse emittance increase per machine* is $\Delta\varepsilon_n$ = 0.1 to 0.2 which is not more stringent than the emittance budget of the SSC and calls for the same precautions (Chap. 3).

(vii) There is little hope that *electron cooling* on the MEB injection front porch (25 GeV/c) would make much sense. With a cooling time constant of some 300 sec, the collider load time would increase from about 2 to 10 hours. The electron beam characteristics[8] which would be required for this job are impressive: an electron beam current of 3 A (MEB proton current ~0.3 A), at 13 MeV.

## 5. Suggestions for R & D

Quite a few of the items touched upon require more R&D effort, most of them with direct benefits for the SSC and LHC as well. Here a list of those issues:

(i) Development of reliable, low-duty-cycle $H^-$ sources with I > 50 mA, pulse length ~ 100 μsec, $\varepsilon_n$ ~ 0.1 μm.

(ii) $H^-$ injection with charge exchange into a synchrotron beyond 1 GeV - are there problems?

(iii) Fast, wide-band transverse dampers for injection imperfections; techniques to eliminate slow steering variations.

(iv) A thorough analysis of coupling impedances and ensuing characteristics of instabilities in the various synchrotrons of the chain.

(v) Investigations into possible structures of the 1.3 GeV (30-40 mA $H^-$) Linac and their optimum choice.

Considering the time scale of a 100 TeV collider and the progress expected during the SSC/LHC construction period, most of these issues are likely to be successfully tackled in time for the ELOISATRON. Thus, *in summary,* there is reasonable hope that an injector chain, filling each ring with some 43000 bunches, with a population of 5 $10^{10}$ p/bunch, and normalized transverse emittances of 0.8 μm, can indeed be built.

## 6. References


1. J.Gareyte, *this Workshop*
2. G.Lambertson, *this Workshop*
3. Helen T. Edwards, *Proc. 2nd European Part.Acc.Conf. Nice, 1990,* Ed. Frontières, Gif-sur-Yvette, France, p. 356
4. R.F.Schwitters, A.Chao, W.Chou, J.M.Peterson, *SPP-Spread-sheets for SSC Parameters, Draft,* April, 1991
5. The LHC Study Group, *Design Study of the Large Hadron Collider(LHC),* CERN 91-03, 1991
6. R.Cappi, R.Garoby, S.Hancock, M.Martini, N.Rasmussen, T.Risselada, J.P.Riunaud, K.Schindl, H.Schönauer, E.J.N.Wilson, The CERN PS Complex as part of the LHC injector chain, *Proc. Part.Acc.Conf. San Francisco, 1991*
7. G.Gelato, L.Magnani, N.Rasmussen, K.Schindl, H.Schönauer, *Proc. Part. Acc. Conf. Washington, DC, 1987,* p.1298
8. J.Bosser, D.Möhl, G.Tranquille, *Private Communication*
9. T.Ellison and S.Y.Lee, Minimum impact electron cooling system for the SSC MEB, *Internal Indiana Univ. Technical Note,* IU/AP/MEBEC/002a, Feb. 1991
10. W.A.Barletta, *this Workshop*
11. K.Johnsen, in *"The ELOISATRON Project"*, edited by A.Zichichi, June 1988 (and updated versions)

# SSC Beam Dynamics Scaled to the Eloisatron*

D. Ritson
*Stanford Linear Accelerator Center*
*Stanford University, Stanford, California 94309*
*and*
*Superconducting Super Collider Laboratory*
*Dallas, Texas 75237*

## Introduction

As crossections drop as $E^{-2}$ a desirable target for a 100 TeV the Eloisatron would be to achieve luminosities $\sim 1.10^{35}\text{cm}^2/\text{sec}$. To understand the impact of such an objective we have compared parameters for the SSC and Eloisatron to differentiate areas which involve considerable extrapolations from current technologies from those which represent more conventional scale-ups. Synchrotron radiation losses per m for the same guide magnetic field associated with such luminosities would be up by $E^2 \times I$ where $E$ is the energy and $I$ is the circulating current. This would result in energy densities of $\sim 250$ times the nominal SSC values. The SSC is already limited by installed refrigeration power and if the circulating current was to be increased would have to use liners at liquid nitrogen temperatures to intercept the radiation as is proposed for the LHC. This issue was the subject of lively discussion at the workshop and is dealt with elsewhere by other authors. This author believed that the radiation could be intercepted by room temperature catchers spaced every 15-25 m around the ring.

Table 1 presents the author's choice for a consistent set of parameters scaled from the SSC current design. To obtain the requisite luminosities it assumes similar bunch spacing but circulating currents an order of magnitude larger than at the SSC. The SSC already uses a bunch spacing as small as 5 m and further reduction does not appear easy. The justification for the choice of bore for the magnets, emittances and attainable luminosities are discussed below. A further section looks into whether seismic ground disturbances might cause unacceptable emittance growth. The conclusion of this section is that careful use of current design practices should be adequate and that it is unlikely that exotic vibration free mounts will be required.

⋆ Work supported by Department of Energy contract DE–AC03–76SF00515.

Table 1

Eloisatron IR Parameters Compared With SSC

| | | Ratio to SSC |
|---|---|---|
| Energy [Tev] | 100. | 5. |
| Circum [Km] | 300. | 4. |
| Bunch separation [m] | 5. | same |
| IPs | 4. | same |
| Magnet bore [cm] | 6. | same |
| IP to first IR quad [m] | 20. | 6. |
| Chromaticity/ip | 100. | 2. |
| $\beta_{max}$ [m] | 40,000 | 4. |
| Stored energy | 200. | |
| IP losses [energy] | 500. | |
| Synchrotron power/m | $/sim300$ | |
| Luminosity[1e35],[nom] | 1. | 100. |
| Current [1e11/bunch],[nom] | 115. | |
| $\beta*_{ip}$ [cm],[nom] | 200. | 4. |
| Emittance [1e-9 cm rad],[nom] | 4. | 1. |
| Crossing angle[microrad],[nom] | 70. | 0.6 |
| $\sigma_{long}$ [cm],[nom] | 12. | 2. |
| Beam overlap [],[nom] | .6 | same |
| Beam size $\sigma_{ip}$ [microns],[nom] | 7. | 1.5 |
| $\delta\nu_{ho}$/ip,[nom] | .002 | 3. |
| $\delta\nu_{lr}$/ip ,[nom] | .004 | 2. |
| Dynamic aperture [$\sigma$][nom] | >10. | same |

### Choice of Aperture

Significant costs or savings are associated with the choice of collider aperture. Typically the machine tune, $\nu$, scales as

$$\nu \propto \sqrt{\frac{E}{B_{max}}}$$

where $E_{max}$ is the maximum energy and $B_{max}$ is the maximum magnetic field.

Thus the betatron tune of the Eloisatron could be expected to scale from the current SSC choice of 100 to around 200 and the $\beta$ values by a comparable factor two.

Invariant brightnesses (particles per bunch/invariant emittance) are largely set at the start of the injector chain though without care degradation can occur during the acceleration cycle. We would propose values $\sim$ 3 times nominal at the SSC. Such values are routinely achieved and used at the Tevatron and SPS. The beam beam induced tune shift parameter is directly proportional to the invariant brightness and higher values are likely to lead to unacceptable high tune shifts using currents 15 times greater than for the SSC and taking into account the adiabatic damping during acceleration gives comparable actual emittances (not normalized emittances) at injection into the SSC and Eloisatron main colliders. Tracking studies made at the SSC show that long term dynamic apertures scale as the magnetic bore to the 1.5 power. Therefore even allowing for less safety margins than presently required for the SSC we can expect magnetic apertures requirements comparable or perhaps slightly less than for the SSC of a 4-5 cm bore for the main Eloisatron collider. Therefore linear costs for the Eloisatron should be comparable to those for the SSC.

### Beam Beam Limits to Luminosities

The currently projected colliders are two ring machines with beams crossing at an angle. This crossing angle is required to be such as to separate long distance crossings by the order of $10\sigma$. If the angle is too small the effects of long distance collisions become prohibitive, if too large the geometric overlap of the beams is poor and there is a loss in luminosity. $10\sigma$ is an estimate of the expected tradeoff between these factors. The author would expect the use of warm bore dipole bends prior to the first IR quad to provide increased separation of the beams 40 or 50 metres downstream from the IPs. If this is done the same considerations apply to the beam beam limitations at the Eloisatron and at the SSC or LHC. On this basis a total (for all IRs) head-on beam-beam tune shift around the machine in the neighborhood of .01-.02 seems in accord with both current experience and extensive computer modeling for the SSC by the author.

**Effects From Ground Motion**

These effects have been discussed for the SSC in a number of reports including the early work of Fischer and Morton 1. The author has investigated aspects of this problem via simulations.

The results of the above work can be summarized as follows. Physical movement of quadrupoles results in moving both the equilibrium orbits and causing motion of the beam centroids about the equilibrium orbit. The motion about the equilibrium orbit will decohere in time and result in filamentation or emittance growth. Such movements are best parameterized in terms of a power spectrum in frequency. The power law spectrum below the betatron rotation frequency ($\sim 250KHz$) causes adiabatic translations of the equilibrium orbits but minimal emittance growth. At the betatron frequency (or modulo the rotation frequency)motion is induced about the equilibrium orbit causing stochastic emittance growth. High frequency components of ground motion are heavily attenuated and the power spectrum falls sharply as a function of frequency so that at these frequencies the power spectrum can be relatively low. However in the absence of adequate "policing" cultural or man made disturbances resulting for instance from traffic, construction or vibration from compressors can dominate the power spectrum at high frequencies. However with care such sources can be controlled.

In general therefore there are two regions that can cause problems. The first is motion of the equilibrium orbits causing the countercirculating beams to miss each other. This is a few Hz effect and can be controlled by feed back to ensure beam centering. The favored method to accomplish this is the "Jostlein feed back" which puts a small circular sweep motion on one of the beams. The relative luminosity is measured with a forward calorimetric detector. If the beams are centered there will be no modulation of the luminosity. The hadronic debris from the IPs is in the neighborhood of 100 KWatts and therefore fast luminosity measurements of high precision are quite feasible. If the beams are off center there will be a modulation signal at the sweep frequency that can be used to provide feedback. The synchrotron betatron coupling effects caused by the sweep frequency are small compared to those produced by the beam crossing angle in association with synchronous motion and therefore will cause only minimal further degradation of beam beam limits. This has been confirmed by simulation studies by the author.

The other region of interest is the few hundred Hz region. Provided cultural noise sources are kept small, careful but conventional support isolation should remove any residual problems. One caveat is that if the radius of the machine should be very substantially increased to cut the guide field and hence the associated synchrotron radiation problems the damaging power spectrum components would be moved into the 50-100 Hz region and this could require more exotic isolation techniques.

**Conclusions**

Over and above the obvious political problems in obtaining funding at a level five times greater than for the SSC and maintaining the requisite level of enthusiasm of physicists attached to such an enormous enterprise the difficulties appear to lie in the engineering domain. The problem of synchrotron radiation power was alluded to in the introduction.

The stored beam energy is up by a factor two hundred from the SSC. Already at the SSC an accidental loss of beam in the accelerator would have catastrophic consequences. Thus the SSC already will require a 100% efficient beam abort system and always provided that this works it should work for the Eloisatron. Of course the engineering of the scrapers, protection collimators and beam dump(s) is certainly not trivial.

The currents required are up by an order of magnitude but the assumed beam-beam limits are those presently observed and coherent instabilities while a function of peak or average circulating currents are also an inverse function of beam energy. This is therefore not a large extrapolation of existing practices.

The required apertures of the magnets will be comparable to those for the SSC.

The tolerances for preventing emittance growth associated with magnet vibrations are more severe for the Eloisatron but do not appear to require exotic technologies, always provided full superconducting guide field are used.

Thus this author concludes that a reasonable luminosity goal for the Eloisatron is indeed in the range of $1.10^{34}$ cm$^2$/sec luminosities up to $1.10^{35}$ cm$^2$/sec.

## REFERENCES

1. Ground motion tolerances for the SSC. G.E.Fischer, P.L.Morton SLAC,SLAC-PUB-3870, Jan 1986.

# Impedance of Many Holes in a Liner within a Beam Pipe*

Robert L. Gluckstern
Physics Department, University of Maryland
College Park, MD 20742

**Abstract**

The longitudinal and transverse impedance of many holes in a thick wall liner have been calculated for hole dimensions which are small compared to the wave length. These are shown to be small compared with the corresponding resistive wall impedances of the liner at the frequency of the lowest transverse coupled bunch mode. Estimates are made of the effect of the propagation of TEM modes in the coaxial region between the liner and beam pipe, and these can be neglected for a sufficiently thick liner. Estimates are also made of the possible coherent effect of uniformly spaced holes.

## 1 Introduction

As one considers the design of proton colliders at higher and higher energies and intensities, it has become clear that the synchrotron radiation emitted by the circulating proton beam is an increasingly serious problem. In fact, both LHC and SSC find it necessary both to remove the heat generated by the synchrotron radiation and to reduce the field generated by the beam at the wall of the beam pipe. The solution currently under scrutiny is a liner between the beam and the beam pipe, with many small holes so as to maintain adequate vacuum. In this report, we review the calculation of the field and impedance of a liner with holes inside a beam pipe.

*Work supported by the U.S. Department of Energy

# 2 Longitudinal Coupling Impedance

The longitudinal coupling impedance is most easily obtained from the expansion of $E_z$, the axial electric field within liner. Specifically we write

$$E_z(r,\theta,z;k) = \sum_n \int_{-\infty}^{\infty} dq \; e^{-jqz-jn\theta} A_n(q) \frac{J_n(Kr)}{J_n(Kb)}, \tag{2.1}$$

where

$$K^2 = k^2 - q^2, \tag{2.2}$$

and where the contour goes below any poles on the negative real q-axis and above any poles on the positive real q-axis. The longitudinal coupling impedance can written in terms of $A_n(q)$ as[1]

$$Z_{\|}(k) = -\frac{2\pi}{Q} A_o(k), \tag{2.3}$$

where $Q$ is the source charge. The inverse transform of Eq. (2.1) leads to

$$Z_{\|}(k) = -\frac{1}{2\pi bQ} \int\int dS e^{jkz} E_z(b,\theta,z;k), \tag{2.4}$$

where the surface integral is now taken over the interior openings of all holes in the beam pipe. Assuming that each hole is small, we may take the unperturbed source fields for $E_r$ and $H_\theta$. Furthermore, if the wavelength is large compared to the hole size, we can use a quasi-static analysis for the induced electric and magnetic dipole moments.[2] We then obtain the same results as for a single hole, but for the present situation we must sum over all $P$ holes. Thus

$$\frac{Z_{\|}(k)}{Z_o} = \frac{jk}{8\pi^2 b^2}(\psi_{in} - \chi_{in})P, \tag{2.5}$$

where $\psi_{in}$ and $\chi_{in}$ are the inside susceptibility (in the $\theta$ direction) and polarizability of the holes.[3] For harmonic $n = kR$, where $R$ is the radius of the ring, one finds

$$\frac{Z_{\|}(k)}{nZ_o} = \frac{j\eta(\psi_{in} - \chi_{in})}{2b\Delta}, \tag{2.6}$$

where $\eta$ is the fraction of the liner surface covered by holes, and where $\Delta$ is the cross sectional area of each hole.

For comparison, we use a similar calculation to obtain the well known longitudinal coupling impedance for the resistive wall.[4] In this case the integral in Eq. (2.4) is evaluated using

$$E_z = -Z_o H_\theta \frac{k\delta}{2}(1+j), \tag{2.7}$$

where $\delta$ is the skin depth of the interior liner surface. Using the source magnetic field

$$Z_o H_\theta = \frac{Q e^{-jkz}}{2\pi b}, \tag{2.8}$$

we find the result for the longitudinal impedance of the resistance wall:

$$\frac{Z_{\parallel}^{RW}(k)}{n Z_o} = \frac{1+j}{2}\frac{\delta}{b}. \tag{2.9}$$

As a practical matter, we want the contributions from the holes to be smaller than that from the resistive wall. Thus we need to restrict the liner surface fraction $\eta$ to a value determined by the condition

$$\frac{\eta(\psi_{in} - \chi_{in})}{\Delta\delta} < 1. \tag{2.10}$$

For circular holes of radius $r_o$, this reduces to[3]

$$\frac{\eta r_o}{4\delta} < 1. \tag{2.11}$$

A similar analysis for the transverse coupling impedance[1] leads to

$$\frac{Z_\perp(k)}{Z_o} = \frac{j\eta(\psi_{in}^{(0)} - \chi_{in})}{2b\Delta} \cdot \frac{2R}{b^2}. \tag{2.12}$$

Since the transverse resistive wall impedance also satisfies

$$Z_\perp = \frac{Z_\parallel}{n}\frac{2R}{b^2}, \tag{2.13}$$

the same considerations as before also apply to the relative contributions of the transverse impedance caused by the holes and by the wall resistance.

The most serious resistive wall effect corresponds to the lowest frequency for the transverse coupled-bunch modes. For this purpose the LHC design assumes $\delta = 0.1$ mm. With a hole radius $r_o = 1$ mm, Eq. (2.11) implies that $\eta < 0.4$. In the current design with $\eta = 0.05$, which is apparently sufficient to maintain a low vacuum within the liner, the contribution of the holes to either the longitudinal or the transverse impedance will be about 1/8 of the corresponding resistive wall coupling impedance at the frequency corresponding to the lowest coupled-bunch mode.

# 3 TEM Modes in the Coaxial Region

In the previous section we ignored the fields in the coaxial region between the outside surface of the liner and the inside surface of the beam pipe of radius $a$. These fields have the potential to be important, since in this region the TEM mode in the forward direction travels at the velocity of light, in synchronism with the source charge.

Analysis of the penetration of the field through the holes in the liner[5] leads to the following estimate for the TEM field level between the liner and beam pipe, relative to the field due to the source particle. Each is proportional to $1/r$, and their ratio is given approximately by

$$F \cong \frac{\psi_{out} - \chi_{out}}{\psi_{in} - \chi_{in}} \tag{3.1}$$

where the subscripts "out" and "in" refer to the "outside" and "inside" values of the polarizability and susceptibility of the hole[3]. In fact the longitudinal impedance is reduced by a factor $1 - F^2$ from the value given in Eq. (2.5).

In Table 1 we list the values of $F$ and $1 - F^2$ for a circular hole of radius $r_o$ in a wall of thickness $t$. The field penetration is strongly reduced for values of $t/r_o > 0.3$ and is less than 15% of the zero thickness value when $t/r_o > 1.5$. Since the outside susceptibility decays as $\exp(-1.84t/r_o)$ for large $t/r_o$, where 1.84 is the first root of $J_1'(x) = 0$, the energy deposited in the beam ($\sim F^2$) decreases by a factor 40 for an increase of 1.0 in $t/r_o$.

Table 1

| $t/r_o$ | $3\psi_{in}/8r_o^3$ | $3\chi_{in}/8r_o^3$ | $3\psi_{out}/8r_o^3$ | $3\chi_{out}/8r_o^3$ | $F$ | $1 - F^2$ |
|---|---|---|---|---|---|---|
| 0 | 1.000 | .500 | 1.000 | .500 | 1.000 | 0 |
| .1 | .871 | .459 | .764 | .360 | .980 | .04 |
| .3 | .778 | .438 | .501 | .210 | .655 | .57 |
| .6 | .732 | .431 | .291 | .100 | .476 | .77 |
| 1.0 | .715 | .430 | .133 | .037 | .337 | .89 |
| 2.0 | .710 | .429 | .021 | .003 | .062 | 1.00 |
| $\infty$ | .710 | .429 | 0 | 0 | 0 | 1.00 |

We make a few observations at this point:

1. The relative level to which the fields build up in the coax is given in Eq. (3.1). It is extremely small because of the exponential decrease of the numerator with increasing wall thickness.

2. The result for the transverse coupling impedance is even simpler, since no transverse TEM mode with $\cos\theta$ dependence exists in the coaxial region. Thus the result in Eqs. (2.12) and (2.13) is not modified by the existence of the coax.

3. In the absence of losses, the impedance depends on the difference between $(\psi_{in} - \chi_{in})^2$ and $(\psi_{out} - \chi_{out})^2$. For zero wall thickness this difference vanishes and the steady state impedance vanishes. In this case the field level in the coax builds up to that within the liner, leading to no steady state contribution to the impedance.

# 4 Uniform Hole Spacing

In the previous sections we have assumed that the holes are spaced such that only modes traveling wth $v = c$ in the forward direction can remain in synchronism with the driving source and build up. In this section we consider the more attractive practical situation of uniform hole spacing and ask if the coherent buildup of other modes will modify our results.

As an example, we consider the propagation of an azimuthally independent $TM_{o\ell}$ mode with the liner in the case where there are $n_\theta$ holes equally spaced in the azimuthal direction, and where $L_z$ is the axial spacing between rings of holes. The steady state values of the fields lead to an impedance[5]

$$\frac{Z_{\|}(k)}{nZ_o} = \frac{\eta}{2\Delta b}\left[j(\psi_{in} - \chi_{in}) - \frac{jn_\theta(kb\psi_{in} - c_\ell\chi_{in})^2}{4\pi b^3 c_\ell(\alpha L_z - j\theta + j\theta_o)}\right], \tag{4.1}$$

where

$$c_\ell = (k^2b^2 - p_\ell^2)^{1/2}. \tag{4.2}$$

$$\theta_o \equiv \frac{n_\theta(k^2b^2\psi_{in} - c_\ell^2\chi_{in})}{4\pi b^3 c_\ell}. \tag{4.3}$$

and where

$$\theta = (kb - c_\ell)L_z/b \qquad (\text{modulus } 2\pi), \tag{4.4}$$

is the phase advance of the mode relative to the beam bunch between adjacent rings of holes. Here the attenuation constant $\alpha$ can be related to the skin depth $\delta$ by

$$\alpha = k^2\delta/2c_\ell. \tag{4.5}$$

Coherence occurs at frequencies given by

$$k = \frac{m\pi}{L_z}\left[1 + \left(\frac{p_\ell L_z}{2m\pi b}\right)^2\right], \quad m = 1, 2, \cdots. \tag{4.6}$$

where $p_\ell$ are the zeros of $J_o(x)$, and the effective $Q_r$ of this "resonance" can be written as

$$Q_r = \frac{4b}{\delta}\left[1 + \left(\frac{L_z p_\ell}{2m\pi b}\right)^2\right]^{-1}. \tag{4.7}$$

For a skin depth $\delta$ corresponding to the GHz frequency range, $Q_r$ is of order $10^5$. Typical parameters suggest that the first resonance occurs at about 75 GHz, and at its peak, contributes several ohms to $Z_{||}(k)/n$.

The narrow width and high frequency of this mode, together with the bunch length of 5-10 cm reduces the effective impedance for the single bunch mode by a factor of 100 or so. But the situation is not really as serious as indicated here since the axial spacing between holes will not be constant to an accuracy of 1 part in $10^5$. Normal fabrication errors (or an intentional spread in the axial spacing) are likely to reduce $Q_r$ to a value more like $10^2$ and the corresponding contribution to the impedance will be of the order of 0.01 ohms or less.

Thus we conclude that the second term in Eq. (4.1) can be neglected and that coherence effects can be kept small. Similar conclusions apply for TE modes, as well as for modes other than TEM in the coaxial region, and also for coherent propagation of deflecting modes.

# 5 Summary

We have obtained the steady state longitudinal and transverse coupling impedance of a liner with holes within a beam pipe for hole dimensions small compared to the wavelength. For a liner with thickness comparable with the hole dimensions the primary results are obtained neglecting all modes within the liner and in the coaxial region between the liner and the beam pipe. These results are compared to the corresponding contributions from the wall resistance and a simple limit is derived for the fraction of the liner surface that can be covered by holes in order that the impedance of the holes be less than the resistive wall impedance.

We then explore the contributions from all modes which may build up within the liner and in the coaxial region. The most serious for the longitudinal coupling impedance is the TEM mode in the coaxial region which is in synchronism with the drive beam. We show that the correction introduced by this mode and all other modes which may be coherent for uniform hole spacing will be smaller than the primary hole impedance result, provided that the axial hole spacing is not uniform to an accuracy of better than 1 part in 100. (To be on the safe side it may be wise to make sure that such a non-unformity in hole spacing is built into the fabrication process.) Thus, the primary results for the impedance in Eqs. (2.6) and (2.12) accurately represent the physical situation.

We also believe that any reflections which may occur in the coaxial region by the mechanical supports for the liner can be negelected since they serve to further attenuate any coherent effects in the coaxial region.

# 6 Acknowledgment

The author would like to thank J. Bisognano, F. Caspers, A. Chao, G. Dôme, J. Gayrete, F. Ruggiero and B. Zotter for helpful comments and suggestions. He is also grateful to SL Division at CERN for its hospitality during the course of part of this work.

# References

1. R.L. Gluckstern, "Coupling Impedance of a Hole in a Thick Wall Beam Pipe", CERN Report SL/92-05 (AP) 1992.

2. See, for example, H.A. Bethe, *Phys. Rev.* **66** (1944) 163 .

3. See, for example, R.L. Gluckstern and J.A. Diamond, *IEEE Transactions on Microwave Theory and Techniques* **39** No. 2, February 1991, p. 274.

4. See, for example, A. Chao, Proceedings of the 1982 Accelerator Summer School.

5. R.L. Gluckstern, "Coupling Impedance of Many Holes in a Liner with a Beam Pipe", CERN Report SL/92-06 (AP) 1992.

# Diffusion phenomenology of transverse beam dynamics and losses in hadronic colliders

A.Gerasimov

Fermi National Accelerator Laboratory*,
P.O.Box 500, Batavia, IL 60505, USA

February 28, 1992

**Abstract**

Methodological aspects of the data analysis of the recent "diffusion" experiment at Fermilab are presented. The data about the beam density evolution over long times (10-30 min.) is shown to fit well into simple phenomenological diffusion models of transverse dynamics of particles. The diffusion intensity depends on the betatron amplitude, and the good quality fit is achieved when using the fast growing functions like $D = D_0 A^n$ with $n > 11$. The fitting is carried out by computer code through the use of the diffusion evolution simulation.

The typical time of beam storage in hadronic colliders corresponds to $10^8 - 10^{10}$ turns of particles around the ring. As the stability of particles under the influence of the nonlinearities of the lattice over such long time presents itself a major operational issue, the design of new large colliders can greatly benefit from a better understanding of long-term stability of particles in existing colliders. The present paper describes the methodological aspects of the data analysis of the recent "diffusion" experiment at Tevatron/1/. The goal of the analysis is to obtain the "microscopic" information about the character of long-term dynamics of particles in the beam from the "macroscopic" measurements of the properties of the beam as a whole through a special fitting procedure with the phenomenological diffusive models of dynamics.

The diffusive phenomenology we are developing is associated with the expectation that the slow dynamical instabilities of transverse nonlinear (due to the nonlinear elements of the lattice) Hamiltonian motion of the particles in the collider should be of diffusive character (dynamical stochasticity). Since the slow processes due to the not too large perturbations (nonlinearities) should be superimposed on top of the fast unperturbed oscillations in linear lattice, the natural variable for the phenomenological description of the "dynamical" diffusion of particles will be the betatron amplitude(s) $A = \sqrt{x^2 + x'^2}$. Thus, we will describe

*Operated by the Universities Research Association, Inc. under contract with the U.S.Dept. of Energy.

the long-time evolution of the beam density by the phenomenological diffusion models of the form:

$$\frac{\partial \rho_J}{\partial t} = \frac{\partial}{\partial J}\left(D(J)\frac{\partial \rho_J}{\partial J}\right) \tag{1}$$

which is the Fokker-Planck equation for the density distribution $\rho_J$ in the action variable $J = A^2$. The choice of the action variable for the Fokker-Planck equation to have a zero drift term can be motivated by the validity of this particular form as the asymptotic regime of (slow) evolution when the small but otherwise arbitrary noise is superimposed on (fast) Hamiltonian motion /2/.

The nonlinearity of the magnetic fields of the storage rings is always increasing with the deviation from the closed orbit, so the dependence of the diffusion intensity $D$ on the amplitude $A$ is expected to be a monotonically increasing function of $A$. The kicked beam distribution, after fast "decoherence", i.e. averaging out all betatron phase dependence due to the tune dependence on amplitude /3/, took a characteristic band shape in the phase space (see Fig.1) and "hollow beam" shape when projected on the real space $x$ coordinate (see below). The amplitude at the center of the band is approximately equal to the kick amplitude, while its width is about the same as the beam size before the kick, as shown in Fig.1. Assuming now the beam density evolution to be due to the diffusion in betatron amplitude with amplitude-dependent diffusion intensity $D(A)$, one can try to extract the function $D(A)$ from the experimental data.

The "diffusion" experiment at Tevatron consisted in applying a fast horizontal deflection to ("kicking") the beam by an impulse on the separator plates and observing the evolution of transverse density distribution afterwards with "flying wire" monitors/1/. The data taken in the experiment consisted of the sequences (over 10-30 min.) of the horizontal and vertical density profiles, measured by the "flying wire" monitors. We used only the horizontal profiles, since the vertical profiles stayed very much constant in all series. The basic feature of observed profiles is that in the course of time they are getting narrower, which conforms well with the expectation that $D(A)$ is the fast-growing function.

As a first step, we developed a computer code to simulate the diffusive evolution (1) with some given dependence $D(J)$. The actual quantities used in the code were the amplitude $A$ and the density distribution $\rho_A = A\rho_J$. The boundary condition used in the simulation was the absorbing wall at $A = A_{max} : \rho(A_{max}) = 0$. The initial condition at $t = 0$ was obtained by taking the "kicked" gaussian distribution $\rho_A = A\exp\left(-\frac{x^2}{2} - \frac{(p+A_k)^2}{2}\right)$, substituting $x = A\cos\theta$, $p = A\sin\theta$, and integrating it over $\theta$. The evolution of the resulting distribution is shown in Fig.2 for some characteristic value of the kick and diffusion intensity dependence $D = D_0A^8$.

In order to allow for the comparison of the simulation results with experimental profiles, we computed the projection of the distribution function(s) in amplitude on the coordinate axis. This was achieved by using the numerical integration in the formula:

$$\rho_x(x) = \int_0^\infty dp\, \frac{\rho_A(\sqrt{x^2+p^2})}{\sqrt{x^2+p^2}} \tag{2}$$

Since the experimental horizontal profiles were measured at the locations with nonzero horizontal dispersion, the dispersion-convoluted profiles $\rho_{xc}$ were also computed through the numerical integration in the formula:

$$\rho_{xc}(x) = \int_0^\infty dx' \, \rho_x(x - x') exp\left(-\frac{x'^2}{2\sigma_d^2}\right) \tag{3}$$

The width $\sigma_d$ was taken to be the same as in the corresponding set of experimental profiles. For the latter, it was calculated from the formula $\sigma_d = \eta_x \frac{\Delta p}{p}$, with the horizontal dispersion function $\eta_x$ at the flying wire location and the energy spread $\frac{\Delta p}{p}$ taken from experimental measurements. The profiles $\rho_{xc}$ were the ones to be actually compared to the flying wire profiles.

The first question we tried to answer was what kind of the diffusion intensity functions $D(A)$ can lead to the qualitatively similar (to the experiment) behaviour of profiles. The class of functions we are limiting ourselves to are the monotonically increasing ones with an additional property $D(0) = 0$. The last feature seems plausible because of extremely slow evolution of unkicked beams observed in the experiment.

We were able to identify two more successfull classes of simple two-parameter functions $D(A)$. The first class is the power-like functions $D = D_0 A^n$ with large exponents $n$. An example of profile evolution for $n = 8$ for some characteristic kick amplitude is shown in Fig.2. The evolution of the profiles with the exponent $n = 8$ has all the qualitative features of the experimental data: the tails are quickly "swept out", the profile widths are decreasing, and at large times the evolution slows down. The beam intensity dependence on time is also quite close to saturation at large time as it should. Overall, we can say that the diffusion phenomenology with power functions with large exponents is close to the familiar concept of dynamic aperture /4/ (for asymptotically large $n$ they are equivalent) and can be viewed as a refinement of the latter.

The second class of functions is:

$$\begin{cases} \mathrm{D(A)=D_0}(A - A_0)^k & A < A_0 \\ \mathrm{D(A)=0} & A \geq A_0 \end{cases} \tag{4}$$

which we call the "edge" functions. The case $k = 1$ and $k = 2$ proved to be not feasible because of the formation of long tails, but all the larger $k$ give a qualitatively correct evolution, as will be shown in some examples later. For practical purposes however the evolution in the "edge" and powerlike models is quite similar.

In the final data analysis /1/, a more complicated four-parameter function $D(I)$ was used,which was the same model (4), with the addition of a constant $D_{00}$ and $k$ being free real parameter. Though the quality of the fit, achieved with that model was higher than what we present below for two-parameter models, was higher, the real advantage is marginal. Indeed, the four-parameter fit is awkward and mostly imparactical while the quality of the fit may not make much difference as long as it exceeds a certain level (20

The next step was to achieve a quantitative agreement between simulated and observed evolution. For this, we developed a special code which optimizes the two parameters of the model aiming to fit the experimental profiles with simulated ones as well as possible. The previously described simulation program was used in this fitting code as an ingredient. In order to have the measure of the goodness of fit, we used not the whole profiles as they depend on time, but rather the dependences on time of two "collective variables": the half-maximum width $\Delta(t)$ and intensity $I(t)$ (area under the profile). The goodness of fit parameter $\epsilon$ was defined as:

$$\epsilon = \sum_n (\Delta_{sim}(t_n) - \Delta_{exp}(t_n))^2 + (I_{sim}(t_n) - I_{exp}(t_n))^2 \qquad (5)$$

where $t_n$ denote the moments of time when the experimental profiles measurements were taken, and both $\Delta$ and $I$ are normalized by the conditions $\Delta(0) = 1, I(0) = 1$. The subscripts *sim* and *exp* stand respectively for simulation and experimental data.

The quality of fitting achieved in such way was quite high and constituted, in considered cases, no worse than 10% error at any time $t_n$ for all quantities. The parameter $A_0$ yielding the best fit for the "edge" model was "well defined", since the dependence of the goodness of fit $\epsilon$ on $A_0$ (for optimized $D_0$) always showed a clear minimum. We discovered also that the values of $k$ from 3 to 5 were always yielding the best fits, so in what follows we fixed $k$ to be equal to 4. When fitting with the power-like models the quality of the fit was always monotonically improving with increasing $n$, with the saturation after $n$'s from 10 to 12. Both "edge" model with $k = 4$ and power-like models with $k > 12$ were giving about the same quality of fit. It should be noted however that we did not make $n$ as high to give the dynamic-aperture type of evolution: the intensity and half-max widths always showed some finite time before saturation even for $n = 14$.

An important feature of experimental data taken was the use of several different kick amplitudes to study each particular configuration of the machine. To make use of this information, we developed another fitting code which minimizes the sum $\epsilon_t$ of goodness-of-fit parameters $\epsilon_1$ and $\epsilon_2$ referring to two different data sets (with 6 and 9 kV kick amplitudes) by varying the two parameters of a single diffusion model. Thus, four experimental curves constituted one data set and were fitted simultaneously: two intensity and two half-max width dependences versus time for 6 and 9 kVs.

We present here the results of fitting for two characteristic examples, while the complete report will be published elsewhere /1/. The first case corresponds to the tunes $\nu_x = .407$, $\nu_y = .45$ and is characterized by rather gradual loss of particles during the observation with the final saturation of intensity at the level of about 20% of initial one for 9 kV data and about 50% for 6 kV data, as shown in Fig.3. The observation time was about 14 min. for 9 kV and 25 min. for 6 kV. The results of the fitting with the "edge" model with $k = 4$ are shown in Figs.3 and 4. The maximum deviation of simulated curves from experimental ones to about 30% is in our opinion within the limits of acceptable for such a phenomenological approach. The edge parameter $A_0$ was found to be $A_0 = 1.7$ , and the graph in Fig.5 shows how the goodness of fit $\epsilon_t$ depends on $A_0$. This graph indicates also that the quantity $A_0$ is not defined within our approach to the precision better than $\pm.15$. In Fig.6, the evolution of density distribution in amplitude $\rho_A$ is shown. The reason for the saturation of both half-max widths and intensities is quite clear from this graph: by the end of the observation time most of the particles initiated beyond $A = A_0$ are lost and distribution density there is nearly zero. In the dispersion-convoluted real-space profiles, shown in Fig.7, these features are concealed, so that it is difficult to understand what is actually happening directly from experimental profiles. Also, it should be noted that the fitting with the power-like models in this case gave very much the same quality of fit for the exponents $n$ larger than 10.

The second set of data corresponded to a different, than in the first two, sextupole configuration and tunes $\nu_x = .391, \nu_y = .45$. The beam intensity dependence on time in this case was rather much different from what was observed in two previous cases: nearly all the loss occured over the first minute. This behaviour was fitted by the edge model with $A_0 = 1.85$ remarkably well, as shown in Figs.8 and 9. The evolution of the density profiles in amplitude is shown in Fig.10, which shows that this situation is quite close to what is understood by dynamical aperture. Like in the previous case, the power-like fitting was yielding very similar results.

Further experiments are necessary in order to further the suggested approach. One upgrade of the experiment, which can be very useful in determining the consistency of the phenomenology, is scanning the kick amplitude with a small step rather than using two or three values. The effect of energy spread and power supply ripples on the diffusion should be also studied, since they are known, the former from theory /5/ and the latter from experiment /6/, to cause stochastic instabilities of dynamics.

An important future supplement to this work can be the use of tracking codes to simulate the experimental data. This can be useful twofold: first, as allowing some understanding for the nature of observed diffusion and providing thus a support for the phenomenology, and second, as a means of checking the reliability of tracking. The last aspect is especially important in view of the ongoing use of tracking for the design of new colliders, e.g. SSC.

Finally, I wish to express my gratitude to Richard Talman and Steve Peggs for a number of usefull discussions.

## References

**1.** T.Chen, A.Gerasimov et al., Phys.Rev.Letters 68 (1992) 33.

**2.** M.Buttiker, E.Harris and R.Landauer, Phys.Rev.A 26 (1982) 2805.

**3.** R.Meller et al., Decoherence of kicked beams, SSC-N-360, May 1987.

**4.** J.Hagel, E.Keil, Eds., Second Advanced Beam Dynamics Workshop on Aperture-related Limitations of the performance and Beam Lifetime in Storage Rings, Geneva 1988.

**5.** A.Gerasimov et al. "Synchrotron sideband overlap in electron positron colliding beams", AIP Conf. Proc. 153, v.2, p.474 (1987).

**6.** D.Brandt et al., Proc. of 2nd European Part. Accel. Conf., V2,1436 (1990).

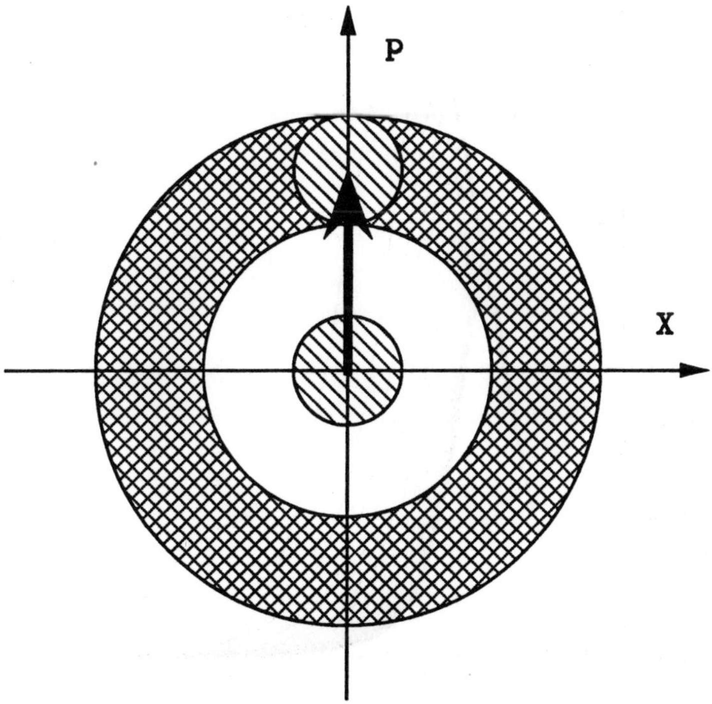


Figure 1: Beam distribution before the kick, immedeately after the kick (both regular hatched) and after decoherence (cross hatched).

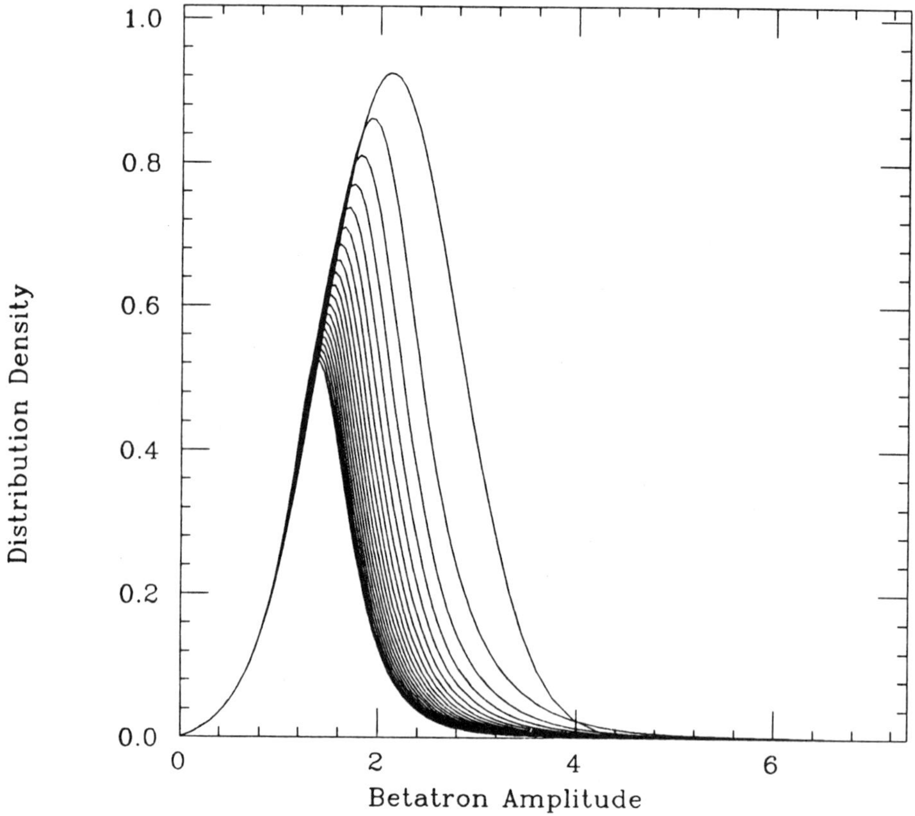


**Figure 2** The evolution of the density distribution $\rho_A$ for the diffusion intensity dependence $D = D_0 A^8$. Successive profiles are equally spaced in time. Betatron amplitude is measured in units of $\sigma$ of the beam distribution before the kick.

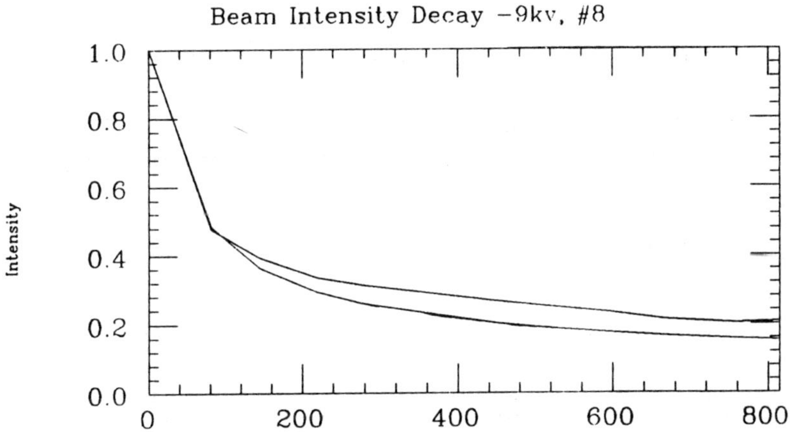


(b)

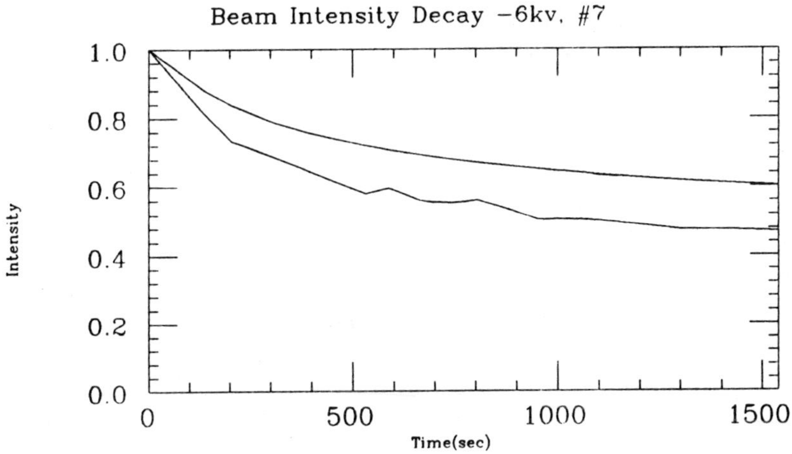


(a)

**Figure 3:** The dependences on time of experimental and simulation beam intensities $I_{exp}$ and $I_{sim}$ for both 6 (graph $a$) and 9kv (graph $b$) kick amplitudes and "edge" diffusion dependence with $A_0 = 1.7$. Time is measured in seconds.

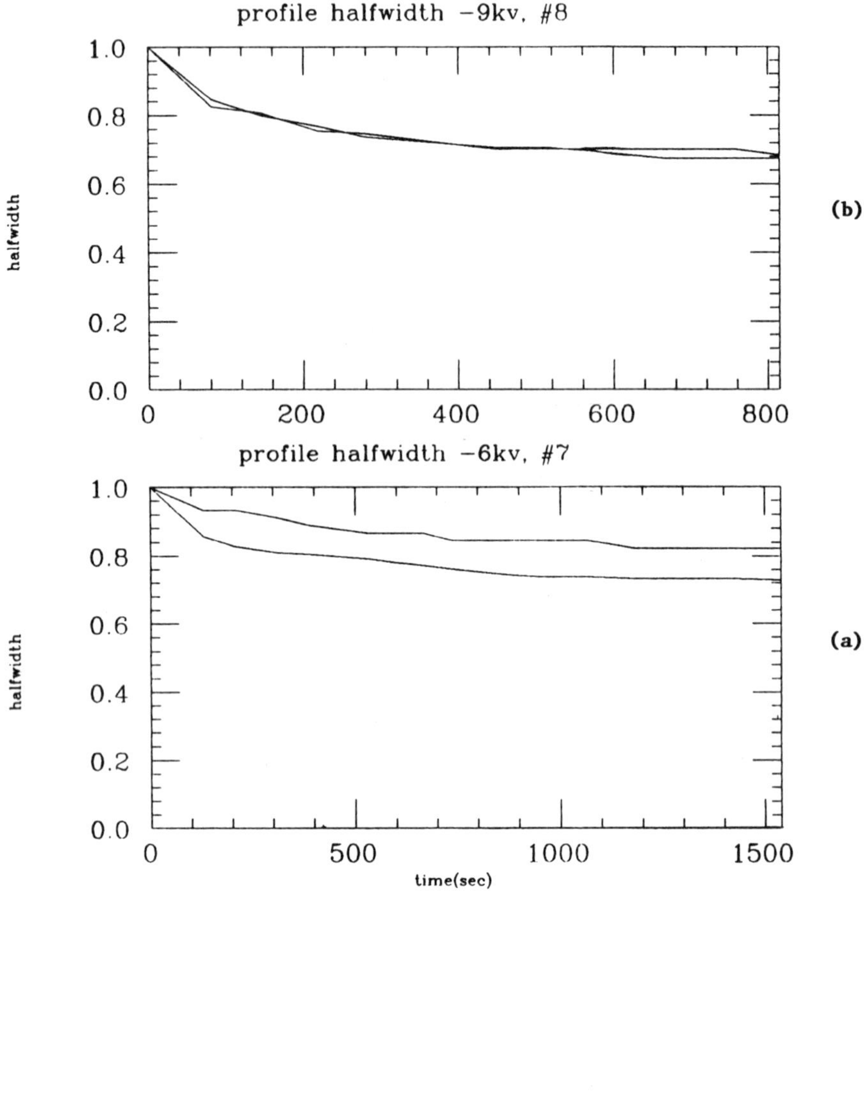


**Figure 4: The dependences on time of experimental and simulation half-max widths $\Delta_{exp}$ and $\Delta_{sim}$ for the case of Fig.3.**

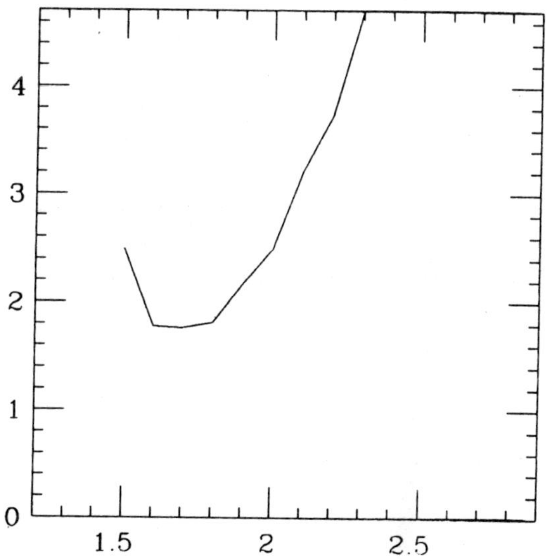


Figure 5: The dependence of the goodness-of-fit parameter $\epsilon$ (5) on the position of the "edge" $A_0$ for the case of Fig.4. Parameter $D_0$ is optimised for each value of $A_0$.

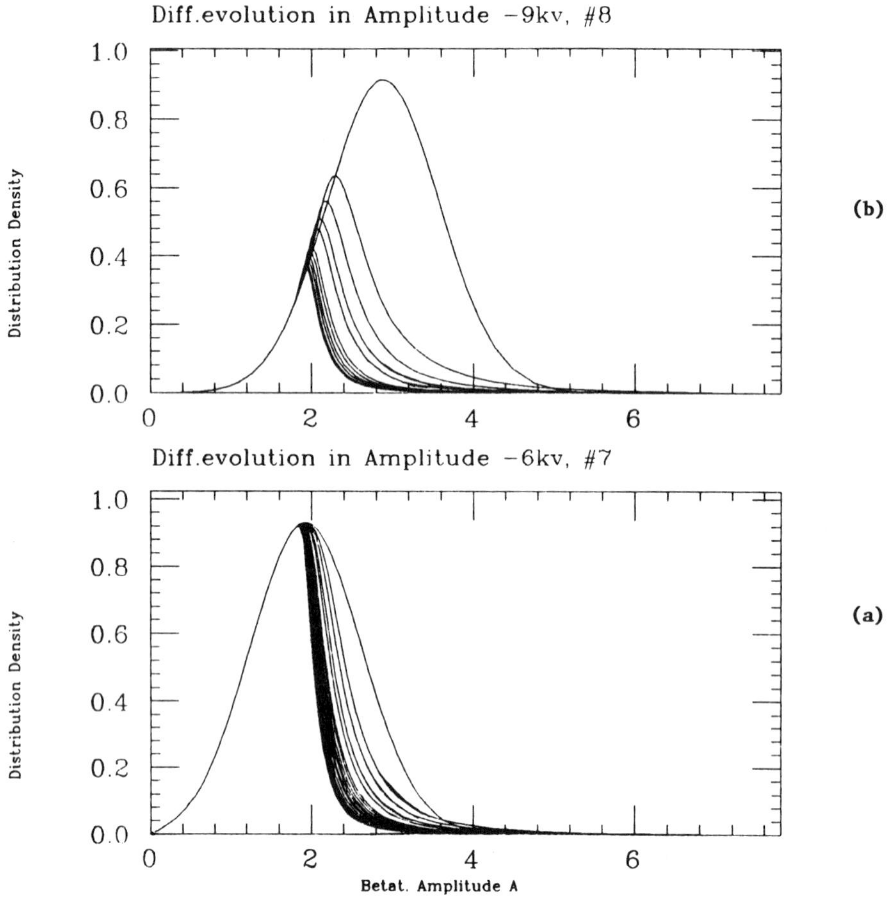


Figure 6: The evolution of the density distribution $\rho_A$ for the case of Fig.4.

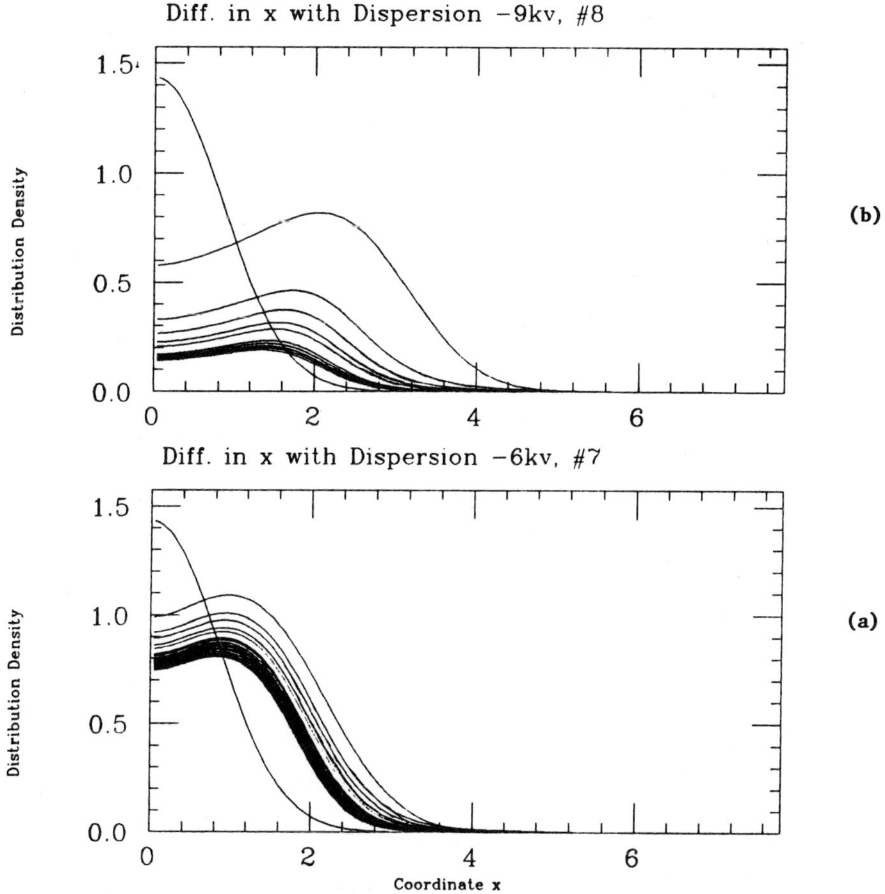


Figure 7: The evolution of dispersion-convoluted density distribution $\rho_{xc}$ for the case of Fig.3.

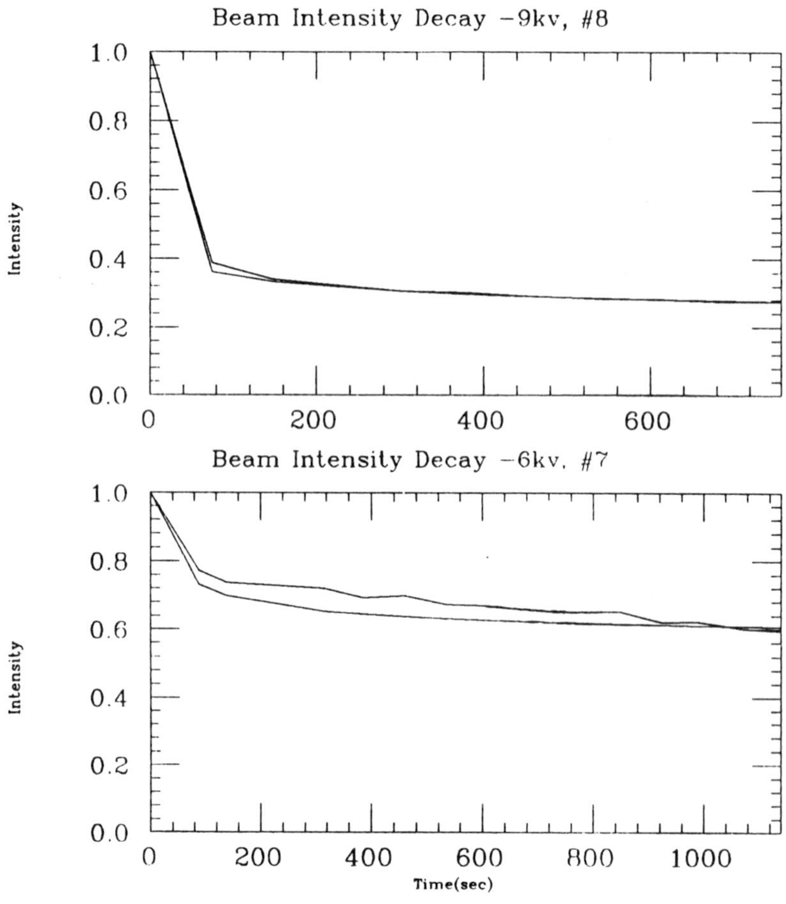


Figure 8: The dependences on time of experimental and simulation beam intensities $I_{exp}$ and $I_{sim}$ for both 6 (graph $a$) and 9kv (graph $b$) kick amplitudes and "edge" diffusion dependence with $A_0 = 1.85$. Time is measured in seconds.

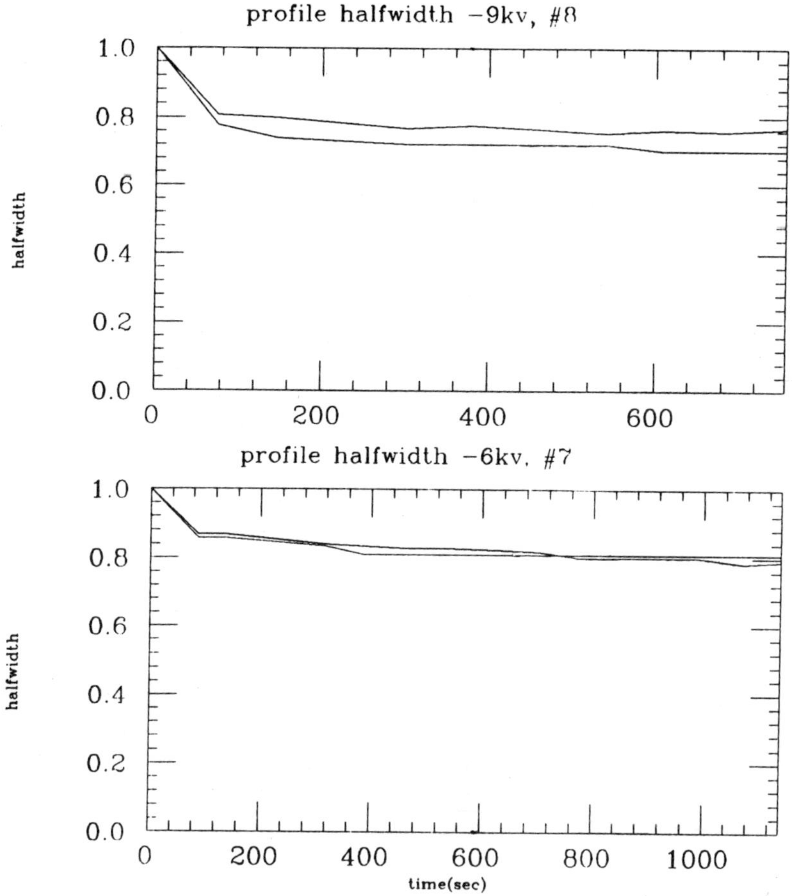


Figure 9: The dependences on time of experimental and simulation half-max widths $\Delta_{exp}$ and $\Delta_{sim}$ for the case of Fig.8.

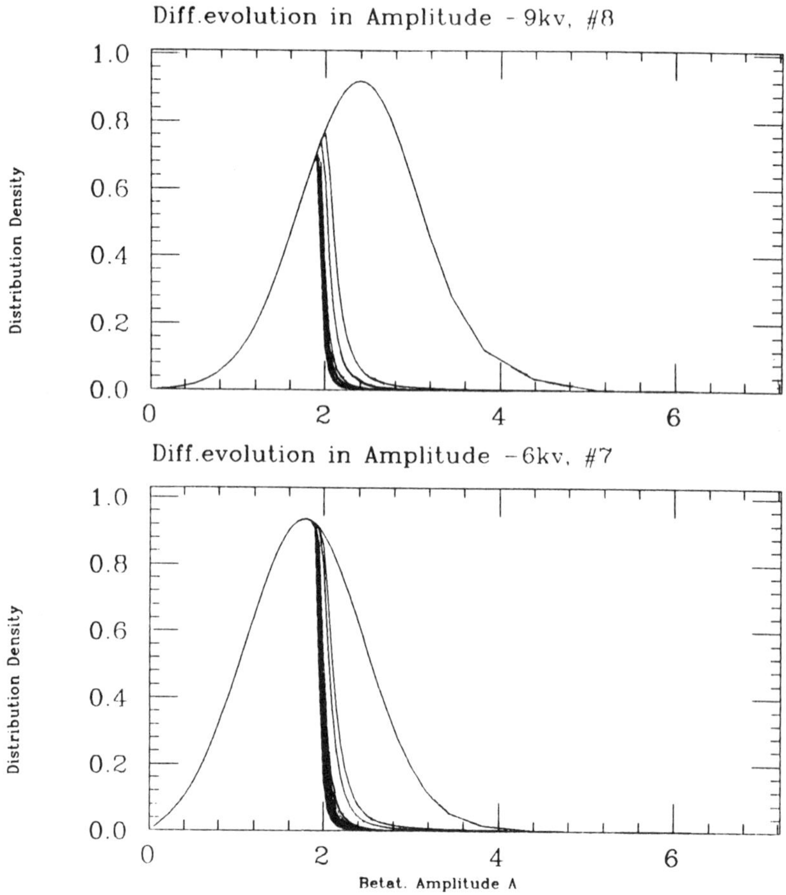


Figure 10: The evolution of the density distribution $\rho_A$ for the case of Fig.8.

# THE EFFECT OF CHROMATIC DECOHERENCE ON TRANSVERSE INJECTION OSCILLATION DAMPING

Gerald P. Jackson

*Fermi National Accelerator Laboratory, MS 341, P.O. Box 500, Batavia, IL 60510*

In order to eliminate or reduce transverse emittance growth during transfers between accelerators, transverse damper systems are used to eliminate residual dipole oscillations before phase space dilution takes place. In transfers where the target accelerator has high chromaticity or the beam has a large momentum spread, phase space dilution due to chromatic decoherence can take place on a scale short compared to the damping time of the transverse injection oscillation damper. The effect of the damper on the beam phase space is not clear while the coherent oscillation is suppressed by this decoherence. The purpose of this paper is to quantify the effectiveness of dampers at eliminating emittance blowup at transfers in the presence of chromatic decoherence.

## 1. Chromatic Decoherence

When a beam is injected with a transverse error, the resultant coherent betatron oscillation can be written as

$$x(n) = x_o \cos(2\pi \nu_x n) \quad , \qquad (1)$$

where n is the turn number after injection, $\nu_x$ is the betatron tune of the target accelerator, and for clarity sake the injection error is assumed to be purely positional of amplitude $x_o$. Since every particle is assumed to have the same betatron tune, this coherent oscillation would continue forever with unchanged amplitude if no damper acted on the beam.

Assuming that the phase space distribution of the beam is Gaussian in all dimensions, and assuming that all particles have the same synchrotron tune $\nu_s$, then in the case of nonzero chromaticity $\xi_x = d\nu_x/(dE/E)$ the betatron motion is described by the equation[1]

$$x(n) = x_o \cos(2\pi \nu_x n) \exp\left[-\frac{2\xi_x^2 \sigma_\delta^2}{\nu_s^2} \sin^2(\pi \nu_s n)\right] \quad , \qquad (2)$$

where $\sigma_\delta$ is the fractional momentum spread of the beam. Of course, due to the sinusoidal nature of the RF voltage, synchrotron tune is a function of longitudinal oscillation amplitude. Therefore, the complete recoherence predicted by this equation each time n is a multiple of $1/\nu_s$ is actually reduced each synchrotron period until the coherent betatron oscillation has completely and permanently decohered into transverse emittance growth.

## 2. Simulation Results

It is possible to simulate chromatic decoherence and the effect of dampers by using a multiparticle simulation computer program. By tracking a large number of test particles through a theoretical accelerator containing the appropriate physics and compiling the average position and rms width of each phase space plane each turn, one can predict the output of actual accelerator diagnostics. The number of test particles used was 1000.

In this paper a consistent set of conditions have been simulated. The only parameters of interest which are modified are the bunch length (to simulate either a long or short bunch), the chromaticity, and the feedback gain. Unless otherwise stated, the conditions simulated are those listed in table 1.

Table 1: Standard parameters used in the computer simulation runs reported in this paper. The values are chosen to approximate injection into the Tevatron Collider.

| Parameter | Value | Units |
|---|---|---|
| Initial Geometric Emittance | $6\times10^{-9}$ | m-rad |
| Initial Beam Size | 0.34 | mm |
| Fractional RMS Energy Width | $0.27\times10^{-3}$ | |
| Synchrotron Tune | $3.28\times10^{-3}$ | |
| Chromaticity | 10 | |
| $\sigma_\delta \, \xi_x / \nu_s$ | 0.82 | |
| Initial Injection Error | 1 | mm |
| Feedback Gain | 0.01 | |
| RF Period | 18.81 | nsec |
| RMS Bunch Length (Short) | 0.78 | nsec |
| RMS Bunch Length (Long) | 3.13 | nsec |

In figures 1 and 2, the evolution of the injection betatron oscillation and the rms beam size are simulated assuming a short bunch.. In cases such as this, where the bunch length is much smaller that the bucket length, the beam particles predominantly sample the linear part of the RF sine wave and the recoherence phenomenon repeats for many synchrotron periods. Particles with larger synchrotron oscillation amplitudes have a synchrotron tune lower than small amplitude particles. If the rms bunch length becomes comparable to the RF period the nonlinearity of the buckets prevents all of the particles from rephasing after a small-amplitude synchrotron oscillation has taken place. Figures 3 and 4 show the evolution of the coherent betatron oscillation and beam size in this situation. Because a large percentage of particles are at larger amplitude, the bulk modulation frequency is also reduced.

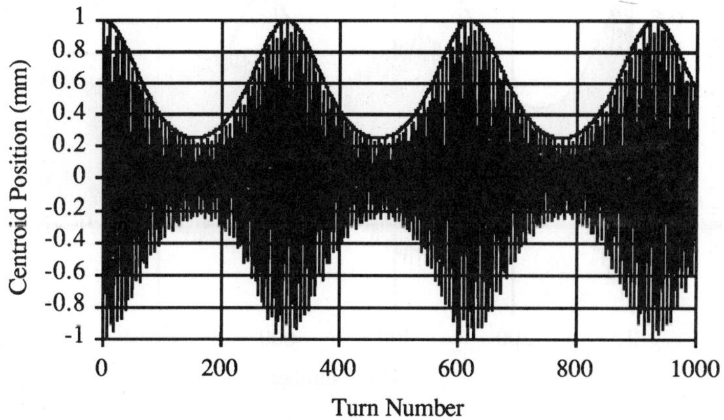


Figure 1: Example of the comparison between the output of a multiparticle simulation program and equation 2 (upper smooth curve) for a short bunch.

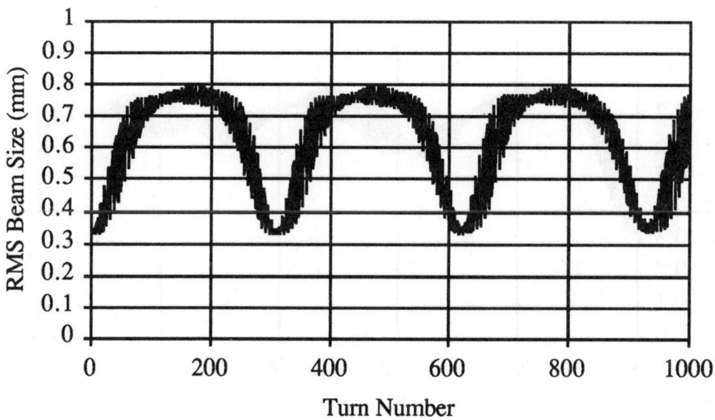


Figure 2: Same run as shown in figure 1, computer program simulation of the rms beam size just after injection in the presence of chromatic decoherence.

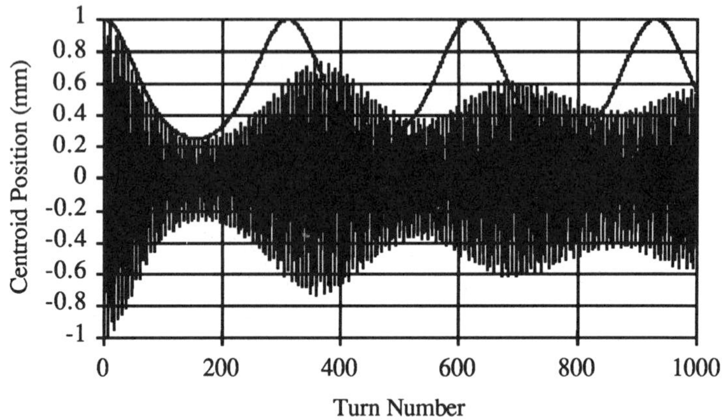


Figure 3: Same conditions as simulated in figure 1, but with a long bunch. Note that the bulk synchrotron frequency is slower due to the dependence of synchrotron tune on amplitude.

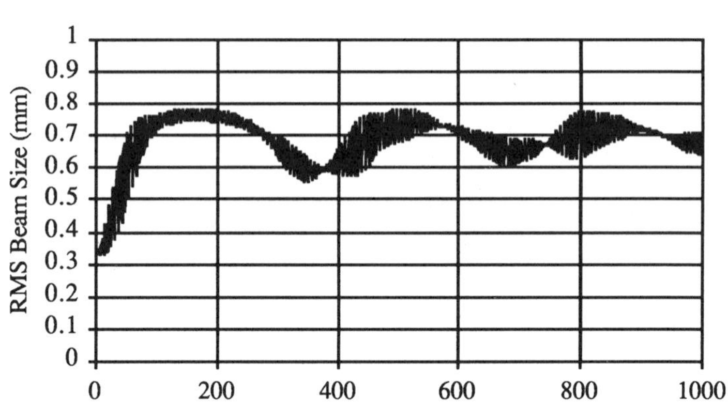


Figure 4: Beam size evolution from the same run displayed in figure 3. Note that with the larger tune spread of the longer bunch, the permanent conversion of coherent amplitude to emittance growth occurs more rapidly.

The dependence of equilibrium emittance as a function of injection error (normalized to the beam size) can be calculated theoretically[2]. Figure 5 is the result of this calculation, where D is the dilution factor defined as the equilibrium emittance divided by the initial emittance and δ is the injection error divided by the beam size. This curve was calculated using a 95% criterion for the emittance and dividing by 6 to generate a rms value (assuming that the distribution was Gaussian). According to the curve, for the standard parameters listed in table 1 the value of d is 0.29. Looking at figure 5, this corresponds to an emittance dilution of a factor of 4, or beam size step increase of a factor of 2. The data in figure 4 confirms this prediction.

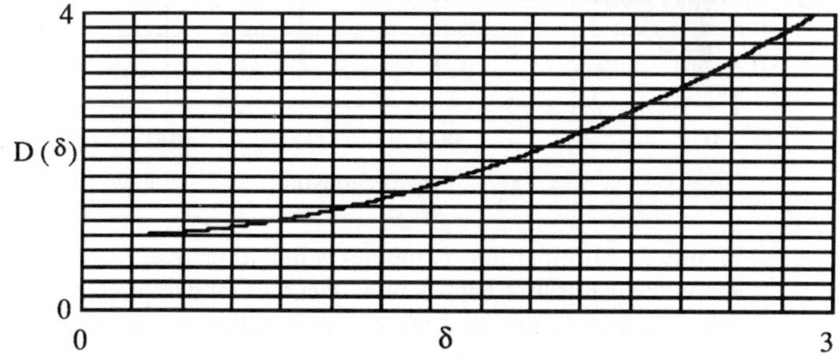


Figure 5: Emittance dilution factor as a function of the transverse injection oscillation amplitude in units of beam size at injection. The horizontal axis is 0.2 per division, the vertical axis is 20% per division.

### 3. Dampers

The effect of a damper is to reduce the oscillation amplitude of an initial betatron oscillation as a exponential function of turn number. A damper is typically composed of a pickup, a preamplifier, a delay, a power amplifier, and a kicker where the distance between the pickup and kicker is an odd multiple of 90° of betatron phase advance apart. Figure 6 contains a sketch of such a system.

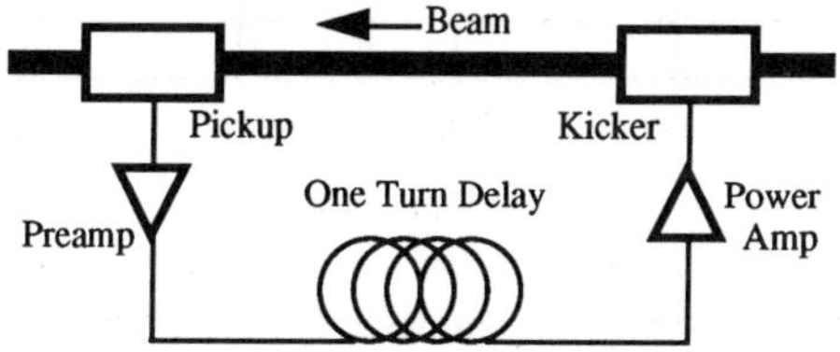


Figure 6: Sketch of a typical transverse injection oscillation damper system.

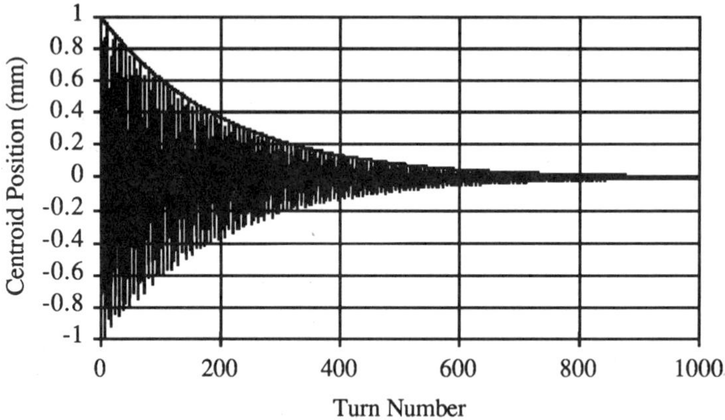


Figure 7: Multiparticle computer simulation result and analytical predication (upper smooth curve) for damping of coherent motion due to a transverse injection oscillation damper system. For this run the chromaticity was set to zero, eliminating the chromatic decoherence shown earlier.

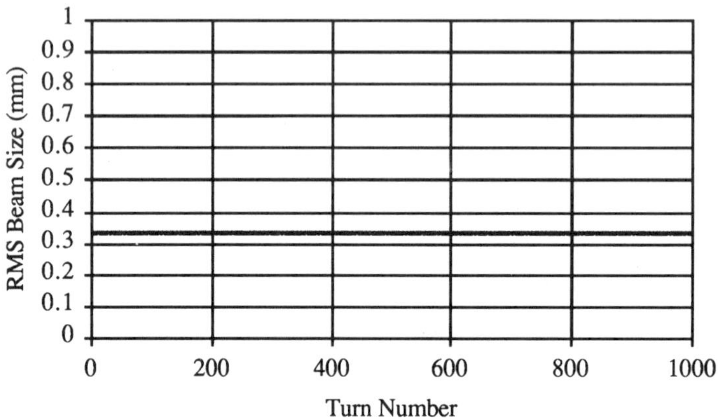


Figure 8: Same run as shown in figure 7, but showing the evolution of the rms beam size when no decoherence mechanism is competing with the damper system.

The gain g of such a system is defined to be the fractional correction the damper makes each turn. A unity gain system therefore completely removes the beam angle at the kicker which corresponded to the measured beam offset at the pickup. For cases where $g \ll 1$, the equation describing the motion of the beam centroid with turn number is an exponential with a characteristic decay time of 2/g

$$x(n) = x_0 \cos(2\pi\nu_x n) \exp\left[ -\frac{n}{\left( \frac{2}{g} \right)} \right] \tag{3}$$

Figures 7 and 8 show the evolution of the centroid oscillation and the beam size of a bunch injected with an initial position error, zero chromaticity, and a damper system active with a gain of 0.01. Note that as in the previous section, the results of multiparticle simulations agree very well with theoretical predictions.

**4. Damping During Chromatic Decoherence**

The point of this paper is to combine the above two phenomena, chromatic decoherence and transverse damping, to assess the utility of injection oscillation dampers as a tool to prevent emittance growth at transfers. Given the fact that a theoretical calculation has not yet been performed, this analysis is limited to multiparticle simulation program results. As a result, in order to make the results presented here applicable to transfers in other accelerators, the worst case scenario of an injection error much larger than the beam size is simulated.

There are only 4 parameters which specify the dynamics of injection damping in the presence of chromatic decoherence. First, $\nu_s$ determines the time scale of the problem. Second, $2\nu_s/g$ determines the damping time (in units of synchrotron periods) in the absence of decoherence. Third, the parameter $\sigma_\delta\xi_x/\nu_s$ determines the depth of the chromatic modulation of the coherent betatron oscillation amplitude. Fourth, bunch length determines the speed (again, in units of synchrotron periods) at which a coherent betatron oscillation is permanently converted into emittance growth.

Since the time scale set by the synchrotron tune is only relevant for extrapolation of results to other accelerators, it can be considered to be an arbitrary parameter in the simulation results presented below. The other three parameters actually determine the effectiveness of the injection damping system. Therefore, though it can be accomplished by another choice of variable, in this paper chromaticity and feedback gain are varied in order to understand the effect of each parameter on the asymptotic emittance after the initial oscillation has damped down. In order to demonstrate the effect of the rate of permanent emittance growth (determined by a finite bunch length), all simulations are performed for both long and short bunch conditions.

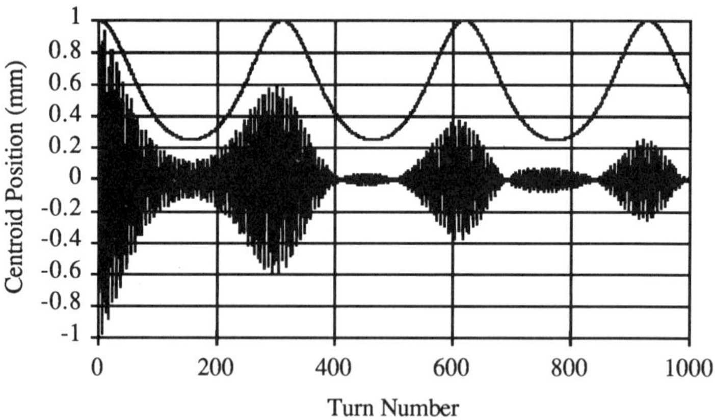


Figure 9: Damping and decoherence with a short bunch. The upper curve is simply the prediction from equation (2) not including the damping.

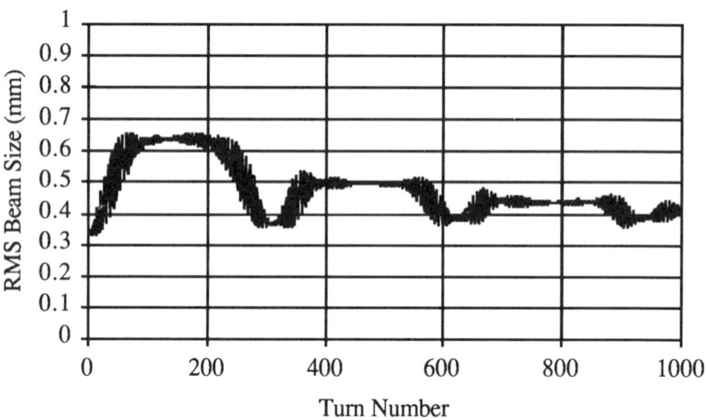


Figure 10: Damping and decoherence with a short bunch for the same simulation run displayed in figure 9. Note that for the standard conditions in table 1, some permanent emittance growth occurs before the coherent oscillation is completely damped.

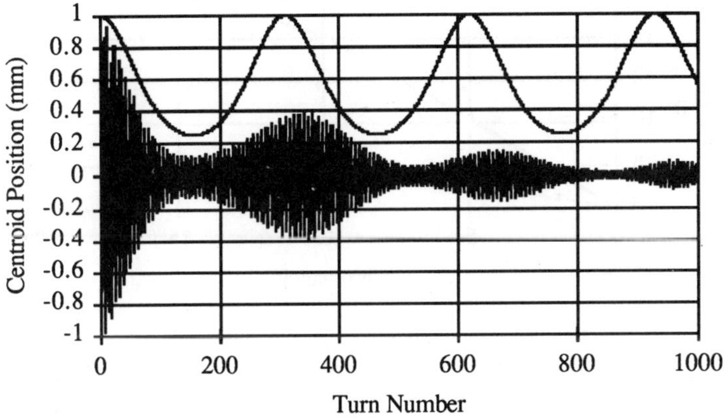


Figure 11: Damping and decoherence with a long bunch. Again, the bulk synchrotron frequency is reduced due to the large synchrotron amplitude particles sampling the nonlinear restoring force of the sinusoidal RF voltage.

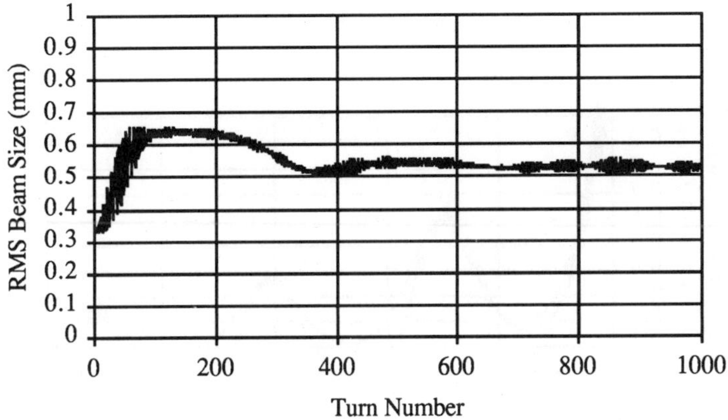


Figure 12: Damping and decoherence with a long bunch. For long bunches the equilibrium emittance is asymptotically approached much quicker than the short bunch case.

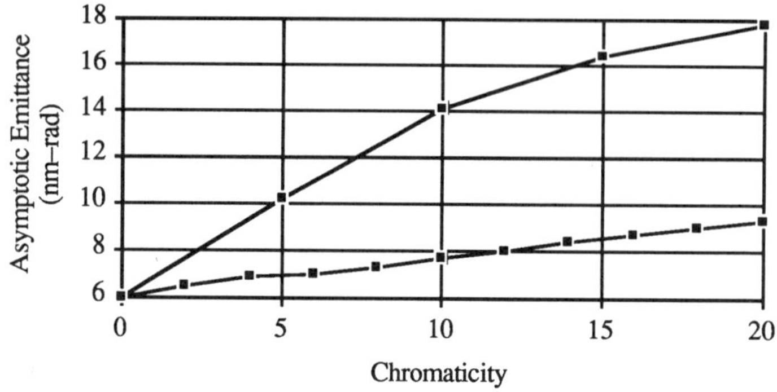


Figure 13: The dependence of the asymptotically attained emittance as a function of chromaticity for a short (bottom) and long (top) bunch.

The effectiveness of injection betatron oscillation damping vs. chromaticity is the first study presented. Since the figure of merit is the increase in emittance due to the injection error during transfer, a plot of equilibrium emittance vs. chromaticity is displayed in figure 13. With both long and short bunch results plotted, the first feature to notice in figure 13 is the fact that the faster the permanent decoherence rate, the larger the emittance growth with increasing chromaticity. In order to keep the emittance growth below 10%, a long bunch must have a chromaticity of less than 1 unit, while for a long bunch a chromaticity of approximately 2 units is tolerable. This is in the case of a damping time approximately two thirds of a synchrotron period (2/g=200 turns).

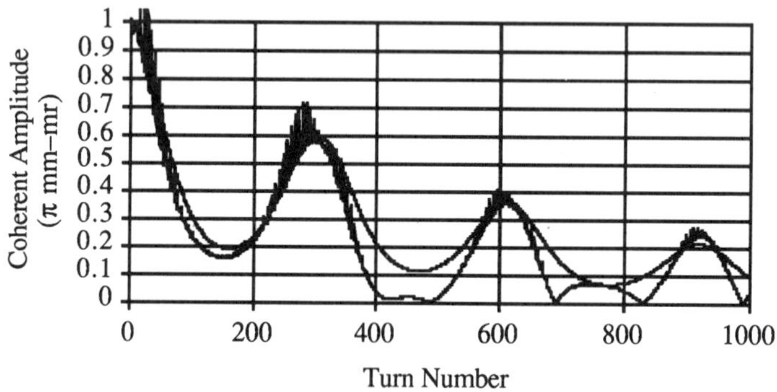


Figure 14: Amplitude of a coherent betatron oscillation vs. turn number for a short bunch. The smooth curve superimposed on the data, and in most cases above the simulated data, is just equation (2) multiplied by an exponential decaying factor.

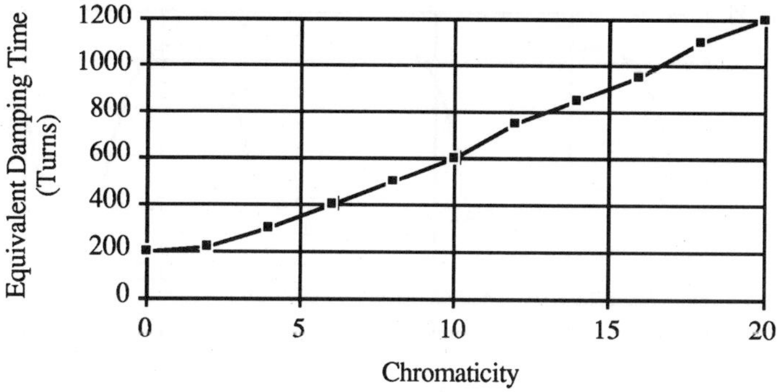


Figure 15: By adjusting the exponential decay time to match the amplitude of the second amplitude recoherence peak, the above plot of equivalent damping time (in turns) vs. chromaticity is generated.

To try and quantify this result a little more, it is necessary to understand the effective damping rate as a function of chromaticity. First, for zero chromaticity the damping time of a g=0.01 feedback system is 200 turns. For a short bunch, the time evolution of the coherent betatron amplitude vs. turn number is fitted with a function which is composed of equation (2) multiplied by an exponential decay factor. By finding the decay time in the exponent which generates the best agreement between this function and the data, an equivalent damping time dependence on chromaticity can be found. Figure 14 shows an example of such a best fit. Figure 15 shows the measured relationship between this equivalent damping time and chromaticity. Oddly enough, this dependence is a straight line with an asymptotic minimum of 200 turns (the value of 2/g). Therefore, for a short bunch and the default machine conditions, the equivalent damping time is 600 turns.

Often the gain of the feedback system is easier to change than the chromaticity. Chromaticity sometimes must be kept high due to instabilities or drifting sextupole moment around the ring due to time dependent persistent currents in superconducting dipole magnets. For fixed chromaticity, the evolution of emittance with turn number as a function of gain is shown in figure 16. Compiling the results of equilibrium emittance for both long and short bunch simulations for a wide range of gains, the plot in figure 17 is generated.

There are a number of interesting features in figure 17. Again, it is clear that a higher gain system is required for long bunches. The surprising feature of these results is the necessity of very short damping times (10 turns for long bunches and 20 turns for short bunches) to keep the emittance growth down to less than 10%.

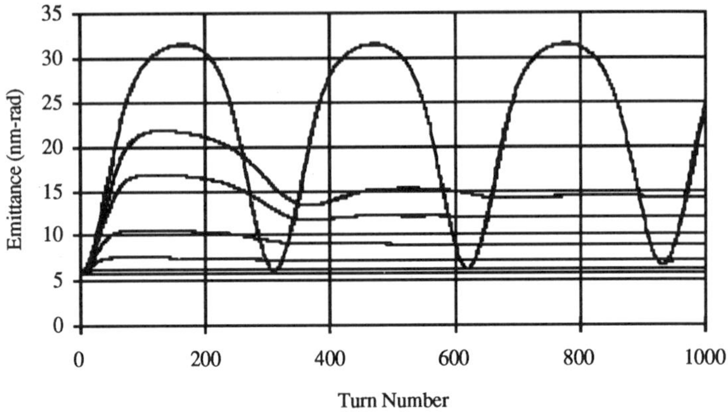


Figure 16: Time evolution of emittance for a number of gains. The top, fully modulated curve is for no damping. The lower curves are for gains of 0.01, 0.02, 0.05, 0.10, 0.20, and 0.50. This data was generated with a long bunch.

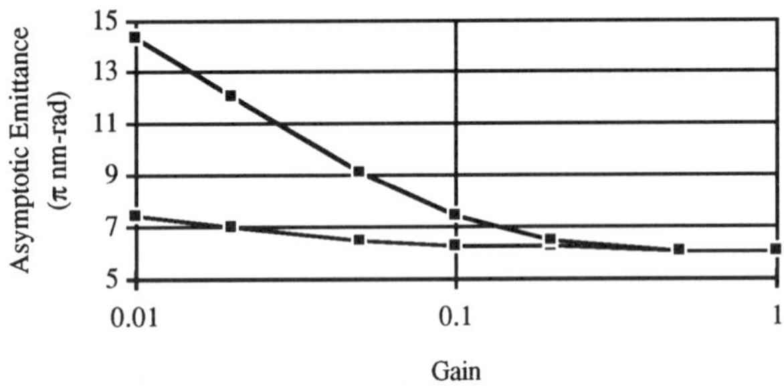


Figure 17: Asymptotically approached emittance as a function of gain for the data displayed in figure 16. The top curve is for a long bunch, the lower curve for a short bunch.

## 5. Extrapolation to SSC and Elosiatron

Tables 2 and 3 contain the SSC[3] and Eloisatron[4] values of the parameters shown in table 1. The initial geometric emittance and beam size are design values. The energy width and synchrotron tune are also design values. The chromaticity was chosen to keep the modulation amplitude $\sigma_\delta \xi_x / \nu_s$ at the value of 0.82. The injection error was chosen to be the same amplitude relative to the beam size as in the Tevatron case. The feedback gain is adjusted so that the damping time is the same number of synchrotron periods as in the Tevatron case. The RF period is calculated from the design RF frequencies, while the bunch lengths are adjusted to be the Tevatron fraction of the RF bucket occupied.

Table 2: Standard parameters used in the computer simulation runs reported in this paper. The values are chosen to approximate injection into the SSC Storage Ring.

| Parameter | Value | Units |
|---|---|---|
| Initial Geometric Emittance | $0.5 \times 10^{-9}$ | m-rad |
| Initial Beam Size | 0.34 | mm |
| Fractional RMS Energy Width | $0.1 \times 10^{-3}$ | |
| Synchrotron Tune | $2.2 \times 10^{-3}$ | |
| Chromaticity | 18 | |
| $\sigma_\delta \xi_x / \nu_s$ | 0.82 | |
| Initial Injection Error | 1 | mm |
| Feedback Gain | 0.0067 | |
| RF Period | 2.8 | nsec |
| RMS Bunch Length (Short) | 0.12 | nsec |
| RMS Bunch Length (Long) | 0.47 | nsec |

Table 3: Standard parameters used in the computer simulation runs reported in this paper. The values are chosen to approximate injection into the Eloisatron.

| Parameter | Value | Units |
|---|---|---|
| Initial Geometric Emittance | $0.1 \times 10^{-9}$ | m-rad |
| Initial Beam Size | 0.1 | mm |
| Fractional RMS Energy Width | $0.03 \times 10^{-3}$ | |
| Synchrotron Tune | $1.0 \times 10^{-3}$ | |
| Chromaticity | 27 | |
| $\sigma_\delta \xi_x / \nu_s$ | 0.82 | |
| Initial Injection Error | 0.3 | mm |
| Feedback Gain | 0.003 | |
| RF Period | 2.2 | nsec |
| RMS Bunch Length (Short) | 0.091 | nsec |
| RMS Bunch Length (Long) | 0.37 | nsec |

Applying the results found for the Tevatron parameters, for the parameters in table 2 the SSC ring must keep their chromaticity below 2 units for long bunches and 4 units for short bunches to restrict the injection emittance growth down to under 10%. Alternatively, the absolute damping time of the injection feedback system must be at most 30 turns for short bunches and 15 turns for long bunches. These damping times correspond to gains of 0.067 and 0.133 respectively, compared with 0.05 and 0.10 in the Tevatron.

In the case of the Eloisatron (a 100x100 Tev proton-proton collider), for the parameters in table 3 the main ring must keep its injection chromaticity below 3 units for a long bunch and 6 units for a short bunch. Varying the damping gain, the absolute damping time of the injection feedback system must be at most 60 turns for short bunches and 30 turns for long bunches (corresponding to damping gains of 0.033 and 0.067).

## 6. Conclusions

In this paper the dependence of emittance growth at injection into a high energy hadron accelerator is explored using a multiparticle computer simulation. Assuming both a short and long bunch, the reduction of emittance growth due to a damper system is shown to be compromised by large chromaticities. The damper gain required for reasonable choices of chromaticities is surprisingly large, meaning that the coherent betatron injection oscillation must be damped with a time constant much shorter than the decoherence time due to chromaticity.

## Acknowledgments

Many thanks to W. Chou and W. Barletta for the invitation to this very interesting and productive workshop, and to Prof. A. Zichichi for hosting the workshop at the Ettore Majorana Center for Scientific Culture in Erice, Sicily. The stimulating discussions with the other participants of this workshop were sincerely appreciated.

## References

1. R. Meller, A. Chao, J. Peterson, S. Peggs, M. Furman, "Decoherence of Kicked Beams", Internal SSCL note SSC-N-360 (1987).
2. M. Syphers, "Injection Mismatch and Phase Space Dilution", Internal Fermilab note FN-458 (1987).
3. SSCL Site-Specific Conceptual Design Report, SSCL-SR-1056 (1990).
4. W. Barletta, Proc. 19th INFN Eloisatron Workshop (1992).

# $\mathcal{Z}$-Transform Description of Feedback Acting on a Coupled Beam

Richard Talman
SSC Laboratory, Dallas, TX, 75237, USA

## ABSTRACT

The power of the formalism of digital control systems for the analysis and cure of accelerator instabilities is demonstrated. The $\mathcal{Z}$-transform can be said to be even more appropriate for accelerators than it is for digital control circuits because a beam position monitor (BPM) is inherently restricted to sampling once per turn and there is no possibility of obtaining more nearly continuous-time performance by increasing the sampling rate. Also interactions of a beam bunch with a fixed structure in the accelerator can most accurately be described as a sampling process. The traditional formalisms of accelerator physics and digital control systems contain numerous results which can be seen to be the same after having deciphered the differences of terminology, but once that is done standard results from the theory of control system can be applied to accelerator problems. An important requirement is to describe the betatron motion by a difference equation—coupled motion requires a fourth order difference equation. These equations are handled easily by the $\mathcal{Z}$-transform. In the analysis of servomechanisms with sampling elements the $\mathcal{Z}$-transform plays the same role as does the Laplace transform $\mathcal{L}$ in the analysis of time-continuous servomechanisms. Just as feedback stabilization of analog circuits against instability can be achieved by adjusting the positions in the complex $s$ plane of the poles of $\mathcal{L}(s)$, so also do the poles in the complex $z$ plane of $\mathcal{Z}(z)$ have a similar interpretation in accelerators. An incidental dividend accompanying this approach is the mathematically satisfactory generalization of "Courant-Snyder formalism" to include damping.

Paper delivered at Workshop on Feedback Control of Multi-Bunch Instabilities, Erice, Italy, 16 November 1992

# 1. Introduction

Except for new results on the description of feedback acting on coupled betatron motion this paper is largely pedagogical. Some valuable references are[1][2][3][4]. Similar work by Zhabitsky[5] was brought to my attention at the workshop by I.N. Ivanov. Except for the material on coupled betatron motion, the present paper is abbreviated from an unpublished, more detailed, report[6].

It has always been clear that there is a close analogy between servomechanism circuits and feedback stabilization of accelerator beams. The discontinuous nature of the signals in the accelerator has tended to vitiate this analogy, with obscure concepts like the "smooth approximation" being introduced to spread the interactions out in time. What is being presented here is a mathematically more sophisticated way of handling the inherently sampled nature of the signals and interactions. The following several paragraphs indicate miscellaneous aspects of this discription.

Mathematically, the $\mathcal{Z}$-transform is the tool of choice for analyzing linear difference equations. To exploit this it is necessary to represent betatron motion by difference equations. Though very simple, being untraditional, this formalism has not been commonly employed in accelerator physics.

In the accelerator formalism called "Courant-Snyder theory" betatron motion is described by transfer matrices. The eigenvalues of those matrices are closely related to "tunes." In the standard circuit analysis of resonators using Laplace transforms, or impedances, or "complex frequencies," poles are important, with the real and imaginary parts being related to damping rates and resonant frequencies. An alternative description of resonators, analogous to the betatron description, is based on transfer maps in which case the real and imaginary parts of the eigenvalues are related to resonant frequencies and damping. This establishes the idea of a close connection existing between eigenvalues of maps and poles of transforms. For circuits containing sampling elements, the poles of the $\mathcal{Z}$-transform play the analogous role. What is especially pleasing in this case is that the poles are more than just connected to the tunes, they are identical (up to a factor of $2\pi$).

If there is damping present, the statement just made must be refined. The poles of the $\mathcal{Z}$-transform can only be identical to the tunes of transfer matrices in the no-damping, Hamiltonian, "symplectic," regime to which the Courant-Snyder theory is restricted. With damping present the $\mathcal{Z}$-transform poles constitute a generalization of that theory. The result that betatron stability requires all eigenvalues to lie on the unit circle generalizes, in the presence of damping, so that all poles must lie inside the unit circle. This result could perhaps be called "obvious" on the basis of "physics"; here it is compactly integrated into the mathematics. Then poles of the $\mathcal{Z}$-transform can reasonably be called "complex tunes". Like eigenvalues of transfer matrices, they come in complex conjugate pairs, but the presence of damping alters the symplectic feature of their occurring in reciprocal pairs. With no damping the seemingly harmless presence of an eigenvalue lying inside the unit circle is still a sure sign of instability, as its reciprocal partner necessarily lies outside. With damping present this inference is not valid as both partners can lie inside the unit circle.

The feedback stabilization of multi-bunch instabilities leads naturally to a digital formalism like that described here, but that aspect will barely be touched in the present paper.

The equations to be derived are rather general in that they describe the possible instability of any two coupled oscillators. Examples of this are: coupling between horizontal and vertical betatron motion; or between longitudinal and horizontal; Robinson instability in the iteraction between a cavity and a beam executing synchrotron oscillations; and interaction between transverse betatron motion and a resonator. Only the last case will be discussed explicitly. First, to illustrate the formalism, the betatron motion will be restricted to a single plane. Then, to illustrate the power of the method, the theory will be generalized to describe the more general situation in which the betatron motion is coupled. This is not just academic since for large proton colliding beams like the Superconducting Super Collider (SSC) or the Large Hadron Collider (LHC) a large degree of coupling will be unavoidable.

**Aside on analog-to-digital conversion and sample-and-hold circuits.** To be processed digitally an analog signal must be converted to a digital signal in an A/D converter. With routinely available digital precision, the least-count inaccuracy of this conversion has no important effect upon the mathematics describing the system. Rather it is the sample-and-hold circuit (necessarily preceding the A/D converter) that complicates things. With some period $T$ a time-varying function $f(\tau)$ is *sampled* to give a sequence of values $f(0), f(T), f(2T)\ldots$. These will be symbolized as $f_0, f_1, f_2\ldots$. This can also be regarded as replacing the function $f(\tau)$ by a "comb" of $\delta$-functions

$$f(\tau) \longrightarrow \sum_{t=0}^{\infty} f_t \delta(\tau - tT) \tag{1.1}$$

(valid only for $t > 0$). In this formula, and for the rest of these notes, the symbol $t$ has been taken over to be the quantized time; *i.e.* an index specifying the sample number (or equivalently the integer part of the time measured in units of $T$). This instantaneously-sampled representation is readily handled mathematically; for signals of limited bandwidth it can even be processed to recover $f(\tau)$ exactly. But, in practice the sampling cannot be instantaneous and a *hold* operation necessarily complicates things since there is another parameter $\Delta T$, the length of time the signal is held constant. Formula (1.1) corresponds to the mathematically simple choice $\Delta T = 0$ but technical considerations dictate a longer value. One natural choice, $\Delta T = T$, converts $f(\tau)$ into a staircase function.

For accelerators with short bunches, the $\Delta T = 0$ choice represents the physics nicely, and makes the formulas simple. At sufficiently high frequencies this will no longer be valid, because of the possibility of shape oscillations within individual bunches.

Topics to be discussed in the remaining sections of the paper are:

(2) Uniform description of resonators and betatron motion by transfer maps and difference equation.

(3) Description of a beam-resonator system with the formalism of digital control systems; the $\mathcal{Z}$-transform and its properties; interpretation of poles; connection with symplectic maps; and instability threshold.

(4) Stability of a coupled beam interacting with a single resonator.

## 2. Uniform description of resonators and betatron motion

### 2.1 A BEAM-RESONATOR SYSTEM

The beam-resonator system to be analyzed consists of a single short bunch of $N$ particles (charge $e$) traveling with transverse displacement $x$ between the electrodes of two long parallel plates (length $\ell_p$, separation $2b_p$) that are connected externally by a leg containing resistance $R$ and inductance $L$, completing an $LRC$ resonator. This can be thought of as a physically realizable configuration containing these actual elements, or as a lumped constant "equivalent" circuit representing an extended resonator present in the ring—for example an cavity "higher mode." While the charge is between the plates its effect on the resonator is equivalent to a charge source proportional to $x$. (On-axis passage of charge causes "common mode" or longitudinal effects which are not being considered.) For simplicity let $x = b_p$ in which case the instantaneous presence of the bunch between the plates is equivalent to opposite sign charges $Ne/2$ and $-Ne/2$ present on the two plates as well as common-mode, same-sign charges $Ne/2$ and $Ne/2$. The opposite-sign charges apply a voltage $Ne/(2C)$ to the external $LR$ leg, where capacity $C$ includes any stray capacity along with the parallel plate capacity. Assuming that the time $\ell_p/c$ that the charge is between the plates is small compared to the period of the resonator, and that the current $\Delta i$ induced in the inductance varies linearly with $x$, one obtains approximately

$$\Delta i = \frac{Ne l_p}{2b_p C L c} x. \tag{2.1}$$

There is also a reciprocal effect of the external resonator on the beam. If charge $q$ resides on the capacitor at the time a particle is between the plates, it suffers an angular deflection given approximately by

$$\Delta p = \frac{\ell_p}{2(pc/e) C b_p} q, \tag{2.2}$$

where $p \equiv dx/ds \equiv x'$ is the slope of the betatron trajectory. These two oscillators, the betatron motion on the one hand, and the $LRC$ circuit on the other, oscillate freely except for these coupling effects. The universal feature of systems to which the analysis of these notes applies is that the excitation in one system varies linearly with the amplitude in the other, and vice versa. The interaction can cause unwelcome instability of circuits that are individually stable but it can also yield welcome stability in the case that one of the circuits would be unstable by itself. The essential intensity dependence of the system follows entirely from the factor of $N$ appearing in Eq. (2.1).

## 2.2 Transfer Map and Difference Equation Representations of an *LRC* Circuit

This section has three purposes: to obtain solutions of the circuit equations needed to describe a resonant circuit by a transfer map; to do this by use of Laplace transform, thereby reviewing manipulations that will be copied for the $\mathcal{Z}$-transform method of solution; and to use the map to convert to a sampled, or difference equation, representation of the resonator.

The time-continuous circuit equations describing the free $LRC$ circuit are

$$Ri + Ldi/dt + q/C = 0; \qquad i = dq/dt. \tag{2.3}$$

Elements of a transfer map are solutions with particular initial conditions, cosine-like with $(q_0 = 1, i_0 = 0)$ and sine-like with $(q_0 = 0, i_0 = 1)$—call them $q_C(t)$, $i_C(t)$, and $q_S(t)$, $i_S(t)$. Laplace transforming the equations, and representing the transform by an overhead bar, as in $\mathcal{L}\{i(t)\} = \bar{\imath}(s)$, the circuit differential equations are transformed into algebraic equations

$$R\bar{\imath} + L(s\bar{\imath} - i_0) + \bar{q}/C = 0; \qquad \bar{\imath} = s\bar{q} + q_0. \tag{2.4}$$

The essential features here are that time derivatives are replaced by multiplicative factors of $s$, and the initial conditions like initial charge $q_0$ appear explicitly. Subsequent formulas are simplified by introducing new parameters, $\tau_q^{-1}$ and $\omega_q$ given by

$$\tau_q^{-1} = \frac{R}{2L}; \qquad \omega_q^2 = \frac{1}{LC} - \frac{R^2}{4L^2}. \tag{2.5}$$

(The subscript $q$ has been added to differentiate from betatron quantities, to be introduced later, that will bear subscripts $x$.) Solving Eq. (2.4) yields

$$\bar{\imath} = \frac{Li_0 - q_0/Cs}{Ls + R + 1/Cs} = \frac{si_0 - q_0/LC}{(s+\tau_q^{-1})^2 + \omega_q^2} = \frac{si_0 - q_0/LC}{(s+\tau_q^{-1} + i\omega_q)(s + \tau_q^{-1} - i\omega_q)}. \tag{2.6}$$

One then looks in a table to find the inverse transform. We require only two transform pairs:

$$\mathcal{L}\{e^{-\alpha t}\sin\beta t\} = \frac{\beta}{(s+\alpha)^2+\beta^2};$$
$$\mathcal{L}\{e^{-\alpha t}\cos\beta t\} = \frac{s+\alpha}{(s+\alpha)^2+\beta^2}. \tag{2.7}$$

For the $\mathcal{Z}$-transform we will require the analogous two pairs. From these formulas the desired solutions can be obtained.

$$\begin{aligned} q_C(t) &= e^{-t/\tau_q}(\cos\omega_q t + \frac{1}{\omega_q\tau_q}\sin\omega_q t), \\ i_C(t) &= -\frac{\tau_q^{-2}+\omega_q^2}{\omega_q}e^{-t/\tau_q}\sin\omega_q t, \\ q_S(t) &= \frac{1}{\omega_q}e^{-t/\tau_q}\sin\omega_q t, \\ i_S(t) &= e^{-t/\tau_q}(\cos\omega_q t - \frac{1}{\omega_q\tau_q}\sin\omega_q t). \end{aligned} \tag{2.8}$$

Before proceeding to generate difference equations it is useful to correlate the positions of the poles (*i.e.* the positions of zeros of the denominator) in Eq. (2.6) with the free oscillation of the circuit. For this purpose the resistance $R$ should be allowed to be negative to permit anti-damping. The poles come in pairs, symmetrically above and below the real axis. Poles in the left/right half-plane yield decaying/growing solutions. Poles on the real-axis correspond to non-oscillatory solutions.

These special solutions, Eqs. (2.8), can be used to propagate general solutions over sampling times of length $T$. For this we introduce new parameters, closely connected to $\omega_q$ and $\tau_q^{-1}$, but specific to time interval $T$:

$$a_q = e^{-T/\tau_q}; \qquad \mu_q = \omega_q T. \tag{2.9}$$

The quantity $\mu_q$ is the phase advance of the resonator during the sampling period which can be arbitrary but in our case is the revolution period for particles to go around the accelerator. That being the case, $\mu_q/(2\pi)$ could be called the "tune" of the resonator. Similarly, $T/\tau_q$ could be called the damping decrement. Since in all practical situations damping is weak, the value of $a_q$ is slightly less than 1.

If, on the $t$ 'th sample, the state of the circuit is given by the vector $(q,i)_t^T$, and there is an impulsive current change $\Delta i_t/2$ just after the sampling, then, at the $t+1$ 'st sampling, the state is given by

$$\begin{aligned} \begin{pmatrix} q \\ i-\Delta i/2 \end{pmatrix}_{t+1} &= \begin{pmatrix} q_C(T) & q_S(T) \\ i_C(T) & i_S(T) \end{pmatrix} \begin{pmatrix} q \\ i+\Delta i/2 \end{pmatrix}_t \\ = \quad & a_q \begin{pmatrix} \cos\mu_q + \frac{1}{\omega_q\tau_q}\sin\mu_q & \frac{1}{\omega_q}\sin\mu_q \\ -\frac{\tau_q^{-2}+\omega_q^2}{\omega_q}\sin\mu_q & \cos\mu_q - \frac{1}{\omega_q\tau_q}\sin\mu_q \end{pmatrix} \begin{pmatrix} q \\ i+\Delta i/2 \end{pmatrix}_t \end{aligned} \tag{2.10}$$

where allowance has been made for another impulse $\Delta i_{t+1}/2$ just before the next sampling. The formula has been written to correspond to sampling at the instant the bunch passes the center of the resonator.

Substituting from Eq. (2.8), the upper row of this matrix equation is

$$a_q^{-1} q_{t+1} = \Big(\cos\mu_q - (\omega_q \tau_q)^{-1} \sin\mu_q\Big) q_t + \omega_q^{-1} \sin\mu_q (i + \Delta i/2)_t. \tag{2.11}$$

Our present purpose is to eliminate $i_t$ and toward that end, we write a similar equation that propagates the solution back to the $(t-1)$'st turn:

$$a_q q_{t-1} = \Big(\cos\mu_q + (\omega_q \tau_q)^{-1} \sin\mu_q\Big) q_t - \omega_q^{-1} \sin\mu_q (i - \Delta i/2)_t. \tag{2.12}$$

Adding these equations yields

$$q_{t+1} - 2a_q \cos\mu_q q_t + a_q^2 q_{t-1} = \Big(\frac{a_q \Delta i_t}{\omega_q}\Big) \sin\mu_q. \tag{2.13}$$

This is the difference equation needed for the $\mathcal{Z}$-transform analysis of the resonator.

## 2.3 Difference equation representation of betatron motion

We now obtain a similar difference equation representation of betatron motion. It is easier because we can begin with the familiar transfer map representation.

If on the $t$'th revolution the phase space coordinates of the particle bunch are given by the vector $(x,p)_t^T$ and there is an impulsive angular deflection $\Delta p/2$ just after the sampling, then at the $t+1$'st revolution, the state is given by

$$\begin{aligned} &\begin{pmatrix} x \\ p - \Delta p/2 \end{pmatrix}_{t+1} \\ &= e^{-\delta_x} \begin{pmatrix} \cos\mu_x + \alpha_x \sin\mu_x & \beta_x \sin\mu_x \\ -\gamma_x \sin\mu_x & \cos\mu_x - \alpha_x \sin\mu_x \end{pmatrix} \begin{pmatrix} x \\ p + \Delta p/2 \end{pmatrix}_t. \end{aligned} \tag{2.14}$$

This is the so-called "Twiss-form" of the transfer matrix, with an *ad hoc* damping factor $e^{-\delta_x}$ included; $\delta_x$ is called the damping decrement; for closer correspondance with the resonator notation we introduce the symbol $a_x = e^{-\delta_x}$. The Twiss parameters $\alpha_x, \beta_x$, and $\gamma_x$, are related by $\gamma_x = (1+\alpha_x^2)/\beta_x$. Again the sampling is taken at the center of the element causing the deflection.

Writing the upper equation explicitly, and also writing the corresponding equation for backwards propagation yields

$$\begin{aligned} a_x^{-1} x_{t+1} &= (\cos\mu_x + \alpha_x \sin\mu_x) x_t + \beta_x \sin\mu_x (p + \Delta p/2)_t, \\ a_x x_{t-1} &= (\cos\mu_x - \alpha_x \sin\mu_x) x_t - \beta_x \sin\mu_x (p - \Delta p/2)_t. \end{aligned} \tag{2.15}$$

Adding these equations yields

$$x_{t+1} - 2a_x \cos\mu_x x_t + a_x^2 x_{t-1} = (a_x \beta_x \Delta p_t) \sin\mu_x. \tag{2.16}$$

This is the difference equation needed for the $\mathcal{Z}$-transform analysis of betatron motion. Neglecting damping, we can also obtain (approximately) $p_t$, the slope at the observation point, by subtracting Eqs. (2.15);

$$p_t = \frac{x_{t+1} - x_{t-1} - 2\alpha_x \sin\mu_x\, x_t}{2\beta_x \sin\mu_x}. \tag{2.17}$$

This equation will not be needed for the "passive stability" analyzed in this paper, but it or a variant would be required for "active feedback". This comment will be expanded upon below.

# 3. Digital Control System Formalism

## 3.1 The $\mathcal{Z}$-transform and its properties

The sort of equation that can be solved by $\mathcal{Z}$-transform is

$$x_{t+2} - 2a\cos\mu x_{t+1} + a^2 x_t = g_{t+1}, \quad t = 0, 1, 2, \ldots \tag{3.1}$$

where $g_t$ is a given sequence of numbers $g_0, g_1, g_2, \ldots$; they constitute the "drive" that leads to a "driven" motion as is familiar for time-continuous oscillators. Eq. (3.1) is a second-order difference equation like Eqs. (2.13) or (2.16). The addition of any solution of the homogeneous equation (*i.e.* a "transient" solution with $g_t = 0$) to a steady-state solution results in another solution. As with the Laplace transform, we explicitly restrict the description to positive time, setting all sequence elements having negative indices to zero, and also setting $g_0 = 0$. If starting values $x_0$ and $x_1$ are given, as we now assume, a unique solution is determined.

The $\mathcal{Z}$-transform of a sequence $f_t$ is symbolized $F^*(z)$ and defined by

$$F^*(z) \equiv \mathcal{Z}\{f_t\} = \sum_{t=0}^{\infty} f_t z^{-t}. \tag{3.2}$$

A Fourier or Laplace transform of a function of time involves an integration over time; here we have an analogous summation over $t$. Operationally the power of the Laplace transform is that it replaces derivatives with respect to time by factors of $s$, thereby converting a differential equation into an algebraic equation. The corresponding virtue of the $\mathcal{Z}$-transform is that it replaces a unit displacement of the $t$ index by a multiplicative factor of $z$, thereby converting a difference equation into an algebraic equation. As with the Laplace transform negative values of $t$ are never allowed to enter.

Though this mathematics is elementary, it is a bit much to give a complete description, and I will list only a few important results; most of them are easy to derive starting from Eq. (3.2). Refer to[2] or other references for more details.

$$
\begin{aligned}
\mathcal{Z}\{f_{t+2}\} &= z^2[F^*(z) - f_0 - f_1 z^{-1}] \\
\mathcal{Z}\{f_{t+1}\} &= z[F^*(z) - f_0] \\
\mathcal{Z}\{f_t\} &= F^*(z) \\
\mathcal{Z}\{f_{t-1}\} &= z^{-1}F^*(z); \quad \text{defining} \quad f_{-1} = 0 \\
\mathcal{Z}\{\sum_{\nu=0}^{t} q_\nu r_{t-\nu}\} &= Q^*(z)R^*(z).
\end{aligned}
\tag{3.3}
$$

The last of these, derived simply by multiplying two series like Eq. (3.2) together and grouping terms, is a convolution result like the well known result that convolution in the time domain yields multiplication in the frequency domain. As with Laplace transforms, the transform of an "output" is given by a "transfer function" multiplied by the transform of an "input." Two transforms more-or-less analogous to Eqs. (2.7) are

$$
\begin{aligned}
\mathcal{Z}\{a^t \sin \mu t\} &= \frac{az \sin \mu}{z^2 - 2az \cos \mu + a^2}, \\
\mathcal{Z}\{a^t \cos \mu t\} &= \frac{z(z - a \cos \mu)}{z^2 - 2az \cos \mu + a^2}.
\end{aligned}
\tag{3.4}
$$

To solve (3.1) , define a displaced sequence $f_t = g_{t+1}$, and $\mathcal{Z}$-transform to get

$$[X^*(z) - x_0 - x_1 z^{-1}]z^2 - 2a \cos \mu [X^*(z) - x_0]z + a^2 X^*(z) = F^*(z). \tag{3.5}$$

Solving for $X^*(z)$ yields

$$X^*(z) = \frac{F^*(z) + x_0 z(z - 2a \cos \mu) + x_1 z}{z^2 - 2az \cos \mu + a^2}. \tag{3.6}$$

Transforming back, using the convolution formula, and substituting for $g_t$ yields

$$x_t = \frac{1}{\sin \mu}[\sum_{\nu=1}^{t-1} g_{t-\nu} a^{\nu-1} \sin \nu\mu - x_0 a^t \sin \mu(t-1) + x_1 a^{t-1} \sin \mu t]. \tag{3.7}$$

The first term gives the driven response, the other terms give the transient response. A powerful feature of the method is the explicit inclusion of transients but here such details will be unimportant.

## 3.2 Solution of coupled oscillator equations. Interpretation of poles. The connection with symplectic maps

Our main task is to solve equations (2.13) and (2.16), after substituting from earlier equations (2.1) and (2.2). The equations are repeated here with indices translated by 1 to bring them into conformity with Eq. (3.1);

$$\begin{aligned} x_{t+2} - 2a_x \cos \mu_x x_{t+1} + a_x^2 x_t &= r_x \, q_{t+1} \\ q_{t+2} - 2a_q \cos \mu_q q_{t+1} + a_q^2 q_t &= r_q \, x_{t+1} \end{aligned} \tag{3.8}$$

where

$$r_x = \sin \mu_x \big( a_x \beta_x \frac{\ell_p}{2(pc/e)Cb_p} \big); \quad r_q = \sin \mu_q \big( \frac{a_q}{\omega_q} \frac{Ne\ell_p}{2b_p CLc} \big). \tag{3.9}$$

These are two coupled, second-order, difference equations. As they are written, neither the betatron motion nor the resonator is externally driven, but the coupling effectively makes the betatron motion drive the resonator and the resonator drive the betatron motion. This can be thought of as a "passive" feedback system, in which an "idler" resonator is excited by, and interacts back on, the beam.

**Aside.** "Active" feedback would be described similarly except that the second of Eqs. (3.8) would describe external circuitry that sensed $x_t$ and generated a deflecting force to be applied on the right-hand side of the first equation. For example, referring to Eq. (2.17), if a feedback kicker situated exactly at the observation point, applies a kick $\Delta p_t$ given by

$$\Delta p_t = G \frac{x_{t+1} - x_{t-1} - 2\alpha_x \sin \mu_x \, x_t}{2\beta_x \sin \mu_x}, \tag{3.10}$$

that kick can reduce the oscillation amplitude. Choosing the value $G = -1$ as the "gain" value would correspond to reducing the oscillation amplitude to zero on a single passage. In practice $G$ would probably be chosen so that $-1 << G < 0$.

After applying $\mathcal{Z}$-transformation to Eqs. (3.8), applying formulas (3.3) to displace the indices, the result is two linear equations for $Q^*(z)$ and $X^*(z)$. After solving these (with arbitrary initial conditions,) and applying the inverse $\mathcal{Z}$-transformation the detailed motion would be obtained. But our primary interest is not in the detailed motion. Rather, the issue of importance is whether the subsequent motion is "stable" or "unstable" or, in other words, whether the transient solutions grow or damp. If there were even one anti-damped solution it would result in unacceptable performance. Jumping right away to the answer, it should be anticipated from the "Courant-Snyder" theory, and/or from the discussion of maps earlier in this paper, that the condition for stability is that all zeros of the following polynomial equation must lie inside (or possibly on) the unit circle in the complex $z$ plane:

$$\det\begin{vmatrix} z^2 - 2a_x \cos\mu_x\, z + a_x^2 & -r_x z \\ -r_q z & z^2 - 2a_q \cos\mu_q\, z + a_q^2 \end{vmatrix} = 0. \tag{3.11}$$

The elements of this determinant are coefficients of Eq. (3.8) after it has been $\mathcal{Z}$-transformed. It is this determinant that appears in the denominator when the transformed equations are solved using Cramer's rule. To complete the argument another transform pair is required:

$$\mathcal{Z}\{\alpha^t\} = \frac{z}{z-\alpha}. \tag{3.12}$$

After performing a partial fraction expansion, $X^*(z)$ and $Q^*(z)$ consist of sums of terms like this. (The factor $z$ in the numerator can be "shifted away," and we ignore the possibility of repeated roots.) Convergence of the resultant sequence after transformation back to the time domain clearly depends on the magnitude $|\alpha|$; it diverges for $|\alpha| > 1$ and converges for $|\alpha| < 1$.

Multiplying out the determinant in Eq. (3.11) yields

$$(z^2 - 2a_x \cos\mu_x\, z + a_x^2)(z^2 - 2a_q \cos\mu_q\, z + a_q^2) - Kz^2 = 0 \tag{3.13}$$

where

$$K = \sin\mu_q \sin\mu_x \left(\frac{a_q}{\omega_q}\frac{Nel_p}{2b_pCLc}\right)\left(a_x\beta_x\frac{\ell_p}{2(pc/e)Cb_p}\right). \tag{3.14}$$

It will be seen shortly that the sign of $K$ strongly influences the stability. This, and the properties of the factor $\sin\mu_q \sin\mu_x$, account for the fact that sum resonances are unstable and difference resonances are stable. A plausible suposition is that instability threatens "when the tunes are equal, or close." But it can be seen from Eq. (3.13) that the tunes enter only in the form of $\cos\mu_x$ and $\cos\mu_q$ (at least for small $K$) and it is actually the equality of cosines that threatens instability. The cosines can be equal in the "sum" situation, $\mu_q = \mu_x$, in which case $\sin\mu_x \sin\mu_q > 0$, or in the "difference" situation, $\mu_q = 2\pi - \mu_x$, in which case $\sin\mu_x \sin\mu_q < 0$.

Before proceeding to determine whether all the roots of a polynomial equation like this lie within the unit circle we obtain an important generalization of a "symplectic" property of the motion. When there is no damping, as in Hamiltonian mechanics, $a_x$ and $a_q$ in Eq. (3.13) are both 1. In that case, though it is a quartic equation in $z$, Eq. (3.13) can, after division by $z^2$, be written as a quadratic equation in $z + z^{-1}$. That means that the roots come in reciprocal pairs, one outside, one inside or both on the unit circle. With damping present (assumed equal for both motions to simplify discussion), the generalization of this property is that Eq. (3.13) is quadratic in $z/a + a/z$, where $a_x = a_q = a$. That means the poles are related to the circle of radius $a$ (one outside, one inside or both on) the way the poles are related to the unit circle in the no-damping case. Because $a < 1$ both can be off the circle of radius $a$ and yet be inside the unit circle. If the damping factors are unequal (*e.g.*, no betatron damping or even slight anti-damping) it is still possible for the resonator to "lend" some of its stability to the beam, keeping both stable.

### 3.3 The Jury-Blanchard stability criteria

It is required to find a set of conditions, necessary and sufficient for the zeros of a polynomial in $z$ to lie within the unit circle. Such conditions resemble the conditions that the poles of a Laplace transform lie in the left half-plane; that is required for stability as was explained above. One such prescription is due to Jury and Blanchard.[8] . Let the polynomial equation be written in the form

$$F(z) = b_0 + b_1 z + b_2 z^2 + b_3 z^3 + b_4 z^4 = 0. \tag{3.15}$$

The coefficients are assumed to be real, with $b_4 > 0$. In that case the zeros are known either to be real or to come in complex conjugate pairs. The method works for arbitrarily high powers, but for simplicity it will be made specific to the problem at hand. Also it is slightly different, depending upon whether the highest power of $z$ is odd or even. One evaluates a set of determinants which, in this case, are given by

$$\begin{aligned} c_0 &= b_0 b_0 - b_4 b_4; \quad c_1 = b_0 b_1 - b_3 b_4; \quad c_2 = b_0 b_2 - b_2 b_4; \quad c_3 = b_0 b_3 - b_1 b_4; \\ d_0 &= c_0 c_0 - c_3 c_3; \quad d_1 = c_0 c_1 - c_2 c_3; \quad d_2 = c_0 c_2 - c_1 c_3. \end{aligned} \tag{3.16}$$

The conditions, all of which must be satisfied for stability, are

$$\begin{aligned} F(1) = b_0 + b_1 + b_2 + b_3 + b_4 &> 0, \quad (1) \\ F(-1) = b_0 - b_1 + b_2 - b_3 + b_4 &> 0, \quad (2) \\ b_4 - b_0 &> 0, \quad (3) \\ |c_0| - |c_3| &> 0, \quad (4) \\ |d_0| - |d_2| &> 0. \quad (5) \end{aligned} \tag{3.17}$$

The number of conditions is one more than the number of poles. Condition 1 is obvious; proceeding out the positive real axis, if the value $F(1)$ is negative and the value $F(\infty)$ is positive, as it is, then there must be a real zero greater than 1. Condition 2 is just as obvious, based on the same argument on the negative real axis. The other conditions are not obvious, but they are easily applied.

For our problem the coefficients are given by

$$\begin{aligned} b_0 &= a_x^2 a_q^2, \\ b_1 &= -\,2a_x a_q(a_q \cos\mu_x + a_x \cos\mu_q), \\ b_2 &= a_x^2 + a_q^2 + 4a_x a_q \cos\mu_x \cos\mu_q - K \equiv b_{20} - K, \\ b_3 &= -\,2(a_x \cos\mu_x + a_q \cos\mu_q), \\ b_4 &= 1. \end{aligned} \tag{3.18}$$

The stability problem can be posed as finding the band of stable values of $K$, running from the most negative stable value to the most positive stable value. With $K = 0$ all five conditions are evidently satisfied, since the uncoupled oscillators are stable by assumption. That means that such a band must exist and it must include the origin.

In the case at hand, $K$ enters only $b_2$ of the original coefficients. That means that it enters only conditions 1, 2, and 5, which can be written

$$\begin{aligned} F(1) &= b_0 + b_1 + (b_{20} - K) + b_3 + b_4 > 0, \quad (1) \\ F(-1) &= b_0 - b_1 + (b_{20} - K) - b_3 + b_4 > 0, \quad (2) \\ |d_0| - |d_{20} &- \{(b_0^2 - b_4^2)(b_0 - b_4)\}K| > 0, \quad (5) \end{aligned} \tag{3.19}$$

where symbols $b_{20}$ and $d_{20}$ refer to the coefficient values with $K = 0$. The expression in curly braces $\{(b_0^2 - b_4^2)(b_0 - b_4)\} \equiv P$ is clearly positive.

The conditions can now be made more explicit. Define $K_{1+}, K_{2+}, K_{5-}$, and $K_{5+}$, as candidate values for the values of $K$ at the ends of the stable interval. The first two are simple; they are given by

$$\begin{aligned} K_{1+} = F_0(1) =& b_0 + b_1 + b_{20} + b_3 + b_4 \\ =& (1 - 2a_x \cos\mu_x + a_x^2)(1 - 2a_q \cos\mu_q + a_q^2) \\ K_{2+} = F_0(-1) =& b_0 - b_1 + b_{20} - b_3 + b_4 \\ =& (1 + 2a_x \cos\mu_x + a_x^2)(1 + 2a_q \cos\mu_q + a_q^2). \end{aligned} \tag{3.20}$$

Both of these are necessarily positive as the subscript is intended to indicate. Condition 5 is more complicated because of the absolute values; it yields

$$K_{5-} = (d_{20} - |d_0|)/P; \quad \text{and} \quad K_{5+} = (|d_0| + d_{20})/P. \tag{3.21}$$

Since there is only one negative candidate, $K_{5-}$, the condition clearly defines the negative stability limit. Finally then, the band of stability can be expressed as

$$K_{5-} < K < \min[K_{1+}, K_{2+}, K_{5+}]. \tag{3.22}$$

When the values of Eq. (3.21) are substituted into these formulas the results, while explicit, are complicated and not easily instructive. For the case of equal damping of both resonators, and exactly on resonance (*i.e.* equal tunes), one obtains for the radius of the outermost pole

$$\begin{aligned} |z|^2 =& a^2 \pm a\sqrt{K}(\sin\mu - \cos\mu) \\ \simeq& 1 - 2\delta \pm \sqrt{K}(\sin\mu - \cos\mu). \end{aligned} \tag{3.23}$$

In that case the instability threshold is given by

$$K(\text{threshold}) \simeq 4\delta^2/(\sin\mu - \cos\mu)^2. \tag{3.24}$$

Because $K$ is proportional to $N$, there will necessarily be a beam current above which the system is unstable.

To flesh out this example we substitute plausible numerical values for the parameters (taken from an experiment performed at CESR[7] ) into Eqs. (3.9), re-expressing beam charge in terms of beam current $I = 0.01\,\text{mA}$ to obtain

$$\begin{aligned} r_x &= -\sin\mu_x(0.9999 \times 40m \frac{2.5m}{2 \times 5.0 \times 10^9 V \times 1.50 \times 10^{-10} f \times 0.06m}), \\ &= -\sin\mu_x \times 1.11 \times 10^3 \quad \text{meters/coulomb} \\ r_q &= -\sin\mu_q(\frac{0.974}{4\pi}\frac{0.01A \times (2.5 \times 10^{-6}s)^2 \times 2.5m}{0.06m \times 1.5 \times 10^{-10} f \times 0.0077h \times 3 \times 10^8 m/s}). \\ &= -\sin\mu_q \times 5.82 \times 10^{-10} \quad \text{coulombs/meter.} \end{aligned} \tag{3.25}$$

Stability is governed by the parameter $K$ that depends on $r_q$ and $r_x$ only through their product;

$$K = 0.65 \times 10^{-6} \sin\mu_q \sin\mu_x \frac{I(\text{amperes})}{0.01}. \tag{3.26}$$

With Eq. (3.24), this yields the maximum stable beam current. This example has been worked through in detail not because of its special importance but in order to illustrate obtaining the dependence of the instability threshold on the parameters of the problem.

## 4. Stability of a coupled beam interacting with a single resonator

The formulas needed for describing coupled betatron motion are contained in reference[9] . That paper, as well as deriving the equations, demonstrates their validity using actual data from the Large Electron Positron Collider (LEP). For our present purposes we mainly need the difference equation describing coupled betatron motion, and for this we just copy a few formulas describing free betatron motion[9] , modifying them to account for external interactions.

To simplify the discussion we will assume the lattice has no intentional coupling, meaning that any coupling present is due to unknown errors. The presence of these errors "perturbs" the actual system, causing the lattice functions to deviate from the "model"—that is, from the functions calculated assuming no errors. We will use the term "model-independent" to describe equations whose coefficients are accurately known (because they have been measured in the perturbed situation) and the term "perturbative" to describe calculations that assume the errors are small enough that quantities calculated using the idealized lattice model constitute good approximations to the perturbed situation.

The column vector of coordinates $X = (x, p, y, q)^T$ represents small transverse deviations from the reference orbit. Evolution of a vector $X$ from longitudinal coordinate $s_0$ to $s$ is described by a transfer matrix $M(s, s_0)$,

$$X(s) = M(s, s_0)X(s_0). \tag{4.1}$$

The fact that $M$ is symplectic is critical to the derivation of the required difference equation. Using the matrix

$$S = \begin{pmatrix} 0 & -1 & 0 & 0 \\ 1 & 0 & 0 & 0 \\ 0 & 0 & 0 & -1 \\ 0 & 0 & 1 & 0 \end{pmatrix}, \tag{4.2}$$

the "symplectic conjugate" of $M$ is defined by

$$\bar{M} = -SM^T S. \tag{4.3}$$

In this equation the order of $M$ can any even number, and $S$ is the obvious generalization of Eq. (4.2). Partitioning the $4 \times 4$ matrix $M$ into $2 \times 2$ elements, it and its symplectic conjugate are defined by

$$M = \begin{pmatrix} A & B \\ C & D \end{pmatrix}; \quad \bar{M} = \begin{pmatrix} \bar{A} & \bar{C} \\ \bar{B} & \bar{D} \end{pmatrix}. \tag{4.4}$$

The condition for $M$ to be symplectic can be expressed by

$$\bar{M} = M^{-1}. \tag{4.5}$$

Especially important for analyzing the state of coupling is the off-diagonal combination $E = C + \bar{B}$ and its determinant $\mathcal{E}$.

$$E = C + \bar{B} = \begin{pmatrix} e & f \\ g & h \end{pmatrix}; \quad \det|E| = eh - fg \equiv \mathcal{E} \ . \tag{4.6}$$

The eigenvalues of $M$ satisfy simple equations

$$\begin{aligned} \Lambda_A + \Lambda_D &= \mathrm{tr}A + \mathrm{tr}D, \\ \Lambda_A\Lambda_D &= \mathrm{tr}A\,\mathrm{tr}D - \mathcal{E} \end{aligned} \tag{4.7}$$

and there are real angles $\mu_A = 2\pi\nu_A$ and $\mu_D = 2\pi\nu_D$ satisfying

$$\begin{aligned} \Lambda_A \equiv \lambda_A + 1/\lambda_A &= \exp(i\mu_A) + \exp(-i\mu_A) = 2\cos\mu_A \\ \Lambda_D \equiv \lambda_D + 1/\lambda_D &= \exp(i\mu_D) + \exp(-i\mu_D) = 2\cos\mu_D. \end{aligned} \tag{4.8}$$

Since the tunes $\mu_A$ and $\mu_D$ are readily and acurately measurable with a spectrum analyzer, the perturbed eigenvalues $\Lambda_A$ and $\Lambda_D$ can be assumed known, no matter how strongly the betatron motion is coupled (*i.e.* they are model independent, as defined above.) The combination

$$M+\bar{M}=M+M^{-1}=\begin{pmatrix}\mathrm{tr}A & 0\\ 0 & \mathrm{tr}D\end{pmatrix}+\begin{pmatrix}0 & \bar{E}\\ E & 0\end{pmatrix} \tag{4.9}$$

turns out to have simpler properties than $M$. Using the fact that $M^{-1}$ can be used to propagate backwards in time, this relation can be used to obtain four third-order, coupled difference equations that relate the coordinates on three successive turns (labeled $-$, 0, $+$):

$$\begin{aligned} x_+ - \mathrm{tr}A\, x_0 + x_- &= hy_0 - fq_0,\\ p_+ - \mathrm{tr}A\, p_0 + p_- &= -\,gy_0 + eq_0,\\ y_+ - \mathrm{tr}D\, y_0 + y_- &= ex_0 + fp_0,\\ q_+ - \mathrm{tr}D\, q_0 + q_- &= gx_0 + hp_0. \end{aligned} \tag{4.10}$$

It is possible to uncouple these equations. Start by squaring Eq. (4.9), subtracting 2I, and using Eqs. (4.7);

$$M^2+M^{-2}=\begin{pmatrix}-2+\mathrm{tr}^2A+\mathcal{E} & 0\\ 0 & -2+\mathrm{tr}^2D+\mathcal{E}\end{pmatrix}+(\Lambda_A+\Lambda_D)\begin{pmatrix}0 & \bar{E}\\ E & 0\end{pmatrix}. \tag{4.11}$$

From Eqs. (4.9) and (4.11) we form the combination that eliminates the off-diagonal, coupling terms and operate on $X_0$ with it. The matrix identity is

$$M^2+M^{-2}-(\Lambda_A+\Lambda_D)(M+M^{-1})+(2+\Lambda_A\Lambda_D)\begin{pmatrix}I & 0\\ 0 & I\end{pmatrix}=0, \tag{4.12}$$

and the resulting difference equation is

$$X_{++}+X_{--}-(\Lambda_A+\Lambda_D)(X_+ + X_-)+(2+\Lambda_A\Lambda_D)X_0=0. \tag{4.13}$$

All components satisfy the same equation; in particular the directly observable quantities $x$ and $y$

$$\begin{aligned} x_{++}+x_{--}-(\Lambda_A+\Lambda_D)(x_+ + x_-)+(2+\Lambda_A\Lambda_D)x_0 &=0,\\ y_{++}+y_{--}-(\Lambda_A+\Lambda_D)(y_+ + y_-)+(2+\Lambda_A\Lambda_D)y_0 &=0. \end{aligned} \tag{4.14}$$

What makes these equations remarkable is that in full generality with arbitrary coupling, with coefficients that are lattice invariants independent of location in the ring, they constrain the possible turn-by-turn sequence of transverse beam location readings coming from any beam position monitor. Since any spurious superimposed signal will not in general satisfy this constraint, these equations can give a measure of the importance of noise or distortion due to nonlinear electronics in the signal processing or distortion due to nonlinear motion. Eqs. (4.14) generalize the difference equation for uncoupled motion, Eq. (2.16). With no coupling Eqs. (4.14) and (2.16) can be shown to be consistent.

To begin to incorporate external feedback deflections into Eqs. (4.13), suppose a deflecting element at the origin applies purely horizontal deflection $\Delta p_0$ on turn number 0. Temporarily we ignore deflections $\Delta p_-$ and $\Delta p_+$. One then obtains the vectors $X_-$ and $X_+$;

$$X_+ = M\begin{pmatrix} x_0 \\ p_0 \\ y_0 \\ q_0 \end{pmatrix} + M\begin{pmatrix} 0 \\ \Delta p_0/2 \\ 0 \\ 0 \end{pmatrix}; \quad X_- = M^{-1}\begin{pmatrix} x_0 \\ p_0 \\ y_0 \\ q_0 \end{pmatrix} + M^{-1}\begin{pmatrix} 0 \\ -\Delta p_0/2 \\ 0 \\ 0 \end{pmatrix} \tag{4.15}$$

where the deflection has been split in two, half just before and half just after the origin. The sum $X_+ + X_-$ can be written

$$X_+ + X_- = (M + M^{-1})X_0 + (M - M^{-1})\begin{pmatrix} 0 \\ \Delta p_0/2 \\ 0 \\ 0 \end{pmatrix} \tag{4.16}$$

where the first term represents "free motion" term and the second "driven." Similarly we obtain

$$X_{++} + X_{--} = (M^2 + M^{-2})X_0 + (M^2 - M^{-2})\begin{pmatrix} 0 \\ \Delta p_0/2 \\ 0 \\ 0 \end{pmatrix}. \tag{4.17}$$

Forming the same combination as in Eq. (4.13), the driven equation becomes

$$X_{++} + X_{--} - (\Lambda_A + \Lambda_D)(X_+ + X_-) + (2 + \Lambda_A\Lambda_D)X_0 =$$

$$\Big(M^2 - M^{-2} - (\Lambda_A + \Lambda_D)(M - M^{-1})\Big)\begin{pmatrix} 0 \\ \Delta p_0/2 \\ 0 \\ 0 \end{pmatrix} \tag{4.18}$$

$$+ \text{ terms proportional to } \Delta p_- \text{ and } \Delta p_+.$$

As well as not yet having the $\Delta p_-$ and $\Delta p_+$ terms explicit, the effects of $\Delta p_{--}$ and $\Delta p_{++}$ need to be understood. These effect $p_{--}$ and $p_{++}$ but not $x_{--}$ and $x_{++}$. Since only the $x$ component of Eq. (4.18) will be used $\Delta p_{--}$ and $\Delta p_{++}$ can be ignored.

As it must be, the left hand side of this equation is the same as the equation of free motion (4.13) and, as emphasized above, (assuming the two tunes have been measured) its coefficients are known in a completely model-independent way without having made any perturbative assumptions. On the other hand the right-hand side is not simple, and depends on poorly-known elements of $M$. But since the right hand side contains only terms proportional to the presumably-small deflections $\Delta p_t$, a perturbative description of the driven motion is obtained by neglecting coupling and using the unperturbed, model values for the elements of $M$. This assumption will not be valid for coupling so strong that the "wrong plane" response after two or three turns, due to impulsive deflection in one plane, is appreciable compared to the "correct plane" response. With this assumption and using Eq. (2.14) with $\delta_x = 0$, the contribution of $\Delta p_0$ to the right hand side of the equation for $x$ becomes

$$\beta_x\Big(\sin 2\mu_x - (\Lambda_A + \Lambda_D)\sin\mu_x\Big)\Delta p_0. \tag{4.19}$$

To the same approximation the influence of deflections $\Delta p_-$ and $\Delta p_+$ can also be included and the complete equation is

$$\begin{aligned}
x_{t+4} - (\Lambda_A + \Lambda_D)x_{t+3} + (2 + \Lambda_A\Lambda_D)x_{t+2} - (\Lambda_A + \Lambda_D)x_{t+1} + x_t = \\
+\beta_x\Delta p_{t+2}\Big(\sin 2\mu_x - (\Lambda_A + \Lambda_D)\sin\mu_x\Big) \\
+\beta_x\frac{\Delta p_{t+1} + \Delta p_{t+3}}{2}\Big(\sin\mu_x + \sin 3\mu_x - (\Lambda_A + \Lambda_D)\sin 2\mu_x + (2 + \Lambda_A\Lambda_D)\sin\mu_x\Big).
\end{aligned} \tag{4.20}$$

This equation generalizes the first of the two Eqs. (3.8), where now we regard $\Delta p_t$ as the "state variable" describing the state of the external feedback system on turn $t$. If, as before, the external system is itself a resonator, it will be described by the second of Eqs. (3.8)

$$\Delta p_{t+2} - 2a\cos\mu\ \Delta p_{t+1} + a^2\Delta p_t = r\,x_{t+1} \tag{4.21}$$

where coefficient $r$, proportional to beam current, sets the strength of the resonator excitation due to the beam.

After $\mathcal{Z}$-transforming Eqs. (4.20) and (4.21) we are left with two linear equations for the $\mathcal{Z}$-transforms $X^*(z)$ and $\Delta P^*(z)$. As in Eq. (3.11), the stability of the whole system is governed by the locations of the zeroes of the determinant of coefficients;

$$\begin{vmatrix} z^4 - (\Lambda_A + \Lambda_D)z^3 + (2 + \Lambda_A\Lambda_D)z^2 - (\Lambda_A + \Lambda_D)z + 1 & \text{Eq. (4.20)}|_{\text{rhs}} \\ -r\,z^2 & z^2 - 2a\cos\mu\,z + a^2 \end{vmatrix}. \tag{4.22}$$

As indicated, the upper-right element still needs to be extracted from Eq. (4.20). The polynomial Eq. (4.22) can be analysed as was Eq. (3.11) in order to obtain the boundary of the region of stability. Since the betatron description has assumed zero damping it is possible there will be no stable region. In practice, the beam will be made up of many particles with (incoherent) tunes spread over a range. In that case, as usual, while the coherent tune remains within the band of incoherent tunes, there will be Landau damping. Naturally the physics of this does not depend on the mathematical formalism being used to describe the oscillators. In particular it will be very natural to incorporate the Landau damping into the $\mathcal{Z}$-transform formalism but that has not been attempted here.

# REFERENCES


1. For Laplace transform: Carslaw and Jaeger, *Operational Methods in Applied Mathematics,* Oxford, 1953.
2. For $\mathcal{Z}$-transform: Doetsch, *Guide to the Applications of the Laplace and $\mathcal{Z}$-transform.* Van Nostrand, 1971.
3. For Digital control systems: Houpis and Lamont, *Digital Control Systems* McGraw-Hill,1985.
4. For control systems in general there are many books, such as: Netushil, *Theory of Automatic Control,* Mir Publishers, Moscow, 1973.
5. V.M. Zhabitsky *Theoretical Treatment of a Classical Transverse Feedback System Using Z-Transform*, JINR E9-91-156, Dubna, 1991.
6. R. Talman, *Analysis of Accelerator Instabilities Using the Formalism of Digital Control Systems*, unpublished CESR report, November, 1990.
7. Lloyd Sakazaki, *Multi-Bunch Instabilities and Current-Dependent Phenomena in the Cornell Electron Storage Ring,* Cornell University PhD thesis, 1985.
8. Jury, *Theory and Application of the $\mathcal{Z}$-Transform Method,* J. Wiley, 1964.
9. G. Bourianoff *et al. Determination of Coupled-Lattice Properties Using Turn-By-Turn Data*, AIP Conference Proceedings, Workshop on the Stability of Particle Motion in Storage Rings, Brookhaven, and SSC Preprint 181, December, 1992.

# TRANSVERSE FEEDBACK IN A 100 TeV STORAGE RING

GLEN LAMBERTSON
Lawrence Berkeley Laboratory
University of California
Berkeley, CA 94720, USA

## Introduction

The general purpose for a transverse feedback system in the proton collider is to stabilize coherent bunch motions and reduce them to a small amplitude that will result in an acceptable diffusion rate into incoherent motion. Strong instabilities are expected to be caused by the resistive-wall impedance of the beam tubes in pulsed magnets at the low revolution frequency. The feedback must stabilize thousands of coupled-bunch modes. It must also damp excursions of the injected beam and suppress excitations from noise in magnetic fields, in support structures, and in the feedback circuits. It may also be important in combating some beam-beam interaction effects. Damping by synchrotron radiation will be important in the long-term evolution of the emittance, but it is too weak to be a factor in the coherent damping.

## Design Issues

With ~50,000 bunches in the ring, there are half that number of coupled-bunch modes that may be unstable and one expects many thousands of these to grow under the influence of various narrow-band beam impedances. A broad-spectrum feedback is needed and most suitable would be a bunch-by-bunch method in which the position of each bunch is measured and a correction directed to that same bunch.

The amount of broad-band power required to kick each successive bunch is a measure of the scale of the feedback system and can become of concern if it is required to damp sizable injection errors in a large fraction of the circulating beam.

Feedback gain is not in itself a problem of hardware or of cost, but is limited by the dynamics of the bunch motion if one tries to damp a growth rate $1/\tau$ that approaches the orbital frequency $f_0$. A practical limit of $1/\tau < f_0/6$ has been assumed here and would be a machine design parameter affecting particularly the bore tube. With a second feedback system in betatron-phase quadrature or with multiple systems acting over less than a full turn between pickup and kicker one might extend this limit. But in what follows, a single system in each transverse direction having one turn delay will be assumed.

Continual noise excitations of the transverse motion will, with feedback operating, result in a steady-state oscillation amplitude. The contribution of feedback-circuit noise will be estimated; the strengths of other noise sources are not yet known but the presence of feedback will reduce their effects.

**Example Parameters**

The parameters of 100-GeV rings, one with low magnetic field and one with strong field, are listed for use in these example feedback calculations.

Parameters for two example rings

| | | |
|---|---|---|
| 2 | 8 | = B [tesla] |
| $10^{35}$ | $10^{35}$ | = $\mathcal{L}$ [$cm^{-2}sec^{-1}$] |
| 219 | 877 | = $f_o$ [Hz] orbital frequency |
| 38 | 38 | = $f_B$[ MHz] bunch rate |
| 10 | 10 | = E [TeV] at injection |
| 0.31 | 0.31 | = I [A] average |
| 0.8 | 0.8 | = $\varepsilon_N$ [mm mrad] |
| 2400 | 2400 | = $\beta_\perp$[m] average |
| 7.1 | 3.5 | = $b_o$ [cm] tube radius |
| 6.0 | 6.0 | = $f_o\tau$ [turns] growth time at injection |

At feedback electrodes

| | | |
|---|---|---|
| 5.3 | 2.6 | = b [cm] tube radius |
| 0.42 | 0.30 | = $\sigma$ [mm] at injection |

**Feedback System Description**

Position data for each bunch would be picked up at the rate $f_B$ = 38 MHz, stored digitally for one turn, and amplified to drive a number of kicker electrodes. Each kicker structure can be a stripline pair. The kicker power scales as $b^2/\beta_\perp$, making a location at large $\beta_\perp$ preferred if b cannot be reduced to follow $\sqrt{\beta_\perp}$ . I assume that b can be so reduced to simplify parameters, but a location at maximum $\beta_\perp$ would be equivalent.

To just stabilize a coupled-bunch oscillation having amplitude A and growth $1/\tau$, the kickers must deliver a peak transverse voltage impulse of

$$V_\perp = \frac{2A\ E/e}{\beta_\perp f_o \tau} . \tag{1}$$

The excursion A depends upon details of the injection process not now known. As a reference amplitude we shall use the rms beam size $\sigma$ to find the following values:

$$V_{\perp} = \frac{(2)(4.2\text{x}\,10^{-4})(10^{13})}{(2400)(6)} = 0.583 \text{ MV/turn, 2 tesla}$$

$$\frac{2(3.0\text{x}\,10^{-4})(10^{13})}{(1200)(6)} = 0.833 \text{ MV/turn, 8 tesla} \tag{2}$$

To address all modes, the kickers must operate over a bandwidth of 1/2 $f_B$ = 19 MHz. The kicker electrode is most efficient at low frequency and the strong R-wall modes are at low frequency, therefore choose the band ~ 20 HZ to 19 MHz. Maximum kicker length is one half the bunch spacing of 8 meters and its stripline impedance may be 50 ohm. Calculate the kicker shunt impedances using

$$R_{\perp}T^3 = 2\,Z_L\left(\frac{\sin kl}{kb}\right)^2 \tag{3}$$

to find per kicker unit

| | B | 2 | 8 | tesla |
|---|---|---|---|---|
| $RT^2$ @ | f = 0 | 0.57 | 2.37 | mΩ |
| | f = 23 MHz | 0.12 | 0.48 | mΩ |

Choosing N=12 such units at 2 tesla and 8 units at 8 tesla, we find using

$$P = \frac{1}{2}\,\frac{(V_{\perp})^2}{NR_{\perp}T^2} \tag{4}$$

the total power for an amplitude of one $\sigma$:

| B | 2 | 8 | tesla |
|---|---|---|---|
| N | 12 | 8 | units |
| P | 24.9 | 18.3 | kW |
| P/N | 2.1 | 2.3 | kW/unit |

We must remember that although these modest power levels are for the most unstable ($f_o\tau$=6) modes, the amplitude allowed was only one $\sigma$. Injected beam may have larger excursions. Only a portion of these injected motions will be rapidly growing coupled bunch modes. A power level can be determined that will damp the total excursions before the growing component becomes uncontrollable. For this technique, the gain is set as needed to stabilize the fastest-growing modes. Then the power output is allowed to respond to that gain value up to and limited at the level needed to bring injection errors into control. This avoids excess power that would be called for by linear response to large excursions of a few injected bunches. This saves power most if only a small fraction of the total charge is injected and that is damped before the next batch.

**Noise and Diffusion**

We certainly want the feedback to damp coherent motions to smaller than the beam size $\sigma$, but more demanding is the requirement that diffusion from coherent motion to emittance be at a tolerable rate. For a very simple model of the complex diffusion process, we could assume that it obeys the relation

$$\frac{d\sigma^2}{dt} = DA^2 \tag{5}$$

in which A is the amplitude of coherent motion. A value for D used at the Workshop was $f_0/100$. The growth time for $\sigma$ is then given by

$$T = \frac{100}{f_o} \frac{\sigma^2}{A^2} . \tag{6}$$

We assume that T is long compared to the time to damp motions down to a steady amplitude A.

The damping action of feedback with gain G per turn is

$$\frac{dx}{dt} = - f_o G x . \tag{7}$$

Averaged over oscillations and in terms of $A^2$ this is

$$\frac{dA^2}{dt} = - f_o G A^2 . \tag{8}$$

Opposing this are growth from electronic noise, coupled-bunch instabilities, and disturbances from the fields and the positions of magnets. Adding these, we have

$$\frac{dA^2}{dt} = - f_o G A^2 + \frac{1}{2} f_o G A^2 \langle x_N^2 \rangle + \frac{2}{\tau} A^2 + M . \tag{9}$$

Here $x_N$ is the rms electronic noise at the pickup expressed as beam position, $\tau$ is the coupled-bunch growth time, and M is the rate of growth of $A^2$ from other sources. To use this simplified relation, we must assume that noises are random, that all modes have the same growth rate, etc. And M is just there as a reminder of how those disturbances enter.

At steady state, $dA^2/dt = 0$ and the amplitude is given by

$$A^2 = \frac{\frac{1}{2} f_o G^2 \langle X_N^2 \rangle + M}{f_o G - 2/\tau} . \tag{10}$$

We see that high feedback gain will reduce the contribution of M but for overall minimum A there is an optimum value of G. G cannot be less than $2/f_0\tau$ for the most unstable mode. To obtain some numerical results I shall use the gain $G = 4/f_0\tau = 2/3$, giving

$$A^2 < \frac{2}{3} \langle X_N^2 \rangle + 3M/f_o . \tag{11}$$

A more detailed modal analysis supports the use of the "<" symbol here.

An estimate of the electronic noise can be made, in this case for a single stripline-pair of length $\lambda/4$ and an amplifier with noise figure 1.5 dB. Input noise in the frequency band W = 19 MHz will be

$$P'W = kT \;\; 10^{NF/10}\, W = \left(0.41 \times 10^{-20}\right) 10^{0.15} \left(19 \times 10^6\right) = 1.1 \times 10^{-13} \text{ watt} \tag{12}$$

The signal power from the pickup for excursion x is

$$P = Z_L (I\, x/2b)^2 . \tag{13}$$

Using this pickup gain we can find the equivalent input noise and emittance lifetime as influenced by this noise alone:

| B | 2 | 8 | tesla |
|---|---|---|---|
| P | $1.5\, x^2$ | $6.4\, x^2$ | KW |
| $x_N$ | 8.6 | 4.1 | nanometer |
| T | 3.3 | 1.8 | $10^4$ hour |

This weak effect on emittance lifetime suggests that other noise sources such as magnetic noise and digital least count will dominate. Large dynamic range and fast data rate are required for the feedback. It may be necessary and practical to switch to a higher sensitivity immediately after the injection disturbances have been damped in order to have digital least count near $10^{-7}$ meter.

**Summary**

Practical transverse feedback systems can be provided if the coupled-bunch growth rates from resistive-wall impedance is limited in magnitude by proper design of the beam tube. Both allowed aperture and materials of the magnet bore tubes will be constrained. Linearity of feedback response is not required above some oscillation amplitude smaller than initial beam excursions but determined by the details of beam injection. The product of charge times transverse error in an injected batch is important. Electronic noise can be negligible compared to feedback digital least count and disturbances delivered to the beam through the magnets. The effects of these latter noises on the growth of emittance need to be evaluated in the presence of the feedback systems.

## Acknowledgments

These considerations of feedback were developed with the participation of members of the workshop on "Maximizing Luminosity of Hadron Colliders at 100 TeV" at Erice, Italy, November 1991. I must thank and note especially the contributions by Jacques Gareyte, Gerry Jackson and Karl Schindl. This work was supported by the Director, Office of Energy Research, Office of High Energy and Nuclear Physics, High Energy Physics Division, of the U.S. Department of Energy under Contract No. DE-AC03-76SF00098.

# PS for LHC- Emittance conservation at low energy

R.Cappi
CERN - PS Division
1211 Geneva 23
Switzerland

## 1. Introduction

Transverse emittance conservation is a major concern in the LHC injector chain composed by LINAC, PS Booster ( PSB ), PS and SPS[1)].

In the PS, for example, the total emittance blow-up must be less than 20 %.

We will investigate in this paper three possible effects of emittance blow-up occurring at injection energy, namely space charge, injection errors and transverse instabilities. For each we will evaluate the issues and propose a possible remedy.

## 2. Space charge

One of the peculiarities of the LHC beam is that the PSB will send two batches of 4 bunches into the PS machine. As the PSB has a repetition rate of 1.2 s, the PS will have to wait for the 2nd batch during this time at injection energy with the 1st batch circulating under a strong space charge regime. With "strong" we mean conditions where the incoherent tune shift $\Delta Q_{x,y}$ is larger than ≈ 0.3. $\Delta Q_{x,y}$ is the tune depression ( space charge is a detuning effect ) of the bunch center compared to the zero intensity value.

In first approximation, it can be shown that, for bunches of parabolic longitudinal line density, Gaussian transverse distributions and dropping the minus sign for simplicity

$$\Delta Q_x = \frac{2\, r_0}{ec}\,\frac{I_p\, R^2}{Q_x\,(\beta\,\gamma)^3}\,\frac{1}{a\,(a+b)} \qquad \text{1a)}$$

$$\Delta Q_y = \frac{2\, r_0}{ec}\,\frac{I_p\, R^2}{Q_y\,(\beta\,\gamma)^3}\,\frac{1}{b\,(a+b)} \qquad \text{1b)}$$

where $r_0$ is the proton radius ( = 1,53 $10^{-18}$m )
c is the speed of light ( = 3 $10^8$ m $s^{-1}$ )
e is the proton charge ( = 1.6 $10^{-19}$ C )
R is the machine radius
$I_p$ is the bunch peak current $= \dfrac{3eN_b}{2T_b}$
$N_b$ is the number of particle in the bunch
$T_b$ is the total bunch length
$Q_x$ and $Q_y$ are the hor. and vert. tunes
$\beta$ and $\gamma$ are the usual relativistic factors

and

$$a = \sqrt{2\left[\varepsilon_x \beta_x + \left(D_x \frac{dp}{p}\right)^2\right]} \qquad \text{2a)}$$

$$b = \sqrt{2\left[\varepsilon_y \beta_y + \left(D_y \frac{dp}{p}\right)^2\right]} \qquad \text{2b)}$$

where $\varepsilon_x$ and $\varepsilon_y$ are the rms hor. and vert. emittances given by

$$\varepsilon_{x,y} = \frac{\sigma^2_{x,y}}{\beta_{x,y}} \qquad \text{2c)}$$

and $\sigma_{x,y}$ is the hor. or vert. rms betatron beam size
$\beta_x$ and $\beta_y$ are the average hor. and vert. beta functions
$D_x$ and $D_y$ are the average hor. and vert. dispersions
dp/p is the rms momentum dispersion

Large $\Delta Q$ leads to crossing betatron resonances which can blow up the beam, however the amount of this blow up is difficult to evaluate. Refined tracking programs have been written [2) 3)] but the PS beam simulation during 1.2s actually requires a too long computer time.

So to answer the questions:

Will these effects blow up the beam during injection time? In which way and by how much?

Several experiments[4)] were done in the PS simulating the LHC beam conditions. These tests were performed at the present injection energy of 1 GeV and the space charge tune shift was modified by varying the bunch length with the RF voltage.

The results of these measurements are summarized in Figs.1 and 2.

It is clear from Fig.1 that the LHC beam, having at 1 GeV a $\Delta Q_y = 0.44$, will suffer a too large blow-up of $\approx 50\%$.

For round beams, equations 1a) or 1b) can be written as

$$\Delta Q_{x,y} \cong \frac{r_0}{2ec} \frac{I_p R}{\varepsilon^*_{x,y} (\beta \gamma)^2} \quad 3)$$

where

$$\varepsilon^*_{x,y} = \varepsilon_{x,y} \beta\gamma \quad 4)$$

Formula 3) shows that, for a given emittance, to decrease the tune spread one has to increase the energy and/or decrease the peak intensity (by increasing for example the bunch length ).

In the PS little margin is left for this last possibility as the injection kicker rise time and the RF bucket length limit the bunch duration to a maximum of $\approx$ 200 ns. Although recent developements in longitudinal blow-up techniques for bunch flattening[5)], provide $\approx 10\%$ reduction of peak intensity, the increase of injection energy to $\approx 1.4$ GeV appears as a necessity.

These modifications should reduce the tune shift to a safe value, smaller than 0.3, producing negligible emittance deterioration.

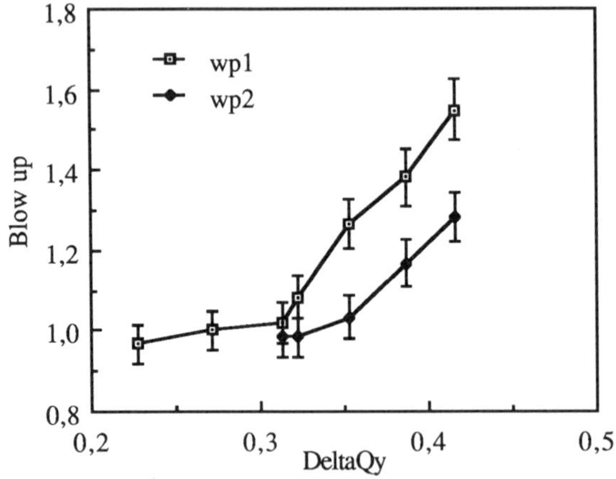


Fig. 1 - Averaged emittance blow-up [ i.e. $1+\Delta<\varepsilon>/<\varepsilon>$ with $<\varepsilon> = (\varepsilon_x+\varepsilon_y) / 2$ ] (calculated after ε b.u.) versus vertical space charge tune shift $\Delta Q_y$ for two different working points : wp1 : $Q_x = 6.22$, $Q_y = 6.22$ and wp2 : $Q_x = 6.22$, $Q_y=6.28$ ( *as expected slightly raising* $Q_y$ *improves the situation* ).

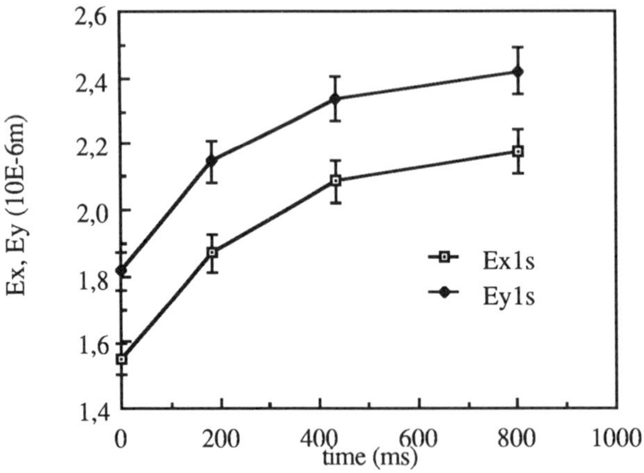


Fig. 2 - Horizontal ( x ) and vertical ( y ) emittances versus time (t = 0 is injection ) in a strong ( $\Delta Q_y \approx 0.45$ ) space charge regime at 1 GeV ( $Q_x = 6.22$, $Q_y = 6.28$).

## 3 Injection errors

### 3.1 Emittance blow-up by injection errors

The emittance blow-up $\frac{\Delta\varepsilon_{x,y}}{\varepsilon_{x,y}} \left( = \frac{\varepsilon_{final} - \varepsilon_{initial}}{\varepsilon_{initial}} \right)$ due to a position error $a_{0x,y}$ (m) of the injected beam is given by[6)]

$$\frac{\Delta\varepsilon_{x,y}}{\varepsilon_{x,y}} = \frac{1}{2}\left(\frac{a_{0x,y}}{\sigma_{x,y}}\right)^2 \qquad \text{5a)}$$

or

$$\frac{\Delta\varepsilon_{x,y}}{\varepsilon_{x,y}} = \frac{1}{2}\,\frac{a_{0x,y}^2}{\beta_{x,y}\,\varepsilon_{x,y}} \qquad \text{5b)}$$

where

$\sigma_{x,y}$ is the r.m.s. transverse betatron beam dimension ( x or y ) at the azimuth where $a_{0x,y}$ is given.

The LHC beam values [7)] at PS injection are foreseen as :

$$\varepsilon_x = \frac{\sigma_x^2}{\beta_x} = 1.53 \quad \mu m$$

and

$$\varepsilon_y = \frac{\sigma_y^2}{\beta_y} = 0.78 \quad \mu m$$

Values of emittance blow-up versus injection errors are shown in Table 1 below .

| $a_{0x}$ (mm) | 0.6 | 1.6 | 2.2 | 3.1 |
|---|---|---|---|---|
| $a_{0y}$ (mm) | 0.5 | 1.1 | 1.6 | 2.2 |
| $\Delta\varepsilon/\varepsilon$ (%) | 1 | 5 | 10 | 20 |

Table 1 : Emittance blow-up ( $\Delta\varepsilon/\varepsilon$ ) corresponding to different injection errors.

It is clear that injection errors will have to be kept below $\approx$ 0.5 mm in both planes.

We know from measurements that, even after a careful injection optimisation, errors of $a_{0x,y} \approx 3$ mm are likely to occur from pulse to pulse and/or from ring to ring of the PSB machine.

At present injection errors are negligible as the beams have three times larger emittances than those discussed above but for the LHC a careful injection optimisation method and eventually a damper system will be necessary.

### 3.2 Injection optimisation method

The description of such a method is outside the scope of this paper so we will mention here only the basic principle[8)].

The objective is to optimise the injection of each bunch, on a pulse to pulse basis, by measuring and eventually correcting its trajectory by controlling the values of the extraction elements of the corresponding PSB ring. This will allow us to minimize short and medium term injection drifts and should decrease the errors below the 1.5 mm value where a transverse damper could then be used for the "final" error minimisation.

## 4. Damper system

Is the present 2 kW transverse feedback[9)] enough powerful to damp the injection oscillations? Is its bandwidth large enough to damp each single bunch?

### 4.1 Required kick, decoherence time and damper gain

Let us calculate the necessary kick ( $V_\perp$/ turn ) to damp an injection error of $a_0$= 1.5 mm to 0.5 mm in a time shorter than the decoherence time $T_c$ .

We assume decoherence effects are only due to the chromatic tune spread given by :

$$dQ = \xi \, Q \, \frac{dp}{p} \qquad 6)$$

where dQ is the rms tune spread

$\xi$ is the machine chromaticity = -1

Q is the tune $\approx$ 6.22

dp/p is the rms momentum spread $\approx 0.5 \; 10^{-3}$

From the condition

$$dQ\,\omega_0\,T_c = \pi / 2$$

where $\omega_0$ is $2\pi$ the revolution frequency, we get

$$T_c = \frac{T_0}{4\,dQ} \qquad 7)$$

where $T_0$ is the revolution time = 2.3 μs.

Substituting the numerical values we finally obtain:

$$\frac{T_c}{T_0} = 80 \quad \text{turns} \quad \text{or} \quad T_c = 0.18\ \text{ms.}$$

In presence of a damper with gain G ( = kick equiv. displ./ displacement ), the evolution of the coherent beam oscillation a(t) is

$$a(t) = a_0 \exp\left(-\frac{G\,t}{2\,T_0}\right) \qquad 8)$$

and from this equation we can compute the necessary gain to achieve a given $\frac{a}{a_0}$ reduction after a time $T_c$ as :

$$G = -2\,\frac{T_0}{T_c}\,\ln\left[\frac{a(T_c)}{a_0}\right] \qquad 9)$$

with

$$a(T_c) = 0.5\ \text{mm} \quad \text{and} \quad a_0 = 1.5\ \text{mm}$$

we finally obtain

$$G = 0.033.$$

The resulting transverse kick expressed as

$$V_\perp = \frac{\Delta p_\perp \beta c}{e} \qquad 10)$$

will be

$$V_{\perp} \approx \frac{G\ a_0\ E/e}{\sqrt{\beta_p\ \beta_k}} \qquad 11)$$

where

$\beta_p$ and $\beta_k$ are the $\beta$ values at the pick-up and kicker position $\approx$16 m.
E is the particle total energy = 2.3 GeV

(*we assume the betatron phase advance between p.u. and kicker to be a odd multiple of π/2*)

the result is

$$V_{\perp} = 4.7 \text{ kV/turn.}$$

### 4.2 Damping kicker

The present strip line kicker has the following dimensions:

| | |
|---|---|
| electrode length | $\ell$ = 0.9 m |
| " width ( total ) | w = 0.03 m |
| hor. electrode half spacing | $b_x$ = 0.07 m |
| vert. " " " | $b_y$ = 0.03 m |

so the kicker coverage factor[10)] given by

$$g_{\perp} = \tanh\frac{\pi w}{4b} \qquad 12)$$

is respectively

$$g_{\perp x} = 0.32$$

and

$$g_{\perp y} = 0.66$$

The response of a stripline kicker at frequency ω is proportional to $(\sin\varphi)/\varphi$ with

$\varphi = \frac{\omega \ell}{2}\left(\frac{1}{v_b}+\frac{1}{v_k}\right)$ where $v_b$ and $v_k$ are beam and line velocities[10] ( both $\approx$ c in our case ).

To be able to treat $k_b$ equally spaced bunches, one has , for the sampling theorem, to satisfy the relation $\omega \geq \pi\, k_b\, f_0$.

Under these conditions, the impedance $R_S$ can be expressed as

$$R_s = 2\,Z_l \left( g_\perp \quad \frac{\ell}{b} \quad \frac{\sin\frac{k_b \ell}{2R}}{\frac{k_b \ell}{2R}} \right)^2 \qquad 13)$$

where $k_b = 8$
$Z_l$ is the characteristic kicker impedance = 50 Ω.

The result is

$$R_{sx} = 1.7\ \mathrm{k\Omega}$$

and

$$R_{sy} = 39\ \mathrm{k\Omega}$$

### 4.3 Kicker power

From the above values one can estimate the required kicker power as

$$P = \frac{V_\perp^{\,2}}{2R_s} \qquad 14)$$

so $P_x = 6{,}4$ kW

and

$P_y = 280$ W

The value for the horizontal plane is more than three times larger than the present 2 kW installed power.

However, rather than increasing the power, it should be possible to solve this problem by a redesign of the kicker electrodes to increase the coverage factor $g_x$. Actually a doubling of the horizontal electrode width from 3 to 7 cm will double the value of $g_x$ providing the required factor 4 in power.

## 5. Transverse instabilities

Resistive wall impedance is the primary source of transverse coupled bunch instabilities at low energy. Nevertheless the LHC bunches will be much longer than the present one (200 instead of 50 ns) as a result only high modes of oscillation ( $m > 5$ ) will be possible, with instability rise time larger than ≈ 50 ms ( compared to the 5 ms of today )

As a consequence the existing transverse feedback is basically adequate to treat these instabilities.

## 6. Conclusions

1) Raising the injection energy from 1 to 1.4 GeV and flattening bunch line density will reduce $\Delta Q_{x,y}$ to < 0.3 preventing emittance blow up due to space charge effects along the 1.2 s injection flat bottom.

2) A careful injection optimisation (*software*) should bring injection errors and drifts to <1.5 mm in position and < 0.1 mrad in angle.

3) The present transverse feedback will have to undergo minor modifications in the horizontal electrodes, p.u. bandwidth and digital fiter in order

a) to increase its efficiency ;

b) to cope with the different number of bunches ( 8 compared to the present 20 ) ;

c) to provide a "bunch by bunch" correction of the injection oscillations ;

d) to cover the entire possible frequency range of transverse instability modes (0.1- 3.7 MHz instead of the present 0.1- 1.5 MHz ).

Hence it could be used as an injection damper to bring the 1.5 mm residual oscillations to less than 0.5 mm, minimizing the emittance blow-up to <1%.

4) The same transverse feedback will damp also the resistive wall instabilities.

## 7. A last remark

This is not all. Other sources of emittance degradation, like betatron or momentum mismatch ( not only at injection but also at extraction ), transition crossing, debunching-rebunching, nonlinearities, etc. are present and they need to be carefully studied in the near future.

## Acknowledgements

The beam studies on space charge effects were performed with M.Martini and J.P.Riunaud of the PS "Beam Studies" team and K.Schindl and H.Schönauer of the PSB group.

I have greatly benefitted from nice and fruitful discussions with G. Lambertson and L. Vos during the workshop.

I am also grateful to K.Hübner and D. Möhl for commenting on the manuscript.

## References

1) R. Cappi, R.Garoby, S.Hancock, M.Martini, N.Rasmussen, T.Risselada, J.P.Riunaud, K.Schindl, H.Schönauer, E.J.N.Wilson, " The CERN PS Complex as part of the LHC injector chain", 1991 IEEE PAC, p.171.

2) S. Machida, "Space Charge Effects in Low Energy Proton Synchrotrons", NIM, A309, pp.43-59, 1991.

3) M. Martini, O.Ponomarev, "Modelling of space charge effects in the CERN Proton Synchrotron", 1991 IEEE PAC, p.1890.

4) R. Cappi, M.Martini, J.P.Riunaud, K.Schindl, H.Schönauer, "Production and measurements in the PS Complex of a high brilliance proton beam in view of the LHC", LHC Note 209, CERN/PS 92-52(PA)(1992).

5) S. Hancock, PS-MD Report, PS-RF-Note 92-17.

6) P.J.Bryant, "Introduction to transfer lines and circular machines", CERN 84-04.

7) The LHC Study Group, "Design Study of the Large Hadron Collider (LHC)", CERN 91-03, p.152.

8) E. Wildner, private communication.

9) J.L.Gonzalez, "Etude et realisation d'un automatisme qui supprime les instabilites transversales des faisceaux de protons, groupes en paquets de haute densite, dans l'accelerateur PS", Memoire, CNAM, to be publ.

10) G. Lambertson, "Dynamic Devices - Pickups and Kickers", AIP Conf. Proceedings 153, V.2, p.1413.

# Requirements for the Transverse Injection Oscillation Dampers

Report from the Transverse Injection Oscillation Damper Working Group
Steven Jackim, Gerry Jackson (chair), and Igor Ivanov

Feedback Control of Multi-Bunch Instabilities in Proton Colliders
at the Highest Energies and Luminosities
Technical Workshop in Conjunction with
INFN Eloisatron Project 25th Workshop: Future Superdetectors and Colliders
at the Frontiers of High Energy and Luminosity

Erice, Sicily
November 13-21, 1992

Transverse injection errors can be caused by

1) Transfer line dipole magnet ripple and drift in power supplies,
2) Insufficient resolution on the settings of transfer line steering magnets,
3) Kicker waveform voltage ripple,
4) Kicker waveform timing jitter,
5) Energy error between the injector and the storage ring in the presence of uncompensated dispersion,
6) Remnant oscillation from an instability in the injector, both longitudinal or transverse.

The goal of this working group was to estimate injection amplitudes and then design transverse damper systems to damp these oscillations before they decohere into emittance growth.

## SSC

After a careful review of the SSC transfer line and magnet geometry, it is apparent that after taking design stage steps to address items 1, 2, 4, and 5, the dominant source of emittance growth during transfers is due to item 3. It is assumed that if item 6 occurs, that already sufficient emittance growth has occurred that the extra amount accumulated during transfer is irrelevant. Let $\delta$ be the size of an injection error normalized to the beam size at the injection point. Figure 1 below shows the dependence of emittance growth (defined as a dilution factor) as a function of $\delta$.

In order to design a transverse injection oscillation damper system, it is necessary to first estimate the amplitude of the oscillation that will have to be suppressed. The scale of the error is set by the distance between the injection and closed orbit trajectories in the Lambertson magnet both at extraction from the HEB and injection into the SSC storage ring. Given that this separation is approximately 20 mm and that the design tolerance on the voltage of each kicker (6 in the HEB, 6 in the SSC) is 1%, if all errors added coherently one would obtain a worst case oscillation amplitude of approximately 2.4 mm. Since the kickers are not all at the same betatron phase and are not all identical, an initial oscillation amplitude of 1.0 mm is considered to be more realistic. The beam size at injection is roughly 0.3 mm, so according to figure 1 the level at which 10% emittance growth would occur is approximately 0.1 mm. These four amplitudes are used to provide a range of numbers for future serious designers.

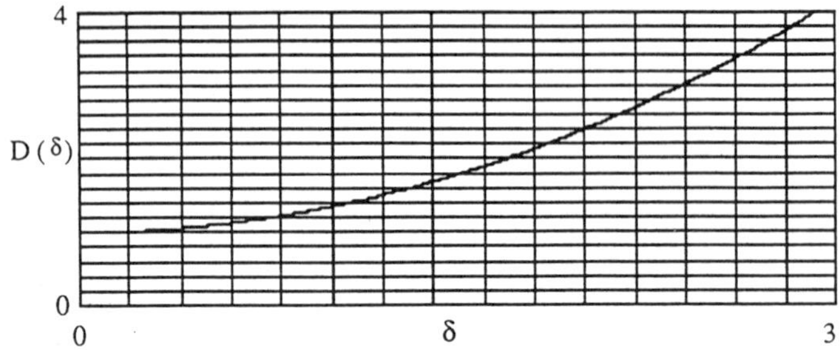


Figure 1: Emittance dilution factor as a function of the transverse injection oscillation amplitude in units of beam size at injection. The horizontal axis is 0.2 per division, the vertical axis is 20% per division.

A beam position detector (BPM) is required to measure the position of the beam each turn. If the voltage measured in the BPM is given by

$$V_{bpm} = \frac{2 I_b Z}{d} A$$

where $I_b$ is the peak beam current, Z is the net impedance of the BPM, d is the distance between the plates of which the BPM is composed, and A is the oscillation amplitude. Table 1 contains reasonable values for these parameters, and also calculates the angle one expects 90° downstream of the BPM assuming an average value of the beta function in either the HEB or the SSC. Note that the higher the value of the beta function at the kicker, the less power that is required in the system. Since the BPM need only be sensitive to oscillation amplitudes on the order of 0.03 mm to damp these oscillations, and noise is not an issue since such a system is only on at injection time, the processing electronics should be rather simple and standard.

| Injection Error Amplitudes | **0.1** | **0.3** | **1.0** | **3.0** | mm |
|---|---|---|---|---|---|
| BPM Impedance | 50 | 50 | 50 | 50 | Ω |
| BPM Electrode Gap | 40 | 40 | 40 | 40 | mm |
| Peak Beam Current | 1 | 1 | 1 | 1 | Amp |
| Peak BPM Voltage | 0.25 | 0.75 | 2.5 | 7.5 | V |
| Average β-Function | 100 | 100 | 100 | 100 | m |
| Equivalent Angle Error | 1 | 3 | 10 | 30 | μrad |

Table 1: BPM voltages and angles at kicker.

| Injection Error Amplitudes | **0.1** | **0.3** | **1.0** | **3.0** | mm |
|---|---|---|---|---|---|
| Kicker Length | 10 | 10 | 10 | 10 | m |
| Maximum Kicker Frequency | 7.5 | 7.5 | 7.5 | 7.5 | MHz |
| Kicker Electrode Gap | 40 | 40 | 40 | 40 | mm |

Table 2: Some reasonable kicker electrode geometry numbers.

The hardest part of this system will be in the kicker electronics. Since the frequency content of the bunch oscillations in a given batch of injected charge is assumed to be comparable or below the inverse rise time of the kicker waveform, a total bandwidth of 1-3 MHz is assumed to be sufficient given a 1.7 μsec risetime. Table 2 contains some reasonable values for the geometry of the kicker electrodes themselves. Note that for a given length of electrostatic kicker, the maximum frequency is set by the requirement that the wavelength be longer than 4 times the length of the kicker.

The kicker can either be capacitive or a 50Ω transmission line. Given these two options, and the fact that the deflection per volt is given by

$$\theta = \frac{2\, V_p\, L}{P_o\, d}$$

for an electrostatic kicker with ±Vp on each plate of length L and separation d for a beam of injection momentum $P_o$. Because of the additional magnetic field contribution, a transmission line kicker has twice the deflection for the same voltage across the plates. Assuming that the kicker will try and eliminate the entire transverse position error measurement each turn made by the BPM (g=1), table 3 contains some parameters relevant to the design of the power amplifier and the system electronics. Given that some of the power estimates are rather high, table 4 contains the same numbers assuming that only 10% corrections are made each turn (g=0.1).

| Injection Error Amplitudes | **0.1** | **0.3** | **1.0** | **3.0** | mm |
|---|---|---|---|---|---|
| Kicker Voltage (Capacitive) | 4 | 12 | 40 | 120 | kV |
| Kicker Voltage (50Ω Trans. Line) | 2 | 6 | 20 | 60 | kV |
| Kicker Power (50Ω Trans. Line) | 0.04 | 0.36 | 4 | 36 | MW |
| Electrical Gain (Capacitive) | 84 | 84 | 84 | 84 | dB |
| Electrical Gain (50Ω Trans. Line) | 78 | 78 | 78 | 78 | dB |
| Dynamic Range (3mm->0.03mm) | 40 | 40 | 40 | 40 | dB |
| Damping 1/e Decay Time | 6 | 6 | 6 | 6 | turns |
| On time (full dynamic range) | 28 | 28 | 28 | 28 | turns |

Table 3: Kicker amplifier and system electronics estimates assuming full corrections each turn (g=1)

| Injection Error Amplitudes | **0.1** | **0.3** | **1.0** | **3.0** | mm |
|---|---|---|---|---|---|
| Kicker Voltage (Capacitive) | 0.4 | 1.2 | 4.0 | 12.0 | kV |
| Kicker Voltage (50Ω Trans. Line) | 0.2 | 0.6 | 2.0 | 6.0 | kV |
| Kicker Power (50Ω Trans. Line) | 0.4 | 3.6 | 40 | 360 | kW |
| Electrical Gain (Capacitive) | 64 | 64 | 64 | 64 | dB |
| Electrical Gain (50Ω Trans. Line) | 58 | 58 | 58 | 58 | dB |
| Dynamic Range (3mm->0.03mm) | 40 | 40 | 40 | 40 | dB |
| Damping 1/e Decay Time | 20 | 20 | 20 | 20 | turns |
| On time (full dynamic range) | 92 | 92 | 92 | 92 | turns |

Table 4: Kicker amplifier and system electronics estimates assuming 10% corrections each turn (g=0.1)

Another kicker option is a resonant system in which the kicker plates (acting as capacitors) are coupled by a inductor magnetically coupled to the power amplifier. Centered at 60 MHz with a ±3 MHz bandwidth, a Q=10 system is needed using plates at most 1 m long to get the same kick for the same power as the 50Ω transmission line case. The utility of this option is the availability of suitable power amplifiers.

Given that the dynamic range is approximately 40 dB, either digital or analog techniques can be utilized. The suggested solution of the working group is an analog system based on optical fiber technology. In the simplest incarnation of such a system, the BPM shown in figure 2 consists of a single 180° hybrid which takes the analog difference of the two plate signals near 60 MHz. This difference signal is then delivered to the power amplifier via an optical fiber delay line. Since the power amplifier need only supply power in a ±3 MHz band around 60 MHz into a matched load (a resonating kicker), there should be plenty of appropriate units commercially available.

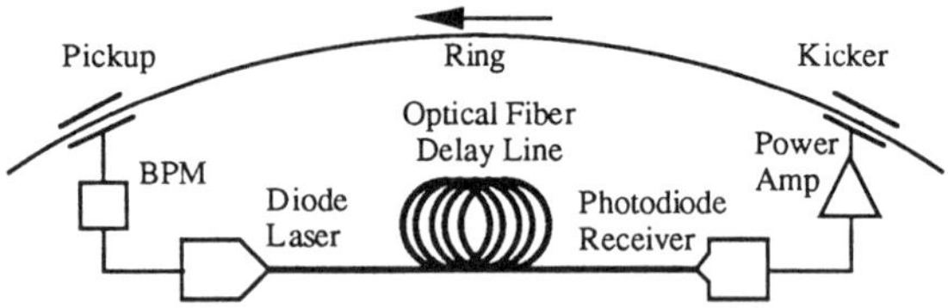


Figure 2: Sketch of a proposed transverse injection oscillation damper for the SSC.

## ELOISATRON

The root reason for the size of the injection errors is the spacing between the closed orbit and injection trajectories in the Lambertson magnets. This distance will not scale down as machine energy is increased. On the other hand, the tolerance for injection errors will decrease due to smaller beam sizes at injection. In the case of the Eloisatron, beam size will be approximately 3 times smaller than in the SSC case. But because of the higher injection energy, more kickers will be required, making the estimated injection error larger. A rough estimate would be a factor of 3 larger injection error over the span of an injected batch (the 3 mm category). See table 5 for some estimates of the voltages and powers required for a g=0.1 system at 10 TeV.

| Injection Error Amplitudes | **0.1** | **0.3** | **1.0** | **3.0** | mm |
|---|---|---|---|---|---|
| Kicker Voltage (Capacitive) | 2 | 6 | 20 | 60 | kV |
| Kicker Voltage (50Ω Trans. Line) | 1 | 3 | 10 | 30 | kV |
| Kicker Power (50Ω Trans. Line) | 10 | 90 | 1000 | 9000 | kW |
| Electrical Gain (Capacitive) | 78 | 78 | 78 | 78 | dB |
| Electrical Gain (50Ω Trans. Line) | 72 | 72 | 72 | 72 | dB |
| Dynamic Range (3mm->0.01mm) | 50 | 50 | 50 | 50 | dB |
| Damping 1/e Decay Time | 20 | 20 | 20 | 20 | turns |
| On time (full dynamic range) | 114 | 114 | 114 | 114 | turns |

Table 5: Kicker amplifier and system electronics estimates assuming 10% corrections each turn (g=0.1) for the Eloisatron. All numbers in tables 1 and 2 are still assumed.

The power levels for a likely oscillation amplitude of 3 mm is approaching 10 MW, which is far too large to construct under reasonable financial and engineering constraints. Alternate solutions like constructing programmable pulse forming network (PFN) kickers designed to counteract the measured ripple in the transfer kickers should be contemplated.

# LONGITUDINAL DAMPER FOR THE SSC

Daniel Boussard (CERN)
Jim Bridges (ANL)
John Fox (SLAC)
Dave McGinnis (FNAL)

November 18, 1992
Erice, Italy

## INTRODUCTION

The purpose of this report is to discuss the design of damper systems that will damp injection oscillations and coupled bunch mode instabilities for the SSC.

RF Parameters

| | | | |
|---|---|---|---|
| RF Frequency | 360 MHz | Harmonic number | 104544 |
| Number of bunches | 17420 | Revolution Frequency | 3.4 kHz |
| Bunch Frequency | 60 MHz | | |

Cavity Parameters (Fundamental mode)

| | | | |
|---|---|---|---|
| Number of Cav. | 32 | R/Q | 150 Ω |
| Shunt Impedance | 3 MΩ | Q | 20,000 |
| Volts/cavity | | Power/cavity | |
| Injection | 206 kV | Injection | 6 kW |
| Storage | 625 kV | Storage | 62 kW |

Beam Parameters

| | | | |
|---|---|---|---|
| Total Current | 72 mA | Transition gamma | 105 |
| Synchrotron freq. | | Synchrotron Spread | |
| Injection | 7 Hz | Injection | 0.8 Hz |
| Storage | 4 Hz | Storage | 0.46 Hz |
| Inject Energy Error | 0.0031 % | Inject Phase Error | 8 degrees |
| Bunch Length (6σ) | 1.2 nS | | |
| Z/n threshold due to inductive wall: | | (injection) | 1 Ω |
| | | (storage) | 3 Ω |

Formulas for Parameters

The spread in synchrotron frequency was calculated by using the approximation:

$$\Delta f_s = f_s \frac{\left(\tau_{bunch} f_{rf} \pi\right)^2}{16}$$

The threshold for Z/n in which Landau Damping ceases to exist due to the broadband inductive impedance of the accelerator is given by:

$$\frac{Z}{n} = \frac{3}{64}\left(\frac{\pi}{h}\right)^2 \frac{N_{cav} V_{cav}}{\left(\frac{I_{DC}}{N_{bunch}}\right)} \left(\tau_{bunch} f_{rf}\right)^5$$

The phase error at injection due to an energy error is:

$$\phi_{inj} = \left(\frac{\Delta E}{E}\right)_{inj} \frac{f_{rf}}{\gamma_t^2 f_s}$$

### THE INJECTION DAMPER

The voltage needed to damp the injection oscillations is:

$$V_d = \frac{1}{\tau} \frac{N_{cav} V_{cav} f_{rf}}{\pi f_s^2 \gamma_t^2} \left(\frac{\Delta E}{E}\right)_{inj}$$

where $\tau$ is the damping time. Using the above parameters:

$$V_d = 43 \frac{\text{kV} - \text{sec}}{\tau}$$

For damping times on the order of 0.1 seconds, this voltage is rather large. To supply this voltage, one can use the acceleration RF cavities as injection damper kickers. The kickers must have at most a 1uS rise time to accommodate the 1.7uS gap between batches during the injection process. This rise time could be accomplished by putting an equalizer upstream of the RF cavities if the RF system has no feedback circuitry. However, one could use RF feedback on the cavities to reduce the Q of the cavities. The main advantage of this solution is that the impedance of the cavity around the fundamental frequency is also reduced, thereby increasing the stability of the coupled bunch modes. The minimum shunt impedance using RF feedback is:

$$R_{min} = 4\frac{R}{Q} f_{rf}\tau_{delay}$$

where $\tau_{delay}$ is the loop delay of the RF feedback.

To provide the fast voltage swing of the injection damper, the RF cavity power supply must have in reserve:

$$P_{reserve} = \frac{V_{cav}\frac{dV_{cav}}{dt}}{2\pi f_{rf} R/Q}$$

The time rate of change of the cavity voltage is equal to the maximum voltage that the injection damper must supply divided by the rise time of the damper.

RF feedback parameters.

| | | | |
|---|---|---|---|
| Damper Volts/cavity | 10 kV | Damping time | 0.13 S |
| Reserve Power | 20 kW | Loop delay | 0.5 uS |
| Loop Gain | 30 | Loaded R | 100 kΩ |
| Q feedback | 720 | Feedback BandWidth | 500 kHz |

Since the cavity voltage is only 6MV at injection compared to 20 MV at storage, the Klystrons should be able to provide the extra 20 kW at injection. If many more than one cavity is fed off a single Klystron, then a loop gain of 30 might be difficult to achieve due to mismatches of the cavities and the cavities' feeds. The loop gain could be reduced, however, the impedance of the cavity around the low order modes will be increased. As will be shown in the next section, this not a strong problem.

There are a few details that should be mentioned concerning the injection damper design. Because a rise time of 1uS will not damp the bunches at the ends of the batch as strongly as in the bunches in the middle of a batch, a sample and hold scheme that measures the phase error at the middle of the batch should be used so that the kicker pulse spreads uniformly over all the bunches in a batch. Also the injection damper could be converted into a coupled bunch mode damper for the lowest order couple bunch modes after the injection process is finished. However, the phase detection scheme for the coupled bunch mode damper will be different than the injection damper.

## COUPLED BUNCH MODE DAMPER

The growth rate of a coupled bunch mode is approximately:

$$\frac{1}{\tau} = \frac{qI_{total} f_{rev} Form(f) N_{cav} Z_{cav}(f)}{4\pi\gamma_t^2 E f_s}$$

where:

$$Form(f) = \frac{12}{\tau_{bunch}} \int_0^1 J_1(\pi f \tau_{bunch} x) x dx$$

Using the ANL cavity as an example for longitudinal modes:

| Frequency (MHz) | Zcav (MΩ) | Growth Time (Sec) |
|---|---|---|
| 360 | 0.10 (Feedback on) | 1.243 |
| 536 | 1.67 | 0.060 |
| 922 | 0.62 | 0.187 |
| 939 | 0.23 | 0.518 |
| 1173 | 0.18 | 1.125 |
| 1210 | 0.49 | 0.468 |
| 1509 | 0.40 | 5.596 |

The growth due to the fundamental mode at 360 MHz is calculated for the cavity being totally mistuned. Since the cavity impedance will spread across the -m mode lines almost as equal as the +m mode lines, the actual growth time will be much longer than 1.2 S.

The above table was given for a cavity with no Higher Order Mode Dampers. We will assume that HOM dampers could be designed that reduce the cavity impedance by a factor of ten for most modes. This will increase the growth times in the above table by a factor of 10.

The most troublesome mode is at 536 MHz. We will propose a bunch by bunch coupled bunch mode damper as shown in Figure 1. The damper system uses a sum mode pickup to detect the phase error of the beam. The system uses a broad band (30 MHz) digital notch filter at baseband to perform a differentiation of the phase signal at the synchrotron sidebands at every harmonic of the revolution frequency in the band. The response of the baseband filter is up-converted using phase signals in quadrature so that the damper system is linear. The frequency of the damper system should be centered where the dipole form factor is largest which is near 1/bunch length (800 MHz). The analog bandwidth of the system should be one half the bunch frequency (30 MHz). The sampling rate of the filter is equal to the bunch frequency (60 MHz at 8 bits). The delay of the digital notch filter should be about 1/6 of a synchrotron period so as to maximize the response at the synchrotron sidebands but still keep a phase shift between bands greater than 90°. To accommodate this delay and sampling rate the digital filter should have enough memory to contain 1.5 Msamples.

Since the damper system is linear, one can think the damper as having a negative impedance equal to:

$$Z_{damper} = -G_{electronics}\sqrt{Z_{pickup}Z_{kicker}}$$

Because the damper system occupies only 30 MHz of bandwidth around 800 MHz, an impedance transforming network with a Q of about 10 could be built so as to raise the pickup and kicker impedance to 500 Ω. With an electronic gain of 80 dB, the damping time would be about 0.6 sec which would eliminate the growth due to the cavity impedance at 536 MHz.

Because of the large pickup impedance, the RF beam signal at the harmonics of 60 MHz will be about 34 dBm. Because this signal will saturate the front end mixers, we propose that an analog notch filter with a notch depth of 30dB be included on the front end of the damper system so as not to degrade the noise figure of the damper system.

Because of the large gain of the damper system, this system will be swamped by the injection oscillations. Thus, this damper should be off during the injection of a batch and turned on when the injection oscillations are reduced by a factor of 10. (Alternatively, the damper could be switched on gradually with the gain of the system reaching a maximum .25 seconds after injection.) This should occur 0.25sec after the bunches have been injected. Since this number is less than the growth rate of the coupled bunch mode, the damper system should easily handle the unstable modes. However, once the coupled bunch mode damper system is turned on the residual injection oscillations will still be the largest signal. The power produced on the damper kicker is:

$$P_{damper} = \frac{\left(\Delta\phi I_{total} G_{electronics} Z_{pickup}\right)^2}{2Z_{kicker}}$$

For a residual phase oscillation of 0.8°, the damper system will develop a power of 25 kW.

Coupled bunch mode parameters

| | | | |
|---|---|---|---|
| Center freq. | 800 MHz | Analog Bandwidth | 30 MHz |
| Digitizing rate | 60 MHz | Memory | 1.5 Meg |
| Digitizer | 8 bits | | |
| Pickup Impedance | 500 Ω | Kicker Impedance | 500 Ω |
| Damping time | 0.6 sec | Electronic Gain | 80 dB |
| Kicker Volts | 5000 V | Kicker Power | 25 kW |

Summary

Injections oscillations are damped using the RF cavities as kickers. The bandwidth of the cavities is increased using RF feedback and the reduction in shunt impedance reduces the impedance of the cavity near the fundamental frequency. The phase detector must include a sample and hold so as to kick all bunches in a batch evenly. After injection oscillations are damped, the RF system is used to damp the low order coupled bunch modes.

A separate wideband coupled bunch mode damper is needed to damp instabilities due to the higher order modes in the cavities. ( It was assumed that higher order modes in the cavities already have at least 20 dB of passive damping.) This system must not be on during the injection of a batch to avoid saturation.

Preliminary estimates of a possible growth of longitudinal emittance due to quantization noise in the coupled bunch mode damper show that this should not be a serious problem.

## APPENDIX

The system uses a sum mode pickup to detect the phase error of each bunch. This phase error signal will be digitized at the bunch crossing rate as the damper requires a bandwidth of at least half the bunch frequency (30 MHz) to faithfully process the highest coupled bunch modes. A correction signal to be applied back unto the bunches can be generated in several ways. At the minimum, the feedback filter must generate a 90 degree phase shift at the synchrotron frequency, and suppress the DC components of the nominal synchronous phase. There may be operational advantages to allow the processing of saturated error signals, or operate the damper in both linear and non-linear feedback modes.

The simplest suitable type of feedback filter would be a two tap FIR filter. Such a circuit would consist of a digital time delay, followed by a digital subtraction circuit. The filter can be operated as a differentiator if the delay line is roughly 1/6 of a synchrotron period, or as a bandpass filter if the delay line tap spacing is 1/2 a synchrotron period long. Both approaches can be tailored to generate the proper 90 degree phase shift, and both can suppress the DC component of a bunch's synchrotron oscillation (equivalently, act as a notch filter at the revolution frequency). The disadvantage to the simple differentiator is the 6 dB loss in amplitude of the oscillation signal, which reduces the resolution of the digital detector by two. In contrast, the bandpass filter does not reduce the resolution of the front end detector. This may or may not be important, depending on the ultimate noise that the system should damp to, and the front end detector resolution.

The SSC design, with a revolution frequency of 3.4 KHz and synchrotron frequencies of 4 - 7 Hz, would result in a system which samples the bunch oscillations at roughly 500 to 850 samples per synchrotron period. This is a grossly oversampled system, and there is no need from a signal processing viewpoint to operate so far above the Nyquist limit. As an example, a brute force digital filter, implementing the simple differentiator filter using a 1/6 synchrotron period delay line requires roughly 1.5 Msamples of memory. An alternative approach would downsample the input data stream, and compute the output signals from roughly 4 - 20 input samples per bunch.

Such downsampled processing could simplify the overall size of the feedback filter to less than 50 - 200 Ksamples, with the tradeoff of more sophisticated signal processing hardware. This downsampled processing, including an interpolating filter to generate a new kick for each bunch on every turn, deserves consideration as a means to implement a flexible system with several operating modes. Such a system, with an output interpolating filter, would not increase the noise spectral density at the synchrotron frequency. Due to the low oscillation frequencies of several Hz, such systems could easily be implemented in one or two commercial DSP microprocessors.

# WIDE-BAND TRANSVERSE FEEDBACK FOR HADRON SUPERCOLLIDERS

*Reported by R. Cappi, G. Lambertson and L. Vos*

## 1. Introduction

A feedback system of sufficient frequency bandwidth can respond to the coherent motion of individual beam bunches and direct corrective kicks to those same bunches. This system is intended to suppress the growth of coupled bunch motion driven by narrow band beam impedances and to damp motions that are produced by external sources such as the motion of magnets or field ripple from power supplies. We assume that the initial excursions of the injected beam have been damped to an amplitude of 0.03 mm by a narrow band feedback. A list of parameters for the three proton rings is shown in Table I. The resistive wall impedance is strong in these large diameter rings because it scales as $R^{3.5}/b^3$; R is the orbit radius and b the radius of the beam tube aperture. While it increases with particle energy, because of magneto resistance in the cold bore, the growth rate it produces and the voltage kicks required to correct large excursions are greatest at injection. The injection values have been the basis for determining the feedback voltage required.

Transverse higher-order modes (HOM) in the rf cavities would, if unmodified, greatly exceed the resistive wall impedance, being in the order of Megohms per meter. But modest efforts to damp these can reduce them by one or two factors of ten, at which level they fall below the resistive wall. Further reductions of the HOM impedances are possible.

Non linearity in the transverse focussing forces that arise from the magnet fields, space charge , or beam-beam forces cause gradual decoherence of coherent bunch motions. The consequent increase in emittance can arise quickly and results in a large loss of luminosity. It is necessary that the feedback reduces the bunch motion to a very small amplitude (e.g. 1 μ) to bring this dilution to an acceptable rate. This function of emittance control can be the greatest demand for gain in the feedback.

## 2. Determination of feedback parameters

The growth rate of motion driven by a transverse impedance $R_\perp$ is found, for relativistic bunches, using

$$\frac{1}{\tau}=\frac{I_0 f_0 \beta_\perp R_\perp}{2E/e}$$

in which

$1/\tau$ = growth rate
$I_0$ =average beam current
E =particle energy in eV
$f_0$ =orbital frequency
$\beta_\perp$ =betatron function where $R_\perp$ influences the beam

To correct this growth requires a feedback gain per turn of $G=2/f_0\tau$ . To provide damping and some reserve gain, we list in Table II twice that gain. Similarly, the minimum transverse voltage kick per turn, defined as $\Delta V = \Delta p_\perp c / e$, is given by

$$\Delta V = I_0 A R_\perp = GA(E/e)/\beta_\perp$$

We also double this value in Table II. The quantity A is the largest coherent amplitude expected, here taken to be 0.03 mm.

The gain required for control of the emittance growth exceeds that for the resistive wall. The main effect to be considered is the quadrupole movement due to low frequency ground motion. Power supply ripple also falls into this category. The power spectrum of these effects scales approximately as $\omega^{-4}$. The increasing size of the supercolliders makes them more and more sensitive to these phenomena. We have estimated that these effects scale roughly like $\gamma^2$. The motion of the quadrupoles at the lowest betatron frequency provokes a constant coherent amplitude of the beam in the presence of tune spread. The number of quadrupoles and the optical properties of the machine cause an important amplification from the quadrupole motion amplitude to the amplitude of the coherent beam oscillation. This coherent oscillation continuously blows up the beam emittance. Simulations for the SSC have demonstrated that the emittance increase is unacceptable. Therefore it is vital to apply feedback during collisions to preserve luminosity lifetime. The gain of the system has to be large enough so that the damping rate by the feedback is much faster than the rate of decoherence. The simulations show that in such a case the emittance growth is reduced by 2 or 3 orders of magnitude. The feedback will reduce the amplitude of the coherent oscillation to an amplitude on the order of the resolution of the system.

For the SSC it has been calculated that a resolution of (0.3-1) μ leads to an emittance growth rate of (60-20) hours. This is longer than the damping caused by synchrotron radiation (13 hours), hence acceptable. For ELN a similar resolution is required. In the case of the LHC it is not clear yet whether the quadrupole movement is troublesome. Indeed, this machine profits from the $\gamma^2$ dependance and from the fact that the emittance is four times larger than in the sister machines.

The importance of the gain parameter can be estimated from the following formula which describes the effect on the emittance of the reduction of the coherent amplitude by feedback :

$$\frac{d\varepsilon}{dt} = \left(\frac{4\pi\Delta\nu}{G}\right)^2 \left(\frac{d\varepsilon}{dt}\right)_0$$

where

$\left(\frac{d\varepsilon}{dt}\right)_0$ = rate without feedback

$\frac{d\varepsilon}{dt}$ = acceptable rate with feedback

$\Delta\nu$ = tune shift at amplitude $\sigma$ (e.g. from beam-beam)

The maximum voltage kick may be limited electronically at $2\Delta V$ as determined above, then the high gain is available when the beams are colliding and the motions are very small.

TABLE I
Ring Parameters

| | LHC | SSC | $10^{35}$ELN |
|---|---|---|---|
| $I_0$ [A] | 0.8 | 0.071 | 0.10 |
| Injection E [TeV] | 0.45 | 2 | 10 |
| $\varepsilon$ [$\pi$ mm.mrad] | 3.75 | 1.0 | 0.7 |
| $f_0$ [kHz] | 11.25 | 3.44 | 0.84 |
| $f_B$ [MHz] | 67 | 60 | 36 |
| $\beta_\perp$ at kick [m] | 170 | 400 | 1200 |
| R-wall [MΩ/m] | | | |
| inject | 100 | 700 | (see |
| collide | 180 | 4200 | Table II) |

TABLE II
Feedback parameters

| | LHC | SSC | $10^{35}$ ELN |
|---|---|---|---|
| $\tau$, R-wall [msec] | 10 | 70 | 18 |
| [turns] | 112 | 240 | 15 |
| 2G, R-wall | 0.036 | 0.016 | 0.26 |
| $2\Delta V$ [kV/turn] | 4.8 | 3.0 | 70 |
| G, $\varepsilon$ growth | 0.02 | 0.2 | 0.2 |
| f-band [MHz] | 67-100 | 60-90 | 36-54 |
| kicker length [m] | 1.12 | 1.25 | 2.1 |
| kicker Rs [kΩ] | 96 | 219 | 4X112 |
| power [Watt] | 120 | 20 | 4X1370 |

## 3. Hardware implementation

The basic scheme of the transverse feedback system is shown in Figure 1. The position of the pick up and the kicker has to be separated by an odd number of quarter betatron wavelengths. The system should damp all the possible transverse rigid bunch modes. This defines a frequency range between $f_B$( the bunch frequency) and 3/2 $f_B$ [60-90 MHz]* . Other frequency bands are possible. A lower band would be impractical for PU and power amplifier response and a higher band implies limitations on the kicker length and possible excitation of higher modes. The excitation of mode m=1 , for example, can be estimated by computing the ratio of the Hermitian modes $h_0(\omega)$ and $h_1(\omega)$.[1]

$$h_m(\omega) = \frac{y^{2m}}{\Gamma\left(m+\frac{1}{2}\right)} e^{(-y^2)}$$

* in [ ] are shown the numerical values for the SSC

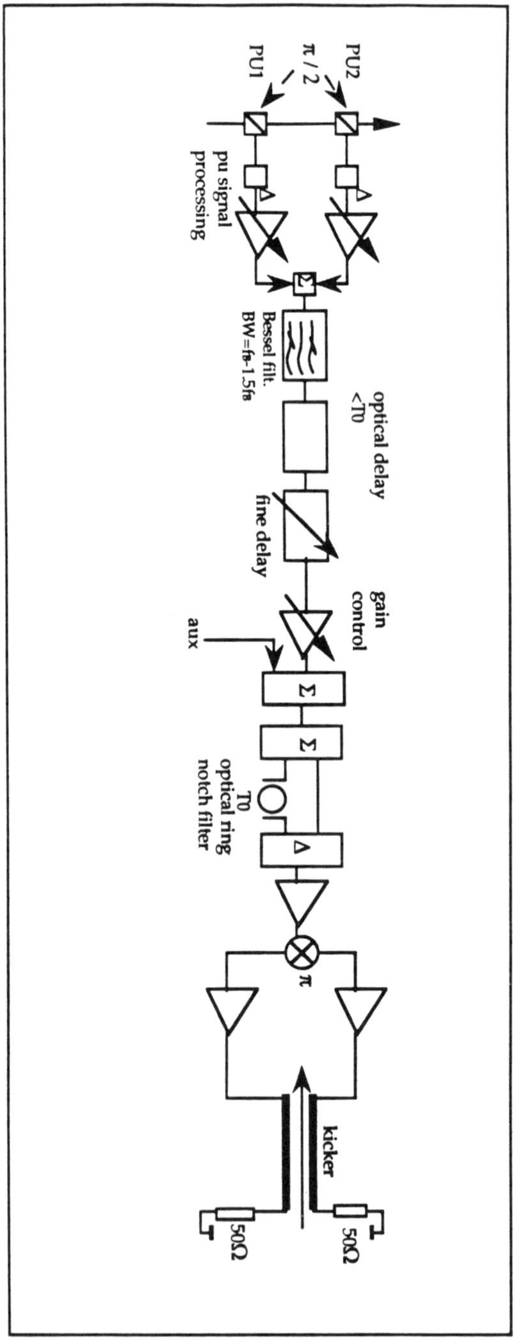
PU1
π / 2
PU2
pu signal
processing
Bessel filt.
optical delay
<T0
fine delay
gain
control
aux
Σ
Σ
T0
optical ring
notch filter
Δ
π
kicker
50Ω
50Ω

Figure 1

with :

$y = \omega\sigma_t$
$\sigma_\tau$=rms bunch length
$\omega = \omega_{FB} - \omega_\xi$
$\omega_{FB}$=feedback frequency
$\omega_\xi = \xi\nu\omega_0/\eta$
$\xi$= chromaticity
$\omega_0 = 2\pi\, f_0$
$\eta = \gamma_t^{-2} - \gamma^{-2}$

It follows that

$$\frac{h_1(\omega)}{h_0(\omega)} = 2(\omega\sigma_t)^2$$

For the SSC if $0<\xi<0.03$ then $h_1/h_0<2\%$ considered to be negligible. It should be recalled that the chromaticity should be kept small and positive to avoid the head-tail instability.

The band-width of the PU stations and the associated analog processing has to be large enough to cover the required frequency range, with high phase linearity, i.e. constant delay. The variation of the delay in the band-width must be less than 1 ns ($30^0$ at 90 MHz). The filter to eliminate the out of band signals should be of Bessel type in order to keep the delays within the required limits.

The resolution of the beam position measuring system is a parameter which is critical for the emittance control by the feedback system. It is determined by the ratio of the noise voltage and the signal voltage. As an example we will compute this ratio for a standard position monitor assuming nominal beam conditions.

The noise voltage will be very much the same for the three machines since the band-widths are similar. The proposed monitors for LHC and SSC are 50Ω striplines shorted at one end. For a low noise first amplifier stage we find :

$$V_n = \sqrt{4kTR\Delta f} = 5\mu V$$

The signal is computed from :

$$V_s = nef_0 Z_T = I_0 Z_T$$

where k is Boltmann's constant, T the absolute temperature, Δf the bandwidth, n the number of particles and $Z_T$ is transfer impedance of the monitor.

The processing frequency, which is not necessarily the working frequency of the feedback system, is choosen where $Z_T$ is maximum. For the LHC this would be 400 MHz. The resolution in position is a function of the distance 2b between the electrodes :

$$\delta x = \frac{V_n}{\sqrt{2}V_s}\frac{b}{2}$$

The following Table III summarises the results.

TABLE III
Resolution of standard position monitors

| | LHC | SSC | $10^{35}$ELN |
|---|---|---|---|
| $I_0$ [Amp] | 0.8 | 0.07 | 0.38 |
| $Z_T$ [Ω] | 6 | 6 | 6 |
| $V_s$ [V] | 4.8 | 0.42 | 2.4 |
| b [mm] | 23 | 17 | 40 |
| δx [μ] | 0.01 | 0.07 | 0.03 |

It can be seen that the resolution is sufficient. The margin is large enough to include some amount of signal loss in the cable and it permits working at a frequency where $Z_T$ is not at its maximum value. This may be convenenient for the signal processing. It is interesting to point out that the resolution margin may be increased by a reduction in the bandwidth if only a part of the feedback spectral width carries usefull information as is the case of the low frequency excitation by ground motion and power supply ripple.

The optical delay line must satisfy the relation :

$$T_{ODL} + T_{FD} + T_{EL} = T_B$$

where

$T_{ODL}$ = time delay optical line
$T_{FD}$ = fine delay
$T_{EL}$ = total delay of electronics (PU to kicker)
$T_B$ =beam time of flight from PU to kicker ($\sim 1/f_0 = T_0$)

As $T_{FD}$ and $T_{EL}$ are in general much smaller than $T_0$, then $T_{ODL}$ is only slightly smaller than $T_0$.

The notch filter based on the optical line can provide a 50 dB reduction in revolution harmonics. [2]. To provide a stable delay it requires temperature stabilisation.

A wide band amplifier with 0-50dB remote gain control allows to adjust the feedback gain as a function of the machine cycle. An auxiliary input is provided. It may be used to excite the beam for tune measurements, beam transfer function or as the output of a dedicated loop for ground motion effects at frequency :

$$f_{gm} = f_B + \nu f_0 \quad [\approx 60.001\text{MHz}]$$

This loop is narrow band to maximise the signal to noise ratio. The bandwidth however has to be larger than $2\pi f_0$ so that the delay does not exceed $T_0$.

## 4. Kicker systems

The kick amplitudes and the band widths of the broad band feedbacks do not call for serious optimization because costs and technical problems are small. Nevertheless, we treated them with what seemed sensible criteria that recognize both aspects. The frequency band, of width 1/2 $f_B$, has been placed in the interval between 1 and 1.5 $f_B$ to avoid requiring power amplifiers at frequencies within a few hundred Hertz of zero. In this band the power demand is greatest near $f_B$ where coupled bunch modes are most strongly driven by the resistive wall impedance.

The kickers consist of opposing pairs of striplines each having characteristic impedance $Z_\ell$ of 50 Ω. For maximum coupling near $f_B$, we call for a kicker length $\ell=\lambda/4$ at $f_B$. Then the shunt impedance $R_S$ is given by :

$$R_s = \left(\frac{\Delta V^2}{2P}\right) = 2Z_\ell\left(g\frac{\sin k\ell}{kb}\right)^2$$

where

$k = 2\pi f/c$
$b$ = half gap between plates
$g$ = electrode geometric factor ~ 1
$Z_\ell$ = 50 Ω

The voltage response (proportional to $\sqrt{R_s}$ ) at the upper frequency 3/2 $f_B$ falls to 0.47 of that at $f_B$ but the kick required is less there. Using $\Delta V$ and $R_S$ , we find the amplifier power as listed in Tablc II. In the ELN we have used four kickers to reduce voltage and power.

## References

[1] BBI , B. Zotter, CERN LEP/TH 89-74
[2] G. Jackson ,this workshop

# THE SSC TRANSVERSE FEEDBACK SYSTEMS

W. Chou and J. Peterson

*SSC Laboratory* *

*2550 Beckleymeade Ave, Dallas, TX 75237, USA*

**Abstract**

The transverse feedback systems are needed at the SSC for several different reasons. The requirements of these systems are analysed and specified. In addition to the general requirements (power, bandwidth and gain, etc.), specific attention is given to the noises in the systems, which need to be controlled in order to keep the emittance growth at a tolerable rate. A quantitative treatment is given to specify the allowable noise level in the feedback systems. A brief discussion of the damping process, beam decoherence, beam heating and coherent beam instabilities are also included.

## 1 Introduction

The SSC transverse feedback systems are designed for the following four different purposes:

(a) Correction of the injection errors:
At injection there are errors in the beam position, angle, energy and phase. These errors can lead to beam emittance growth if not corrected. Because these errors are relatively large compared with other types of perturbations as decribed below, the requirement of the feedback system is that it should have enough power to kick the beam quickly back to the desired orbit before any significant decoherence occurs.

(b) Damping of the resistive wall instability:
The transverse impedance of the wall of the beam tube increases as the frequency decreases. At low frequencies (around several hundred hertz), the wall impedance of the SSC is in the order of gigaohm per meter. It can lead to fast beam blowup in the transverse planes. A feedback system with a large gain is needed.

(c) Damping of the coupled bunch instability driven by the rf cavities:
The higher order modes (HOM) of the rf cavities, if not properly damped and/or staggering-tuned, may drive the beam unstable. The feedback system needs a wide bandwidth to cure this type of instability.

(d) Control of possible emittance growth from the above and/or other mechanisms (e.g., ground motion, coolant flow, power supply ripples, etc.):
Because of the long memory of the proton beam (the radiation damping time in the SSC is about 13 hours), any external perturbation will be remembered by the beam and may lead to eventual emittance growth. A feedback system that keeps the coherent motion of the beam below certain allowable amplitude can effectively reduce the emittance growth rate.

It is seen that each purpose imposes different requirements on the feedback systems. This, of course, does not mean that a total of four systems would be required. One system may well

* Operated by the Universities Research Association, Inc., for the U.S. Department of Energy under Contract No. DE-AC35-89ER40486.

serve several different purposes.

A special concern of the SSC is the preservation of the required small transverse beam emittance. Its design value is 1 mm-mrad (rms, normalized), which is just a fraction of the existing proton storage rings. A high noise level in the feedback system or a poor pickup resolution will lead to a fast emittance dilution as will be discussed in detail later. As a matter of fact, this is believed to be the possible source of the emittance blowup that has been observed both at the Tevatron and SPS when the transverse feedback system is turned on. Therefore, in addition to the general requirements of a feedback system (such as the power or kicker voltage, the bandwidth, the gain, etc.), one needs in particular to specify the noise level or, equivalently, the pickup resolution of the SSC feedback systems for the purpose of controlling the emittance growth.

This paper is organized in the following way. Section 2 is an analysis of the injection errors and their corrections by means of a feedback system. Section 3 discusses the continuous emittance growth due to external excitations in the absence and in the presence of a feedback system, respectively. It is followed by a discussion of the resistive wall instability in Section 4. The instabilities driven by the HOM are the contents of Section 5. The description of the SSC feedback systems and their specifications are given in Section 6. Finally, Section 7 is the summary and conclusions.

## 2 Analysis of injection error corrections

### 2.1 Injection process at the SSC

In the present design, there will be 8 batches injected from the High Energy Booster (HEB) into the Collider. Each batch has about 2000 bunches. The bunch spacing is 16.7 ns, the batch spacing 1.7 $\mu$s, and the abort gap 4.2 $\mu$s. It is the batch spacing (1.7 $\mu$s) and the injection kicker waveform ripples that determine the bandwidth requirement of the feedback system used to correct the injection errors.

### 2.2 Emittance growth and error tolerances

#### 2.2.1 Injection errors and emittance dilution

At injection into the Collider several types of errors can lead to growth in the transverse emittances $\epsilon_x$ and $\epsilon_y$. Errors in beam position ($\delta x$, $\delta y$), errors in angle ($\delta x'$, $\delta y'$), and errors in beam energy ($\delta E$) all produce coherent, dipole-type betatron oscillations, which, if not immediately corrected, will decohere (i.e., smear out in phase space) and so increase the emittance. The function of the damper† is to reduce these coherent dipole oscillations before the emittance growth through decoherence becomes appreciable. (The decoherence mechanisms and rates will be considered in Sections 2.4 and 2.5.)

If the beam optics of the injection transport line are not matched to the optics of the Collider, coherent quadrupole-type betatron oscillations are induced, which also eventually smear out and produce emittance growth [1]. However, the normal (dipole field) damper can not correct the quadrupole type of coherent oscillation. Therefore, in this paper we shall not

† The terms "damper" and "feedback system" are used exchangeably in this paper.

consider any further the effects of optical mismatches.

The magnitude of the coherent amplitude $x_c$ due to injection position and angle errors in, say, the horizontal plane is:

$$|x_c|^2 = \delta x^2(1+\alpha_x^2) + 2\alpha_x\beta_x\delta x\delta x' + \beta_x^2\delta x'^2 \tag{1}$$

where the lattice functions $\alpha_x$ and $\beta_x$ refer to the lattice position where $\delta x$ and $\delta x'$ are measured. The eventual fractional emittance increase produced by decoherence is, for the case of small coherent amplitudes, relative to the rms beam width $\sigma_x$ $(=\sqrt{\epsilon_x\beta_x})$ :

$$\frac{\delta\epsilon_x}{\epsilon_x} = \frac{|x_c|^2}{2\sigma_x^2} \tag{2}$$

For the case of an injection energy error $\delta E/E$, the resultant coherent amplitude is:

$$|x_c|^2 = (\frac{\delta E}{E})^2\left(D_x^2(1+\alpha_x^2) + 2\alpha_x\beta_x D_x D_x' + \beta_x^2(D_x')^2\right) \tag{3}$$

where $D_x$ and $D_x'$ are the dispersion function and its slope at the point of injection. Equation (3) points out the advantage of injecting into a region of zero dispersion.

### 2.2.2 Tolerable injection errors

It is of interest to estimate the magnitude of injection errors that could be considered "tolerable" in that the corresponding emittance growths are, say, less than 10%, in which case an injection damper might not be required. We will compute the injection position, angle, and energy errors that *singly* would produce a ten percent emittance dilution. Consider a beam-monitor point with $\beta_x = 300$ meters and $\alpha_x = 2.4$. At the Collider injection energy of 2 TeV and a normalized emittance of 1 mm-mrad, the rms beam size $\sigma_x$ is 0.38 mm, and the rms beam angle $\sigma_x'$ is 1.27 $\mu$rad. The tolerable coherent amplitude $x_c$ is 0.17 mm, which could be produced either by an injection-position error $\delta x$ of 0.065 mm *or* by an injection-angle error $\delta x'$ of 0.57 $\mu$rad. Since the injection area is nominally dispersion-free, let us estimate the tolerable injection energy error on the basis of an assumed spurious dispersion magnitude $D_x$ of 1 meter and a slope $D_x'$ of 0.01. In this case the corresponding fractional injection energy error $\frac{\delta E}{E}$ that would produce a 10% emittance dilution is $0.31\times10^{-4}$, which is comparable to the nominal rms fractional energy width $\frac{\sigma_E}{E}$ of $1\times10^{-4}$. Since each of these three "tolerable" errors separately produces a 10% dilution, the actual tolerance level of each when all three are present is certainly lower, possibly by a factor of $1/\sqrt{3}$. These estimates are listed in Table 1.

Table 1. Injection Error Tolerances for $\delta\epsilon/\epsilon = 10\%$

| $\beta_x$ (m) | $\alpha_x$ | $\sigma_x$ (mm) | $\sigma_x'$ ($\mu$rad) | $x_c$ (mm) | $\delta x$ (mm) | $\delta x'$ ($\mu$rad) | $D_x$ (m) | $D_x'$ | $\delta E/E$ |
|---|---|---|---|---|---|---|---|---|---|
| 300 | 2.4 | 0.38 | 1.27 | 0.17 | 0.065 | 0.57 | 1 | 0.01 | $0.31\times10^{-4}$ |

### 2.2.3 Transfer magnet tolerances

Errors in the strengths of the dipole magnets in the extraction system of the HEB and in the injection system of the Collider all contribute to the injection errors that we have been considering. The Lambertson septum magnets and the C-magnets (which functionally are extensions of the Lambertsons) are dc-powered, so that it is their degree of regulation with which we are concerned. The kicker magnets are pulsed, and it is the ripple on the flat-top that is of concern. In Table 2 the number of each magnet type $N_m$, the length of each magnet $L$, the nominal field strength $B$, the total bend angle $\theta$ for each magnet group, the average beta-function value $\beta$ for each magnet group (horizontal $\beta$ for the kickers, vertical $\beta$ for the Lambertsons and the C-magnets), and the fractional error $\delta\theta/\theta$ that will produce a 10% emittance dilution due to that group of magnets alone. The transfer energy is 2 TeV (which corresponds to a magnetic rigidity of 6671.3 T-m), and the normalized emittance is 1 mm-mrad.

Table 2. Transfer Magnet Tolerances for $\delta\epsilon/\epsilon = 10\%$

| MAGNET TYPE | $N_m$ | $L$ (m) | $B$ (T) | $N_mBL$ (T-m) | $\theta$ (mrad) | $\beta$ (m) | $\delta\theta/\theta$ |
|---|---|---|---|---|---|---|---|
| HEB kickers | 6 | 2.625 | 0.125 | 1.97 | 0.30 | 59 | $4.3 \times 10^{-3}$ |
| HEB Lambertson | 7 | 4.86 | 1.1 | 37.5 | 5.6 | 65 | $2.1 \times 10^{-4}$ |
| HEB C-magnets | 3 | 5.0 | 1.1 | 16.5 | 2.5 | 100 | $3.9 \times 10^{-4}$ |
| Coll. C-magnets | 3 | 5.0 | 1.1 | 16.5 | 2.5 | 50 | $5.5 \times 10^{-4}$ |
| Coll. Lambertson | 7 | 4.86 | 1.1 | 37.5 | 5.6 | 100 | $1.7 \times 10^{-4}$ |
| Coll. kickers | 5 | 2.0 | 0.06 | 0.60 | 0.09 | 300 | $6.2 \times 10^{-3}$ |

For a given set of errors in the HEB extraction and the Collider injection magnets, this table can be used to calculate the resultant emittance dilution in the Collider if no injection damper is used. The beam transfer system from the HEB to the Collider is being redesigned, but the design is not yet complete. The magnet system listed above was taken from the SSC Site-Specific Conceptual Design [2]. Although the final magnet system will undoubtedly be different in detail, the main features (numbers and strengths of magnets) are likely to be similar [3].

If only the dc-supplies of the Lambertson septum magnets and the C-magnets exceed the tolerable levels, a low-frequency, narrow-band damper will suffice. If there is excessive wave-shape ripple on the kicker-magnet pulses, the damper bandwidth must accommodate the frequencies observed in that ripple. The dominant ripple frequency in the kicker waveform will probably be about $(2\pi\tau_r)^{-1}$ where $\tau_r$ is the rise time. For the HEB extraction kicker and the Collider injection kicker, $\tau_r = 1.1$ $\mu$sec, so that the damper bandwidth must include frequencies up to 500 kHz.

## 2.3 The damping process

Feedback damping of collective betatron oscillations proceeds as follows. The amplitude $x_1$ of the collective oscillation is measured by a beam position monitor at point 1. A kicker magnet at point 2, chosen to be, say, an integer-plus-one-quarter wavelengths downstream from point 1, is arranged to deflect the beam by the angle

$$\Delta x_2' = -gx_1/(\beta_1\beta_2)^{1/2} \tag{4}$$

where $\beta_1$ and $\beta_2$ are the beta functions at points 1 and 2, and $g$ (the "gain") is typically much less than 1. This deflection produces a change in amplitude $\Delta x_1$ at point 1 on the next turn of

$$\begin{aligned} \Delta x_1 &= \Delta x_2'(\beta_1\beta_2)^{1/2}\sin\phi_{21} \\ &= -gx_1\sin\phi_{21} \end{aligned} \tag{5}$$

where $\phi_{21}$ is the relative phase between points 2 and 1. Thus, the collective amplitude is decreased in proportion to its amplitude.

The damping process can be described in terms of the eigenvalues of the matrix representing these procedures, as has been done by Lebedev et al. [4], but it is also of interest to derive an analytic description from the difference equations. It is convenient to use normalized coordinates, in which the collective amplitude and phase are represented by the vector $\eta$

$$\begin{aligned} \eta &\equiv \frac{x}{\sqrt{\beta}} + i(\frac{\alpha x}{\sqrt{\beta}} + \sqrt{\beta}x') \\ &\equiv \eta_x + i\eta_y \end{aligned} \tag{6}$$

The feedback algorithm can be expressed as

$$\begin{aligned} \Delta\eta_{y_2} &= \sqrt{\beta_2}\Delta x_2' \\ &= -gx_1/\sqrt{\beta_1} \\ &= -g\eta_{x_1} \\ &= -g\eta_{y_2} \end{aligned} \tag{7}$$

The amplitude $|\eta|$ is decreased in one turn by $\Delta|\eta|$:

$$\begin{aligned} \Delta|\eta| &= \Delta\eta_{y_2}\cos\phi_1 \\ &= -g\,|\eta|\cos^2\phi_1 \end{aligned} \tag{8}$$

where $\phi_1$ is the betatron phase at the beam-position monitor. Also, the phase angle is changed by $\Delta\phi$:

$$\begin{aligned} \Delta\phi &= \frac{\Delta\eta_{y2}\sin\phi_1}{|\eta|} \\ &= -g\sin\phi_1\cos\phi_1 \end{aligned} \tag{9}$$

Note that the fractional decrement in the amplitude due to the kicker is monotonic and on average is equal to $g/2$, while the change in phase angle oscillates and has a zero average.

Thus, the evolution of the collective betatron amplitude can be represented by the difference equation

$$\begin{aligned} \frac{\Delta|\eta|}{\Delta n} &= -g\,|\eta|\cos^2\phi_1 \\ &= -g\,|\eta|\cos^2(2\pi\nu n + \phi_0) \end{aligned} \tag{10}$$

where $n$ is the turn number, and $\phi_0$ is the phase at the beam position monitor on the first turn. Integrating equation (10) up to turn $N$, we get

$$\begin{aligned} \ln\left(\frac{|\eta(N)|}{|\eta_0|}\right) &= -g\sum_{n=0}^{N}\cos^2(2\pi\nu n + \phi_0) \\ |\eta(N)| &= |\eta_0|\exp\left(-g\left(\frac{N}{2} + \frac{\sin(2\pi\nu(2N+1)) - \sin 2\phi_0}{4\sin 2\pi\nu}\right)\right) \end{aligned} \quad (11)$$

Thus, the collective amplitude damps as an exponential with a characteristic period of $2/g$ turns, but the exponential also has some minor wiggles.

This description of the damping process, which has ignored the effects of noise in the feedback system, is appropriate to damping of injection errors, where the amplitudes that require correction are much larger than the low noise levels required for acceptable long-time emittance growth.

## 2.4 Decoherence due to chromaticity and momentum spread

The rms tune spread $\sigma_\nu$ due to chromaticity $\zeta$ and rms momentum spread $\sigma_p/p$ is

$$\sigma_\nu = \zeta\frac{\sigma_p}{p} \quad (12)$$

Because of this tune spread each pencil element in a beam tends to stretch out in the betatron-phase coordinate of phase space. After $N$ turns a beam element with a Gaussian momentum spread develops a Gaussian distribution in betatron phase $\phi$ [5]:

$$\rho(\phi, N) = \frac{1}{\sqrt{2\pi}\,|\sigma_\phi|}\exp(-\phi^2/2\sigma_\phi^2) \quad (13)$$

where

$$\sigma_\phi = 2\zeta\frac{\sigma_p}{p}\nu_s^{-1}\sin(\pi\nu_s N) \quad (14)$$

is the rms phase width. Note that the spreading in phase is periodic in the synchrotron period $\nu_s^{-1}$ in the linear approximation. That is, if $\nu_s$ is constant, independent of synchrotron amplitude, each pencil element decoheres and then perfectly re-coheres every synchrotron period. Note also that this action is independent of betatron amplitude, so that the whole beam bunch decoheres and re-coheres as a unit in the linear approximation.

The centroid of the bunch with an initial betatron amplitude $a_0$ has after $N$ turns the amplitude $a(N)$:

$$a(N) = a_0 A(N) \quad (15)$$

where the decoherence factor $A(N)$ is given by [5]

$$A(N) = \exp(-\sigma_\phi^2/2) \quad (16)$$

For the Collider at 2 TeV injection, $\sigma_p/p = 1\times10^{-4}$, $\nu_s = 2.2\times10^{-3}$, and for a residual chromaticity of 5, the linear decoherence factor oscillates between 1.0 and 0.90 at the synchrotron period of 455 turns. At 20 TeV, with $\sigma_p/p = 0.6\times10^{-4}$ and $\nu_s = 1.2\times10^{-3}$ it oscillates betwen 1.0 and 0.88 with a period of 830 turns. Thus, for the Collider the linear decoherence due to chromaticity varies over a range of only about 10% and so does not significantly affect the feedback requirements.

However, the perfect re-coherence of the linear case is spoiled by the non-linear variation of the synchrotron tune with synchrotron amplitude. A particle with synchrotron phase amplitude $\phi_s$ has a synchrotron tune given approximately by

$$\nu_s(\phi_s) = \nu_{s0}(1 - \frac{\phi_s^2}{16}) \tag{17}$$

where $\nu_{s0}$ is the synchrotron tune at zero synchrotron amplitude. For a Gaussian distribution in synchrotron phase space, the rms synchrotron tune spread is $2\nu_{s0}(\frac{\sigma_\phi^2}{16})$. For the Collider at injection, where $\sigma_\phi = 0.41$ radian, this spread in synchrotron tune is $4.7 \times 10^{-5}$ (synchrotron oscillations per revolution), giving a mixing time in synchrotron phase space of about $2.2 \times 10^4$ turns (6.3 seconds). This mixing in synchrotron phase space damps out the oscillations in the linear betatron decoherence process so that the decoherence factor settles down to its average value. At 20 TeV this non-linear synchrotron mixing time is about $4.6 \times 10^4$ turns (13.4 seconds). Thus, in the Collider the non-linear synchrotron tune is not a significant factor in the decoherence process produced by chromaticity effects.

## 2.5 Decoherence due to non-linear magnetic fields

In the Collider the betatron tune varies with amplitude because of non-linear fields in the magnet system. Simulations have shown for typical magnetic error distributions in the lattice that the horizontal tune is well represented by

$$\nu_x = \nu_0 - \mu x^2 \tag{18}$$

where $\nu_0$ is the tune at zero betatron amplitude, $x$ is the betatron amplitude at a focussing quadrupole in the arcs, and $\mu$ is about $1.4 \times 10^{-4}$ mm$^{-2}$ [6]. The elementary smearing action in phase space is that a radial locus of particles transforms monotonically into a spiral.

For a Gaussian betatron distribution Meller et al. [5] derived a decoherence factor, which for small coherent oscillations is:

$$A(N) = (1 + (4\pi\mu\sigma_x^2 N)^2)^{-1} \tag{19}$$

where $\sigma_x$ is the rms betatron width.

Note that the number of turns for the decoherence factor to equal 0.5 is $(4\pi\mu\sigma_x^2)^{-1}$. This is consistent with the expectation that the characteristic decoherence time $\tau_d$ is equal to the product of the revolution time $T_0$ and the inverse of the rms tune spread $\sigma_\nu$, which is about equal to $2\mu\sigma_x^2$.

$$\tau_d = \frac{T_0}{\sigma_\nu} \tag{20}$$

$$\sigma_\nu = 2\mu\sigma_x^2 \tag{21}$$

For the Collider at 2 TeV, $(2\mu\sigma_x^2)^{-1}$ is $1.8 \times 10^4$ turns (5.1 seconds), and at 20 TeV, $1.8 \times 10^5$ turns (51 seconds).

# 3 Analysis of continuous emittance growth (beam heating)

Many external disturbances (ground motion, helium flow, air flow, cooling water flow, power supply ripples, etc.) may cause coherent betatron motion of the beam, which, in turn, would lead to transverse emittance growth. The growth time is usually long (in the order of hours) if these disturbances are kept under certain level. The beam oscillation amplitude is small and the frequency low. A narrow-band, low power feedback system is considered to control the coherent betatron oscillation to an allowable level. (Details will be discussed in the following sections.)

## 3.1 Continuous emittance growth in the absence of feedback

Consider the beam in a storage ring in which there is a continuous, small emittance growth ($\dot{\epsilon}_0$) due to small and random dynamic disturbances, such as, quadrupole motion or power-supply jitter. These disturbances continually produce small-amplitude collective betatron oscillations, which continually smear out through decoherence and so transform into emittance growth. Let the average incremental collective amplitude be $a$, referred to some convenient point such as a focussing quadrupole magnet in the FODO lattice of the arcs. The corresponding incremental increase in beam emittance $\delta\epsilon_0$ is (without feedback)

$$\delta\epsilon_0 = \frac{a^2}{2\beta} \tag{22}$$

where $\beta$ is the value of the betatron function at the reference point. For an average random disturbance rate $\dot{n}$, the average rate of emittance growth $\dot{\epsilon}_0$ is

$$\dot{\epsilon}_0 = \dot{n}\,\delta\epsilon_0 = \dot{n}\,\frac{a^2}{2\beta} \tag{23}$$

Because of the random character of the disturbances, the elemental collective oscillations and their decoherence simply superpose without mutual interference. Note that in terms of these parameters, the characteristic decoherence time $\tau_d$ does *not* affect the outcome (if feedback is not applied).

This emittance growth can be expressed also in terms of an *average* collective amplitude $x_{av}$ and the decoherence time. Since the elementary amplitudes are random, the average collective amplitude is

$$x_{av}^2 \equiv \dot{n}a^2\tau_d \tag{24}$$

so that the total emittance growth can be expressed also as

$$\dot{\epsilon}_0 = \frac{x_{av}^2}{2\beta}\cdot\frac{1}{\tau_d} \tag{25}$$

## 3.2 Decoherence due to beam-beam interaction

A beam bunch undergoing a coherent betatron oscillation tends to smear out in betatron phase space and thus increase its emittance. The smearing action is due to the spread in betatron tune which causes different particles to rotate at different rates so that the coherent motion decoheres.

In the Collider there are three basic sources of betatron tune spread: (1) chromaticity and momentum spread, (2) non-linear betatron motion due to non-linear magnetic fields, and (3) non-linear betatron motion due to beam-beam forces. The first two sources have been studied in the Sections 2.4 and 2.5, respectively. The third one, namely, the beam-beam tune shift is qualitatively similar to the non-linear betatron tune shift discussed in the preceding section in that the betatron tune shift falls non-linearly with betatron amplitude over the principal part of the distribution.

There are two types of beam-beam tune shifts — the head-on and the long range. We will consider the head-on type only in this paper. The total head-on beam-beam tune shift per collision point is

$$\xi = \frac{N_B r_p}{4\pi \epsilon_N} \tag{26}$$

where $N_B$ is the number of protons per bunch, $r_p$ is the classical proton radius, and $\epsilon_N$ the normalized rms beam emittance. Eq. (26) gives the tune shift at zero amplitude. The dependence of the tune shift on the betatron amplitude $a$ is [7]

$$\begin{aligned} \frac{\Delta\nu}{\xi} &= \int_0^1 du\, e^{-\alpha u} \Big( I_0(\alpha u) - I_1(\alpha u) \Big) \\ &= \frac{1}{\alpha} \Big( 1 - e^{-\alpha}\, I_0(\alpha) \Big) \end{aligned} \tag{27}$$

in which $I_0$ and $I_1$ are the modified Bessel functions, $\alpha$ the normalized amplitude

$$\alpha = \frac{a^2}{4\sigma^2} \tag{28}$$

The density distribution of a Gaussian bunch is

$$\rho(a) = \frac{1}{\sqrt{2\pi}\sigma} e^{-\frac{a^2}{2\sigma^2}} \tag{29}$$

The rms tune spread of this Gaussian bunch can be obtained from the integral

$$\left(\frac{\sigma_\nu}{\xi}\right)^2 = \int_{-\infty}^{\infty} \left(\frac{\Delta\nu}{\xi}\right)^2 \rho(a)\, da \tag{30}$$

A numerical integration has been carried out by Lebedev et al. [4] and the result is

$$\sigma_\nu \approx 0.2\, \xi \tag{31}$$

for each collision point. For the Collider at its nominal intensity ($0.8 \times 10^{10}$ protons per bunch), nominal normalized emittance (1 mm-mrad), 20 TeV beam energy, and with four collision points in operation, the total head-on beam-beam tune shift is $3.8 \times 10^{-3}$. The corresponding rms tune spread is $7.6 \times 10^{-4}$, giving a typical decoherence time of $1.3 \times 10^3$ turns (0.38 second).

Thus, we see that at 20 TeV, the tune spread due to the beam-beam effect dominates the decoherence process and determines the gain required of the feedback system.

### 3.3 Continuous emittance growth in the presence of feedback

Next consider the effects of applying feedback to damp the collective oscillation before decoherence has progressed very far. If on each turn the feedback system reduces the angle of the collective oscillation at the feedback kicker by the fraction $g$ ($g$ is the "gain", $<< 1$), the collective amplitude decreases approximately as

$$x = x_0 \exp(-\frac{gN}{2}) \tag{32}$$

Thus, the characteristic feedback damping period $N_f$ is

$$N_f = \frac{2}{g} \quad \text{turns} \tag{33}$$

and the corresponding characteristic feedback damping time $\tau_f$ is

$$\tau_f = N_f\, T_0 = \frac{2}{g}\, T_0 \tag{34}$$

where $T_0$ is the revolution time.

If the damping time $\tau_f$ is considerably smaller than the decoherence time $\tau_d$, the amount of smearing, and thus the amount of emittance growth per elementary disturbance, is reduced. To estimate the amount of this reduction, we will repeatedly use two results from the paper of Meller et al [5]. One is Eq. (19) in Section 2.5, which gives the decoherence factor in a non-linear magnetic field. For small $N$ it can be represented by

$$A(N) \approx 1 - (4\pi\mu\sigma_x^2 N)^2 \tag{35}$$

where $\sigma_x$ is the rms beam width. The other Meller result is Eqs. (14) and (16) in Section 2.4, which give the association of the decoherence factor with the characteristic mean-square phase spread $\sigma_\phi$ at turn $N$ due to chromaticity decoherence. For small $N$, the mean-square phase spread is approximately

$$\begin{aligned} \sigma_\phi &\approx 2\pi\zeta\frac{\sigma_p}{p}N \\ &= 2\pi\,\sigma_\nu N \end{aligned} \tag{36}$$

and the decoherence factor is approximately

$$\begin{aligned} A(N) &\approx 1 - \sigma_\phi^2(N)/2 \\ &\approx 1 - (2\pi\zeta\frac{\sigma_p}{p}N)^2/2 \end{aligned} \tag{37}$$

Thus, for small $N$ and to first order, equations (35) and (37) have the same dependence on $N$. Therefore, it is reasonable to generalize slightly and relate the mean-square decoherence phase spread $\sigma_\phi$ to the rms tune spread $\sigma_\nu$ generally. In other words, for the case of tune spread due to a non-linear betatron tune, we can also estimate the rms betatron phase spread similar to Eq. (36)

$$\begin{aligned} \sigma_\phi(N) &= \sqrt{2}\,4\pi\mu\sigma_x^2 N \\ &= \sqrt{2}\,2\pi\sigma_\nu N \end{aligned} \tag{38}$$

in which $\sigma_\nu$ is the rms tune spread expressed in Eq. (21). There is a $\sqrt{2}$ factor difference between Eqs. (36) and (38), which is considered not important in this rough estimation. This rms spread in betatron phase is a measure of how far the decoherence process has evolved. Decoherence is practically complete when the typical spread is $\pm\pi$.

The principal question at this point is to determine the increase in the emittance for a given rms spread in betatron phase. Consider a one-dimensional analogy: Suppose that in an initial Gaussian distribution $\rho(x) = A\exp\left(-x^2/2\sigma_0^2\right)$, each point at $x$ is spread into another Gaussian distribution $B\exp\left(-(x-x_1)^2/2\sigma_1^2\right)$. The resulting new distribution is found by folding the spreading distribution over the initial distribution:

$$\begin{aligned}\rho(x_1) &= B\int_{-\infty}^{\infty}\rho(x)\exp\left(-(x-x_1)^2/2\sigma_1^2\right)dx\\ &= C\exp\left(-x_1^2/2(\sigma_0^2+\sigma_1^2)\right)\end{aligned}\tag{39}$$

$C$ being the normalizing constant involved. Thus, due to the spreading process, a new Gaussian distribution is formed with a mean-square width equal to the *sum* of the initial and spreading widths. In our two-dimensional beam distribution, it is only the phase coordinate that is being spread in the decoherence process. Therefore, we can estimate that the incremental emittance growth due to the incremental oscillation amplitude $a$ when feedback is applied is, for a perfect (noise-free) system

$$\begin{aligned}\delta\epsilon_f &= \delta\epsilon_0\left(\sigma_\phi(N_f)/\pi\right)^2\\ &= (a^2/2\beta)\,(2\sqrt{2}\sigma_\nu N_f)^2\\ &= (a^2/2\beta)\left(2\sqrt{2}\sigma_\nu\frac{2}{g}\right)^2\\ &= \frac{16\,a^2}{\beta}\left(\frac{T_0}{\tau_d\,g}\right)^2\\ &= \frac{4\,a^2}{\beta}\left(\frac{\tau_f}{\tau_d}\right)^2\end{aligned}\tag{40}$$

in which $N_f = 2/g$ is the feedback damping period, $\tau_f = 2T_0/g$ the damping time, and $\tau_d = T_0/\sigma_\nu$ the decoherence time. The total emittance growth rate with feedback is

$$\begin{aligned}\dot\epsilon_f &= \dot n\,\delta\epsilon_f\\ &= \dot n\,\frac{4\,a^2}{\beta}\left(\frac{\tau_f}{\tau_d}\right)^2\\ &= \dot\epsilon_0\cdot\left(\frac{2\sqrt{2}\tau_f}{\tau_d}\right)^2\\ &= \dot\epsilon_0\cdot\left(\frac{4\sqrt{2}\sigma_\nu}{g}\right)^2\end{aligned}\tag{41}$$

in which $\dot\epsilon_0$ is defined in Eq. (23). Therefore, if the feedback gain is big enough such that

$$g > 4\sqrt{2}\sigma_\nu \tag{42}$$

then we will have

$$\dot\epsilon_f < \dot\epsilon_0 \tag{43}$$

i.e., the feedback will reduce the emittance growth rate.

### 3.4 Noises in the feedback system

If there is noise in the feedback system equivalent to a beam amplitude $x_N$, then in addition to to collective-amplitude contributions $\dot{n}a^2$ due to machine disturbances, there is the contribution $f_0(gx_N)^2$ due to the noise in the feedback system, where $f_0$ is the revolution frequency. Thus, Eq. (41) has to be modified and takes the form

$$\dot{\epsilon}_f = \left(\dot{\epsilon}_0 + \frac{f_0 g^2 x_N^2}{2\beta}\right) \cdot \left(\frac{4\sqrt{2}\sigma_\nu}{g}\right)^2 \tag{44}$$

Let $\epsilon_0$ be the initial emittance, then one may also define the emittance growth rate as

$$\begin{aligned} \frac{1}{\tau_\epsilon} &\equiv \frac{\dot{\epsilon}_f}{\epsilon_0} \\ &\equiv \frac{1}{\tau_{ext}} + \frac{1}{\tau_{noise}} \end{aligned} \tag{45}$$

in which

$$\frac{1}{\tau_{ext}} = \frac{\dot{\epsilon}_0}{\epsilon_0} \cdot \left(\frac{4\sqrt{2}\sigma_\nu}{g}\right)^2 \tag{46}$$

is the emittance growth rate due to external sources, and

$$\begin{aligned} \frac{1}{\tau_{noise}} &= \frac{1}{\epsilon_0} \cdot \frac{f_0 g^2 x_N^2}{2\beta} \cdot \left(\frac{4\sqrt{2}\sigma_\nu}{g}\right)^2 \\ &= 16\, f_0 \left(\frac{x_N}{\sigma_x}\right)^2 \cdot \sigma_\nu^2 \\ &= 0.64\, f_0 \left(\frac{x_N}{\sigma_x}\right)^2 \cdot \Delta\nu^2 \end{aligned} \tag{47}$$

where we have converted the rms tune spread $\sigma_\nu$ to the total tune shift $\Delta\nu$ (which equals $\xi$ times total number of interaction points) using Eq. (31). It is interesting to note that the emittance growth rate due to the noises in the feedback system is independent of the gain. It is also interesting to compare this result with that presented in Ref. [8], which was obtained by Boussard using the well established stochastic cooling theory. Eq. (47) above differs from Eq. (8) in [8] only by approximately a factor of 2.

## 4 Damping of resistive wall instability

The wall resistivity of the beam tube may cause beam instability, which is usually of long correlation lengths and, therefore, shows no discrimination against the bunched beams from the coasting beams. The growth rate is

$$\begin{aligned} \frac{1}{\tau_w} &= \frac{N_{tot}\, c\, r_p}{2\pi\gamma\, \nu_\beta} \frac{Re\, Z_\perp}{R Z_0} \\ &= \frac{N_{tot}\, c\, r_p}{2\pi\gamma\, \nu_\beta} \frac{1}{b^3}\, \delta_s \end{aligned} \tag{48}$$

where $\tau_w$ is the growth time, $N_{tot}$ the total number of particles, $c$ the velocity of light, $r_p$ the classical radius of proton, $\gamma$ the relative energy of the particle, $R$ the average machine radius,

$\nu_\beta$ the betatron tune, $Z_0$ the vacuum impedance, $\delta_s$ the skin depth, $b$ the beam tube radius and $Re\ Z_\perp$ the real part of the transverse impedance of the wall,which is equal to

$$Z_\perp = (1+i)\,\frac{Z_0\,R\,\delta_s}{b^3} \tag{49}$$

The skin depth is a function of the wall conductivity $\sigma_e$ and frequency $\omega$:

$$\delta_s = \sqrt{\frac{2}{\mu\sigma_e\omega}} \tag{50}$$

It is seen that as the frequency is decreased, $\delta_s$ is increased as $\omega^{-1/2}$. So is the growth rate $1/\tau_w$. The frequency range of interest is $\omega = (n-\nu_\beta)\omega_0$, in which $n$ is an integer and $\omega_0 = c/R$ is the angular revolution frequency, which equals to $2\pi \times 3.441$ kHz for the Collider. During machine operation, the fractional part of $\nu_\beta$ would be scanned in order to find the optimal working point. Therefore, the lowest frequency is taken to be $0.1\omega_0$ in the estimates of the fastest growth time.

In the present design, the beam tube of the Collider is made of stainless steel, which is coated in its inner surface with a thin copper layer in order to have low electrical resistivity. In this case, the thickness of the copper layer $\Delta$ is smaller than the skin depth $\delta_s$ in the frequency range mentioned above, and Eq. (48) needs to be modified [9]. With some approximations, the modified equation can be written as

$$\frac{1}{\tau_w} = \frac{N_{tot}\,c\,r_p}{2\pi\gamma\,\nu_\beta}\,\frac{1}{b^3}\,\frac{\delta_s^2}{\Delta} \tag{51}$$

Plugging in Eq. (50) and expressing the instability in terms of the growth time, one has

$$\tau_w = \left(\frac{2\pi\gamma\,\nu_\beta\,b^3}{N_{tot}\,c\,r_p}\,\frac{\mu\omega}{2}\right)\,\sigma_e\Delta \tag{52}$$

In other words, the growth time is proportional to the product of $\sigma_e\Delta$. The specs of the copper layer is

$$\sigma_e\Delta \geq 2\times 10^5 \quad 1/\Omega \quad , \tag{53}$$

in which $\sigma_e$ is the conductivity of the copper layer at 4 K and 0.66 Tesla. The nominal values are $\sigma_e = 2\times 10^9\ \Omega^{-1}m^{-1}$ (corresponding to RRR = 30), $\Delta = 0.1$ mm. This would give a resistive wall instability growth time of 34 ms, or 117 turns during the about one hour injection period [10]. At full energy (20 TeV), the growth rate would be slower due to the beam rigidity. In addition, the large betatron tune spread resulting from the beam-beam collisions may provide significant Landau damping.

This type of instability appears to be the rigid dipole mode oscillation and can be effectively damped by means of a transverse damper. The oscillation amplitude is small (a fraction of the beam size, which is 0.38 mm, rms, at injection). This means the power requirement on the damper is small. The instability frequency is very low, ranging from several hundred hertz to tens of kHz. The bandwidth requirement on the feedback will mainly be determined by the batch spacing (1.7 $\mu$s). The damper must be fast enough in order to act on each individual batch. This implies a bandwidth of about 500 kHz.

Table 3. Coupled Bunch Instability Growth Time

| | Bunch Mode | At 2 TeV | At 20 TeV |
|---|---|---|---|
| Longitudinal | m = 1 | 0.9 s | 5.0 s |
| | m = 2 | 2.2 s | 16 s |
| Transverse | m = 0 | 8.6 s | 91 s |
| | m = 1 | 16 s | 150 s |

# 5 Damping of coupled bunch instabilities driven by HOM

The original candidate of the Collider rf cavity is the PEP 5-cell cavity [2]. The calculation shows that the HOM of this type of cavity would cause multiple bunch instabilities, both longitudinal and transverse. The fastest growth time would be in the order of one to several seconds as listed in Table 3. The instability frequency could be as high as 30 MHz, which is one half of the bunch repetition frequency. The amplitude is small (a fraction of the beam size).

This kind of instability may be avoided if the single-cell type rf cavity is to be used [11]. Because the experiences on other machines have demonstrated that it is feasible to damp the HOM by a factor of 10 to 100 in a single-cell cavity. Thus the instability would be largely suppressed and the need of a wide bandwidth feedback system could be eliminated. However, because the final design of the rf cavities of the Collider has not yet been done, it is prudent to include the discussion of a wide-band feedback system that aims at damping the instabilities driven by the HOM.

# 6 Discription and specifications of the SSC feedback systems

## 6.1 Requirements of the feedback systems

Table 4 summarizes the requirements of the transverse dampers for the four different cases as analysed above.

Table 4. Requirements of the Feedback Systems

| Case | Power | Bandwidth | Gain | Noise |
|---|---|---|---|---|
| (a) Injection errors | Large | Small | Large | (not critical) |
| (b) Resistive wall instability | Small | Small | Large | Low |
| (c) Instability driven by HOM | Small | Large | Small | Low |
| (d) Emittance growth control | Small | Small | Small | Low |

## 6.2 Types of the feedback systems

In the SCDR [2], it is suggested to use two feedback systems in each transverse plane. One is of high power and narrow bandwidth. It would be used during the injection for correcting the injection errors and for damping the resistive wall instability. Another is of low power and wide bandwidth for the purpose of curing the instabilities driven by the HOM.

However, a third system may be required if the noise in the feedback would indeed become a problem and lead to an untolerable emittance growth. This system will have low power and narrow bandwidth for control of (b) and (d) as listed in Table 4. Because the thermal noise power is proportional to the bandwidth, a narrow band system can significantly lower its noise level.

## 6.3 Components of the feedback systems

In each transverse plane, the feedback systems consist of two pickups, three signal processors and two kickers. The two pickups are shared by all the three feedback systems as discussed above. Each feedback system has its own signal processor. One of the kickers is of high power, while another is of low power.

The advantage of the 2-pickup scheme is that its performance is independent of the betatron tune of the machine. The signals from the two pickups give both $x$ and $x'$ (or $y$ and $y'$), from which the position of the bunch center at the kicker can be obtained.

The pickup signal contains both the coherent betatron oscillation (which is needed) and the closed orbit signal. The latter is harmful, in particular, if a digital processor is to be used. Because the quantized noise problem would greatly limit the resolution of the pickups (see Section 6.7.3). One way to get rid of the closed orbit signal is to take the signals from two succesive turns and subtract one from the other before the pre-amplifier.

## 6.4 Algorithm of the 2-pickup, 2-turn scheme

Using two pickups and their readings on two consecutive turns, one can extract the closed-orbit amplitudes at the two pickups and the amplitude and phase of the collective oscillation in the beam. Let $x_1$ and $x_2$ be the two readings on turn 1, and $x_3$ and $x_4$ the readings on turn 2. If $c_1$ and $c_2$ are the amplitudes of the closed orbits at the two monitors, and $\epsilon$ the invariant emittance of the collective oscillation, the readings can be expressed as

$$\begin{aligned} x_1 &= c_1 + \sqrt{\beta_1 \epsilon} \cos \phi_1 \\ x_2 &= c_2 + \sqrt{\beta_2 \epsilon} \cos (\phi_1 + \delta_{12}) \\ x_3 &= c_1 + \sqrt{\beta_1 \epsilon} \cos (\phi_1 + 2\pi\nu) \\ x_4 &= c_2 + \sqrt{\beta_2 \epsilon} \cos (\phi_1 + 2\pi\nu + \delta_{12}) \end{aligned} \tag{54}$$

where $\beta_1$ and $\beta_2$ are the beta functions, $\phi_1$ is the phase of the oscillation at the first pickup on turn 1, $\nu$ is the betatron tune, and $\delta_{12}$ is the phase difference between the two monitors. Eliminate $c_1$ and $c_2$ by forming

$$\begin{aligned} (x_3 - x_1)/\sqrt{\beta_1 \epsilon} &= \cos (\phi_1 + 2\pi\nu) - \cos \phi_1 \\ (x_4 - x_2)/\sqrt{\beta_2 \epsilon} &= \cos (\phi_1 + 2\pi\nu + \delta_{12}) - \cos (\phi_1 + \delta_{12}) \end{aligned} \tag{55}$$

Eliminate $\epsilon$ by forming

$$\frac{x_4 - x_2}{x_3 - x_1}\sqrt{\frac{\beta_1}{\beta_2}} = \frac{\cos(\phi_1 + 2\pi\nu + \delta_{12}) - \cos(\phi_1 + \delta_{12})}{\cos(\phi_1 + 2\pi\nu) - \cos\phi_1} \tag{56}$$

Equation (56) can be solved for $\phi_1$. Then equation (55) can be solved for $\epsilon$. Then solve (54) for $c_1$ and $c_2$.

This analysis has ignored the small effects of the feedback kicker between turns 1 and 2. These effects can easily be included by including the drop in $\epsilon$ and the change in phase produced by the kick:

$$\Delta\epsilon = -g\sin^2\phi_k \tag{57}$$

in which

$$\phi_k = \phi_1 + \delta_{1k} \tag{58}$$

is the phase at the kicker on turn 1, $\delta_{1k}$ being the phase difference between the first pickup and the kicker.

The corresponding change in phase due to the kicker is

$$\Delta\phi = -(g/2)\sin^2\phi_k \tag{59}$$

These changes $\Delta\epsilon$ and $\Delta\phi$ can be included in the second-turn equations (54), and the analysis proceeds as indicated above. In most cases of interest, the gain $g$ is relatively small so that these changes are not important.

## 6.5 Locations of the feedback systems

There are several possible locations for the feedback systems in the Collider — the west or east utility region, the missing dipole locations in the arcs and the IR regions. At this moment, the feedback systems are planned to be installed in the west utility region, in which the injection kickers, abort kickers, Lambertson magnets and rf cavities etc., are located. This makes the maintenance easier. (At the moment, the east utility region is not used.) The proposed locations of the two pickups and two kickers are at the points where the $\beta$-functions have relatively high values. The pickups should avoid the dispersive region. Otherwise the beam loading induced coherent synchrotron oscillation may cause coherent betatron motion through dispersion. A rough estimate shows that the amplitude of this kind of betatron oscillation can be at the level of several tenths of one micron, which is comparable with the required pickup resolution (see Section 6.7.2).

## 6.6 Power, bandwidth and gain

The requirements of the power, bandwidth and gain of the three feedback systems are listed in Table 5. The gain $g$ and damping time $2/g$ are defined in Sections 2.3 and 3.3. $X_p$ is the maximum betatron oscillation amplitude at the pickup, which determines the kick angle.

$$\theta = \frac{g \cdot X_p}{\sqrt{\beta_p \beta_k}} \tag{60}$$

Table 5. General Specs of the SSC Transverse Dampers

| | DAMPER A | DAMPER B | DAMPER C |
|---|---|---|---|
| Purpose | Injection errors | Resistive wall; Emittance control | Cavity driven insta. |
| Gain $g$ | 0.04 | 0.1 | 0.02 |
| Damping time $2/g$ | 50 turns | 20 turns | 100 turns |
| $X_p$ | $\pm$ 2 mm | $\pm$ 100 $\mu$m | $\pm$ 40 $\mu$m |
| Kick angle $\theta$ | 0.27 $\mu$rad | 0.04 $\mu$rad | 0.003 $\mu$rad |
| Kicker length $L$ | 4 m | 4 m | 4 m |
| Kicker voltage $V$ | 1 kV | 150 V | 150 V |
| Kicker power $P_k$ | 40 kW | 0.9 kW | 0.9 kW |
| Bandwidth $\Delta f$ | 500 kHz | 500 kHz | $\geq$ 30 MHz |

in which $\beta_p$ and $\beta_k$ are the $\beta$-function at the pickup and kicker, respectively. The kick angle then determines the kicker voltage $V$ and the kicker length $L$:

$$V \cdot L = \frac{\theta \cdot (\frac{E}{e}) \cdot d}{4} \tag{61}$$

in which $E$ is the energy of the particles and $d$ the separation of the two kicker electrodes (taken to be 3 cm). The voltage on the two electrodes is +V and −V, respectively. The power is calculated by assuming the kicker be a 50 $\Omega$ transmission line.

## 6.7 Noise level requirements

### 6.7.1 Allowable coherent betatron oscillation amplitude

Assume that the allowable emittance growth rate $\dot{\epsilon}_0$ is specified, then the allowable coherent beam oscillation amplitude $x_{av}$ can be derived from Eq. (25) in the case no feedback system is employed. For the SSC, the initial rms emittance at 20 TeV is

$$\begin{aligned} \epsilon_0 &= \epsilon_N/\gamma \\ &= 4.7 \times 10^{-11} \ \text{m}-\text{rad} \end{aligned} \tag{62}$$

where $\epsilon_N$ is 1 mm-mrad, $\gamma$ the relativistic factor. Assume the allowable emittance doubling time is 24 hours, then the allowable average emittance growth rate is

$$\dot{\epsilon}_0 = \epsilon_0 \,/\, (24 \text{ hours})$$

$$= 5.4 \times 10^{-16} \text{ m-rad/s} \tag{63}$$

The total beam-beam tune shift for four interaction points is

$$\begin{aligned} \Delta\nu &= 4\,\xi \\ &= 0.0038 \end{aligned} \tag{64}$$

The rms tune spread is

$$\begin{aligned} \sigma_\nu &= 0.2\,\Delta\nu \\ &= 0.00076 \end{aligned} \tag{65}$$

The decoherence time is

$$\begin{aligned} \tau_d &= T_0/\sigma_\nu \\ &= 0.38 \text{ s} \end{aligned} \tag{66}$$

in which the revolution time $T_0$ is 290 $\mu$s. Thus, from Eq. (25) we get the allowable coherent oscillation amplitude

$$\begin{aligned} x_{av} &= (\dot{\epsilon}_0 \times 2\beta \times \tau_d)^{1/2} \\ &= 0.29 \ \mu\text{m} \end{aligned} \tag{67}$$

in which the average beta-function is 200 m.

### 6.7.2 Required pickup resolution of the feedback system

In the presence of a feedback system, the emittance growth rate will be reduced by a factor of $(4\sqrt{2}\sigma_\nu/g)^2$ as indicated by Eq. (41). But any feedback system will inevitably contain certain level of noises, which will "heat" the beam up, i.e., blow up the emittance. In order to keep the emittance growth under control, one needs to specify the allowable noise level of the feedback system or, equivalently, to specify the pickup resolution.

Again, assume the allowable emittance growth rate is given as before, let us estimate the required pickup resolution by using Eq. (44).

(a) Assume no external excitations on the beam, i.e., $\dot{\epsilon}_0 = 0$:
In this case, the emittance growth is solely due to the noises in the feedback system. Eq. (44) becomes

$$\begin{aligned} \dot{\epsilon}_f &= \frac{f_0 g^2 x_N^2}{2\beta} \cdot \left(\frac{4\sqrt{2}\sigma_\nu}{g}\right)^2 \\ &= 16\, f_0 \frac{x_N^2}{\beta} \cdot \sigma_\nu^2 \end{aligned} \tag{68}$$

in which the revolution frequency $f_0$ is 3.441 kHz, $\beta$ at the pickup is assumed to be 300 m. Solve Eq. (68) for $x_N$, we get

$$x_N = 2.4 \ \mu\text{m} \tag{69}$$

It is independent of the gain of the feedback system.

(b) Now assume there be external excitations which gives an emittance doubling time of 0.1 hour, i.e., $\dot{\epsilon}_0 = \epsilon_0 \ / \ (0.1 \text{ hour})$:
Let the gain $g = 0.1$, solve Eq. (44) for $x_N$ we obtain

$$x_N = 1.8 \ \mu\text{m} \tag{70}$$

### 6.7.3 Theoretical limit of the pickup resolution

The pickup resolution is limited by the thermal noises and electronic noises. The thermal noises are determined by the product of the temperature and bandwidth. The electronic noise of an amplifier is described by a noise factor (NF).

Assuming the amplifiers have a temperature $T$ (°K), a bandwidth $\Delta f$ (Hz) and a noise factor $NF$ (dB), then the noise power when expressed at the pickup is given by

$$P_N\,(Watts) = k_B T \times \Delta f \times 10^{NF/10} \quad , \tag{71}$$

in which $k_B$ is the Boltzmann constant. Sometimes the unit $dBm$ is used. The conversion is as follows:

$$P_N\,(dBm) = 10\,\log_{10} P_N\,(mW) \quad . \tag{72}$$

The signal power of a pair of pickup electrodes is

$$P_S = \frac{1}{4}\,I^2\,(\frac{x}{b})^2\,Z\,\sin^2(\frac{\omega\ell}{c})\,F \quad , \tag{73}$$

in which $x$ is the beam displacement from the axis, $b$ the half distance between the two electrodes, $Z$ the characteristic impedance of the pickup, $\ell$ the length of the electrodes, $c$ the velocity of light, $F$ the geometrical factor (which is about equal to 1 for an electrode subtended angle of 70°). If the amplifiers have an infinite wide bandwidth, then $I$ in (73) should be the peak current. But in a real feedback system, the bandwidth is always finite. Therefore, the value of $I$ is approximated by the average current $I_{ave}$ in our estimation.

By combining Eqs. (71) and (73), the theoretical limit of the pickup resolution can be obtained from

$$\Delta x = \frac{2b}{I_{ave}\sin(\frac{\omega\ell}{c})}\sqrt{\frac{k_B T \times \Delta f \times 10^{NF/10}}{Z}} \quad . \tag{74}$$

This value must be smaller than $x_N$ given by Eq. (69) or (70) in order to avoid the emittance growth problem.

The noise level calculated in the previous sections for the SSC are summarized in Table 6.

There are several other error sources that are not included in the above analysis but may also put a limit to the pickup resolution. These include the least significant bit (LSB) error if a digital system is used, the mechanical vibration of the pickup, the noises from the kicker power amplifier and the cable (which are usually not included in the given noise factor), etc.

When a digital system is used, the LSB error may become significant. As an example, the Tevatron Super-damper utilizes an 8-bit digital system for signal processing. The full scale is about 5 mm, which is determined by the residual orbit error [12]. Therefore, the maximum LSB error is

$$\Delta x_{LSB} = \frac{5\ \text{mm}}{256} \approx 20\ \mu\text{m} \quad . \tag{75}$$

It is much larger than the theoretical limit $\Delta x$ and can result in significant emittance growth. In this case, it is necessary to reduce the orbit error or to compensate the orbit error signal in order to improve the pickup resolution.

Table 6. Specs of Noise Level of the Feedback Systems

| | |
|---|---|
| Initial emittance $\epsilon_0$ | $4.7 \times 10^{-11}$ m-rad |
| Emittance doubling time | 24 h |
| Average emittance growth rate $\dot{\epsilon}$ | $5.4 \times 10^{-16}$ m-rad/s |
| Beam-beam tune shift $\Delta\nu$ | 0.0038 |
| rms tune spread $\sigma_\nu$ | 0.00076 |
| Decoherence time $\tau$ | 0.38 s |
| *Allowed coher oscil amp (w/o feedback)* $x_{av}$ | 0.29 $\mu$m |
| Damping time $2/g$ | 20 turns |
| *Allowed noise level* $x_N$ | 2.4 $\mu$m ($\dot{\epsilon}_{ext} = 0$) |
| | 1.8 $\mu$m ($\dot{\epsilon}_{ext} = \epsilon_0/0.1$ h) |
| Noise power $P_N$<br>(NF = 17 dB) | -82 dBm (bandwidth 30 MHz)<br>($6.2 \times 10^{-12}$ Watts) |
| | -100 dBm (bandwidth 500 kHz)<br>($1.0 \times 10^{-13}$ Watts) |
| *Pickup resolution limit* $\Delta x$ | 0.16 $\mu$m (30 MHz, $\ell = 1.25$ m, $I = 72$ mA) |
| | 0.02 $\mu$m (500 kHz, $\ell = 1.25$ m, $I = 72$ mA) |

# 7 Conclusions

It is important to understand the underlying physics in designing feedback systems. In this paper, special attentions are given to the analysis of the beam emittance growth and coupled bunch instabilities. The instabilities can be driven either by the wall impedance (which has a fast growth rate and a relatively narrow bandwidth) or by the HOM of the rf cavities (which usually grows slowly whereas crosses a broad bandwidth). The emittance growth can be caused either by the injection errors (which is a one-time event) or by continuous external excitations (which is a heating process). In the injection case, the decoherence time is determined by the chromaticity and by the nonlinear parts of the magnetic fields. In the collision case, it is determined by the beam-beam tune spread. The emittance growth time (or growth rate) is studied for both with and without feedback systems.

Based on the above analysis, the requirements of the power, bandwidth, gain and noise level of the feedback systems are specified. Although preliminary, these specs will serve useful references to guide the engineering designs of these systems.

# References

[1] M. Syphers, FN-458, FNAL (June 1987).

[2] Site Specific Conceptual Design, p. 128, SSCL-SR-1056, SSC Laboratory (July 1990).

[3] D. Johnson, private communication.

[4] V. A. Lebedev et al., Preprint 91-120, INP, Novosibirsk (1991).

[5] R. E. Meller et al., SSC-N-360, SSC Laboratory (May 1987)

[6] G. Bourianoff, SSC Laboratory, private communication.

[7] G. López, SSCL-442, SSC Laboratory (May 1991).

[8] D. Boussard, SL/Note 92-79 (RFS), CERN (1992).

[9] J. D. Jackson, SSC-N-110, SSC Laboratory (January 1986).

[10] H. T. Edwards, SSCL-N-771, SSC Laboratory (August 1991). See Appendix 7 therein.

[11] W. Chou and G. Schaffer, PMTN-036C, SSC Laboratory (January 1992).

[12] D. Martin, SSC Laboratory, private communication.

# RADIATION DAMAGE CONSIDERATIONS IN A HIGH LUMINOSITY COLLIDER

## THE INTERACTION REGION

DAVID M. LEE
Los Alamos National Laboratory
P.O. Box 1663
Mail Stop H846
Los Alamos, New Mexico 87545, USA

ABSTRACT

The interaction region in a high luminosity collider will be a source of radiation for all componants in the vicinity and will place stringent requirements on their design. The major componants in the vicinity of the interaction region are the physics detectors that surround the beam pipe and the focusing quadrupole magnets nearby. We will present the radiation levels in such a physics detector and the power in the forward direction that will be deposited in the forward calorimeters and quad magnets. The implications of the levels on a variety of detector componants and electronics will be presented. The calculational techniques and limitations will be reviewed.

## 1. The Interaction Region

The interaction region in present and future colliders consists of a large detector for the physics experiments that surrounds the interaction point (IP), final focusing quadrupole magnets, the beam pipe and assorted vaccum pumps, and shielding. A typical physics detector is shown in figure 1. The detector is designed to cover as much of the solid angle as possible. The region closest to the beam pipe normally contains a particle tracker consisting of silicon microstrip detectors followed by gaseous tracking detectors. Depending on the radiation levels from the interaction point the silicon microstrip detectors begin at a radius of 5-10 cm and the outer radius of the gaseous tracker is 75-150 cm. In the region outside of the tracker volume are the calorimeter systems, electromagnetic calorimeter and hadron calorimeter. The electromagnetic calorimeter may consist of 25 radiation lengths thick of $BaF_2$ or liquid Argon, for example, and the hadron calorimeter may consist of 10-12 interaction lengths thick of uranium or copper. Muon chambers follow the calorimeter system and then the magnet. The diameter of the entire detector can range from 15 to 22 meters and the length up to 40 meters.

In the forward regions are the forward detector componants, calorimeters and muon systems, the final focusing quadrupoles, collimaters, vaccuum pumps, beam pipes etc. The radiation levels in the forward regions are the most severe and it is in these regions that high luminosity machines will present the most difficult problems.

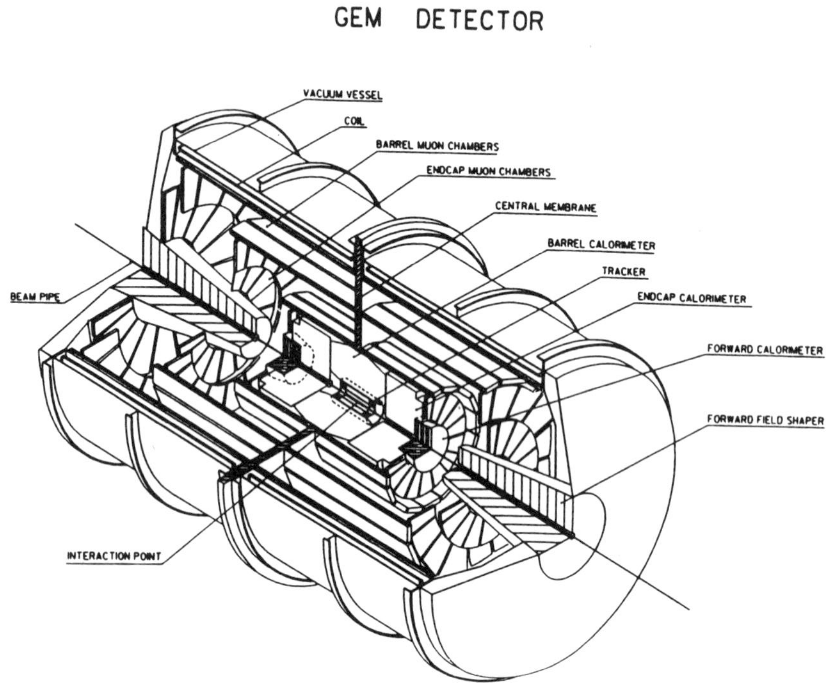


Figure 1. A typical physics detector for a high intensity collider.

## 2. Radiation Levels

The radiation environment in the interaction region comes from a variety of sources. In the central tracking volume the radiation levels are due to minimum ionizing particles, conversions in the beam pipe and other structural materials, and albedo neutrons and photons from the calorimetry and shield. In the calorimeters the radiation is due to electromagnetic and hadronic showers from the particles entering the calorimetry. The very forward regions (beyond eta =6) the primary radiation environment is due to forward beam jets.

### *2.1 The pp Interaction*

The primary pp interaction at the SSC is shown in figure 2 as a side view and figure 3 from the end ( as the beam sees it) with a simulated central tracker visible as concentric circles. Shown are all charged particle tracks emanating from the iteraction vertex. Several features are very noticeable. Conversions of photons in the material of the central tracker are seen as small circles which are low energy electrons. These types of particles are commonly referred to as loopers because they tend to loop in tight circles and cross detector elements many times. Loopers can double the energy deposition in

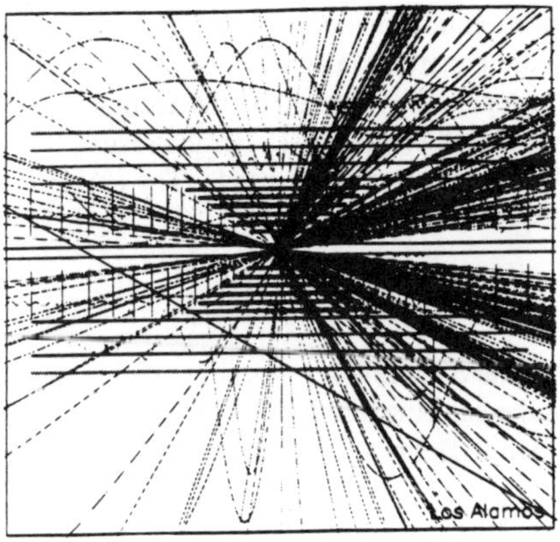


Figure 2. A pp interaction at the SSC, side view.

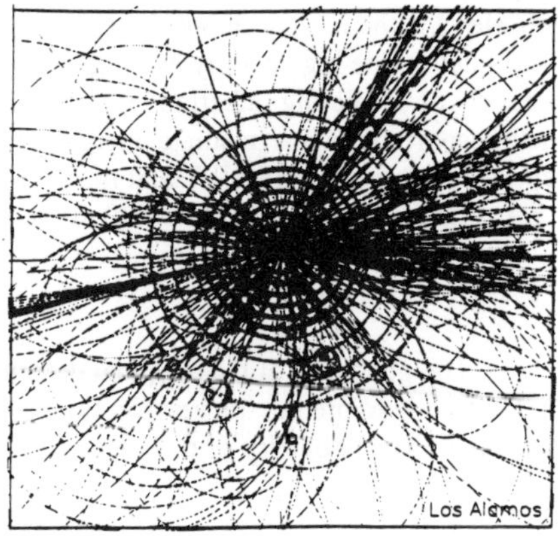


Figure 3. A pp interaction at the SSC, end view.

detector elements. One also notices that the density of tracks is greatest in the center of the detector and therefore one expects that the radiation levels will be greatest in that region. Determination of the radiation levels in the interaction region is generally accomplished with large Monte Carlo codes augmented by empirical data. For a typical ELIOSITRON event, the number of tracks in the primary pp interaction is increased by a factor of 16.8.

### *2.2 Calculational methods*

A variety of calculational tools are available to determine the radiation levels in the interaction region. These tools consist of event generators, high energy transport codes, and low energy transport codes. Table 1 list some of the more commonly used codes.

Table 1

CALCULATIONAL TOOLS

| EVENT GENERATORS | H. E. TRANSPORT | L. E. TRANSPORT |
|---|---|---|
| ISAJET[1] | HETC[2] | MCNP[3] |
| PYTHIA[4] | FLUKA[5] | MORSE[6] |
| DTUJET[7] | GEISHA[8] | ANISN[9] |

Generally the low energy transport codes are quite accurate ( 10 to 20%) while the transport by the high energy codes in the intermediate energy region (5 GeV to 50 -

100 MeV) is the least accurate since the cross section information in this region is incomplete. The high energy event generators are generally in agreement. The overall accuracy of predicting radiation levels is therefore primarily limited by the accuracy of the transport in the intermediate energy region is about a factor of 2. For the results reported in this paper ISAJET and the LAHET[10] code system were used. The LAHET code system uses HETC and MCNP as the transport codes.

*2.3 Neutron Backgrounds*

The geometry of the a typical detector at the SSC collider was simulated in the LAHET code to predict the neutron fluences in the calorimeters and the central volume[11]. The geometry of this detector is shown in figure 4. The neutron fluence in

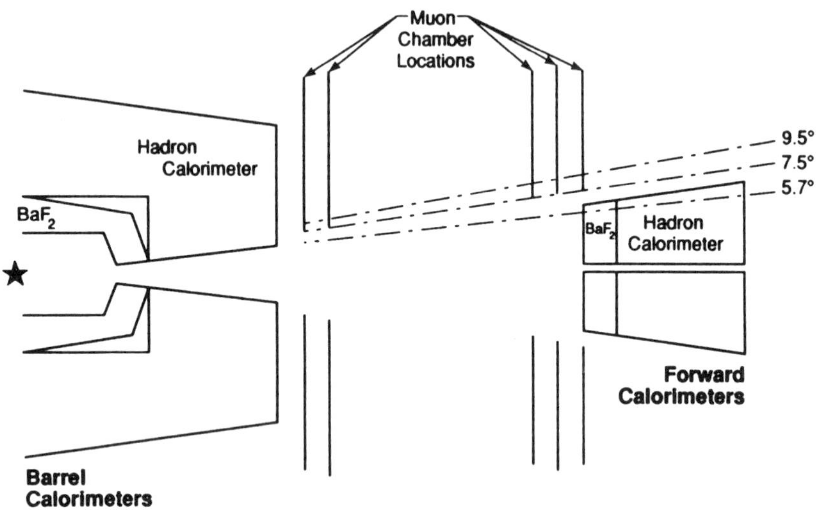


Figure 4. Detector Geometry for simulations. View is one half of the detector without Magnet.

the central volume is shown Table 2 . Two cases were studied, one with a 5 cm borated polyethylene shield in front of the electromagnetic calorimeter ($BaF_2$) and one without a shield. Clearly the polyethylene shield is very beneficial in reducing the neutron fluence. At a luminosity of $10^{34}$ and 200 TeV the ELIOSITRON will have fluences that are 16.8 times the above values. These fluences demonstrate the need for radiation hardened electronics.

The neutron fluences in the calorimeters of this SSC detector is shown in figure 5. In this calculation the barrel calorimeter face was at 1.6 meters from the IP and the forward calorimeter was at 10.53 meters from the IP. The fluence in the forward detectors is very high, approaching $10^{15}$ per SSC year. Corresponding fluences for the ELIOSITRON are 16.8 times higher. This level of neutron radiation will place extreme demands on the electronics as well as the material composition of forward detection systems.

Table 2

EQUILIBRIUM NEUTRON FLUENCE IN THE CENTRAL VOLUME
(neutrons/cm$^2$ - SSC year)

| ENERGY | POLY SHIELD | NO SHIELD |
|---|---|---|
| E > 1 eV | $3.4 \times 10^{12}$ | $4.4 \times 10^{13}$ |
| E > 100 keV | $1.7 \times 10^{12}$ | $8.7 \times 10^{12}$ |

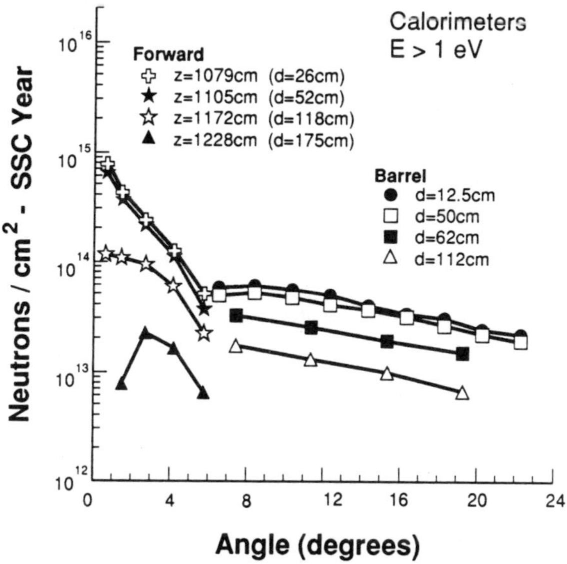


Figure 5. Neutron fluence in the calorimeters.

*2.4 Charged Particle Fluence*

The major constraint that determines the smallest inner radius of the central tracker and of electronic componants in general is the charged particle fluence from the pp interactions and from low energy charged particles (called loopers) that are trapped in the central volume. The dose due to charged hadrons from pp interactions can be calculated as follows,

$$\text{Dose} \propto L \cdot \sigma \cdot \frac{dN_{charged}}{d\eta} \cdot \frac{dE}{dx} \cdot K \qquad (1)$$

where L is the luminosity, $dN_{charged}/d\eta$ is the charged multiplicity in the interval $d\eta$ per collision, $\sigma$ is the inelastic cross section, and K is the looper factor to take into account the secondaries produced from the interactions in the tracker material and beam pipe. The charged multipicity per event, $dN_{charged}/d\eta$, is shown in Figure 6 and the inelastic cross section is shown in figure 7. [12] Incorporating the fit to the data from ref. 12. we get a simple relation for the dose due to hadrons from the interaction,

$$\text{Dose (Rads)/ year} = 2.3 \times 10^{-2} \cdot \frac{s^{0.105} \cdot L \cdot \sigma \cdot K}{r^2 \cdot \sin^2\theta} \qquad (2)$$

where $s = 4E^2$ ( $GeV^2$), r is distance from the interaction point (cm), and $\theta$ is the polar angle. The charged particle fluence can be expressed in a similar manner as,

$$\underline{\text{Charged fluence/year}} = 1.09\times10^{6} \cdot \frac{s^{0.105} \cdot L \cdot \sigma \cdot K}{r^2 \cdot \sin^2\theta} \qquad (3)$$

From the above the dose and charged particle fluence at a cylindrical radius of 10 cm for the ELIOSITRON is,

Charged fluence/ELIOSITRON year =33.5 x $10^{13}$ @ 10 cm radius and K=2

Dose/ELIOSITRON year ( Rads) = 70.6 x $10^{5}$ @ 10 cm radius and K=2

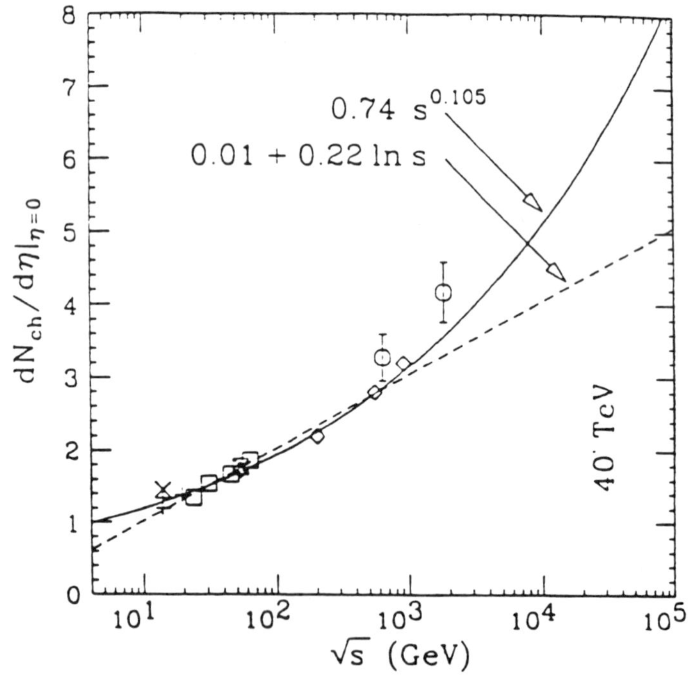


Figure 6. Charged Multiplicity Distribution[12]

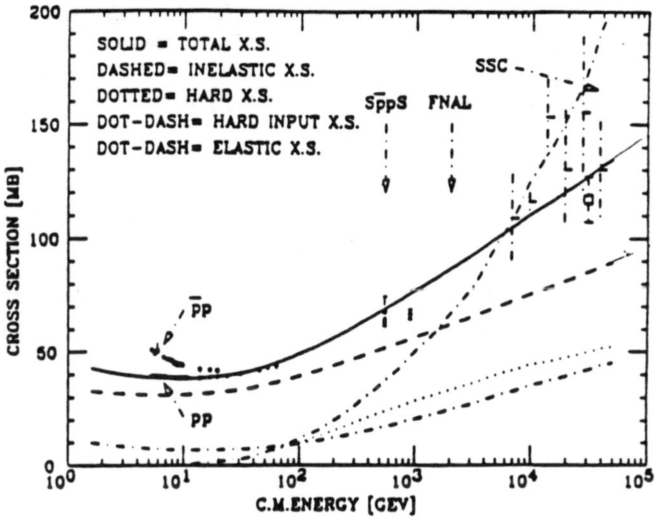


Figure 7. pp cross sections[12]

### 3. Detector Componants and Electronics

The ability of each detector subsystem to survive in a high luminosity environment depends primarily on the material in each detector and the radiation levels present. Some detectors that contain hydrogenous material are sensitive to neutrons with energies over the entire energy range, 1 eV and higher, while other materials have their greatest sensitivity to higher energy neutrons. Silicon, for example, is primarily damaged by neutrons with energies greater than 100 keV because the displacement cross section exhibits a sharp increase above 100 kev. The displacement cross section for silicon is shown in figure 8. The displacement cross section increases by over an order of magnitude above 100 keV.

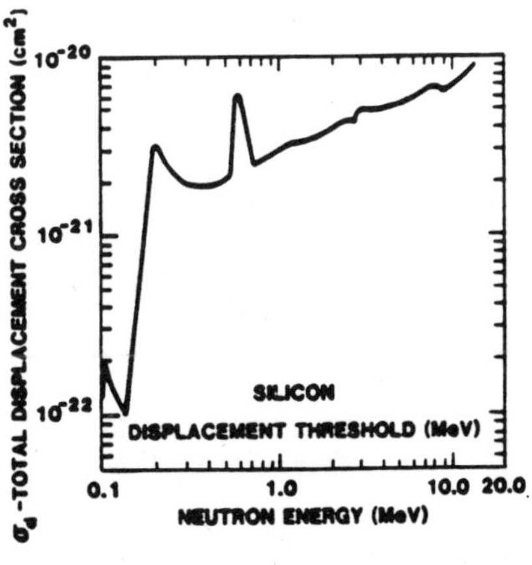


Figure 8. Displacement cross section in Silicon.[14]

Any detector subsystem must contend with a number of concerns to be useful in a collider environment. The central tracker must contend with occupancy, speed and radiation damage. Occupancy is the fraction of time that a detector element is busy and depends on its readout speed, its active area, and the particle flux and it is desireable to keep the occupancy in any detector system less than 10%. For accelerators like the SSC it requires over $10^5$ channels to keep the occupancy to tolerable levels. The particle fluence for the ELIOSITRON will increase by 16.8 times. To keep the occupancy low one can either reduce the size of the active area or increase the distance from the IP. In either case the number of detector elements will increase by 16.8.

### *3.1 Detector Technologies*

The survivability of the various detector technologies available is dependent on the maximum radiation levels that the detector can withstand before the detector is no longer able to provide good information. For scintillating type detectors the criterion is typically light attenuation, for solid state detectors like silicon the criterion may be an increase in the depletion voltage or when the leakage current becomes to large, and for electronics it could be speed deterioration, gain deterioration, or noise increase. Listed in Table 3 is a summary of the hardness levels of a variety of standard technologies that are presently being contemplated for the SSC.

Table 3

DETECTOR TECHNOLOGIES

| | HARDNESS LEVELS | |
|---|---|---|
| TYPE | NEUTRON($n/cm^2$) | GAMMA(rads) |
| SILICON(PADS OR STRIPS) | ~ $10^{14}$ | ~ 5 MRADS |
| SCINTILLATOR(FIBERS OR PADS) | > $10^{14}$ | < 10 MRADS |
| CRYSTALS ($BaF_2$, CsI, etc) | > $10^{14}$ | > 10 MRADS |
| LIQUID SCINTILLATORS | $10^{14}$- $10^{15}$ | > 20 MRADS |
| CRYOGENIC LIQUIDS | > $10^{15}$ | > 20 MRADS |

THE ABOVE NUMBERS ARE SUBJECT TO ORDER OF MAGNITUDE FLUCTUATIONS DEPENDING ON PROCESSING AND PURITY.

Since radiation damage is cummulative, the lifetime of the detector systems depends on how well they are hardened against radiation damage. From the above the various technologies are adequate to give a 5-10 year lifetime for the SSC but will allow for only less than a year at the ELIOSITRON. Clearly, improved hardness capability is needed or alternate technologies are needed to provide an adequate lifetime for the ELIOSITRON.

### *3.2 Electronics*

In most cases the detector technologies require that the electronics be placed close to the detectors. Therefore, the electronics close to the detectors will be in these high radiation environments and the lifetime will depend on their survivability. An indication of the radiation tolerance of different electronic technologies is shown in figure 9.[13] The figure indicates that several technologies are able to satisfactorily operate at fluences above $10^{14}$ - $10^{15}$ and 10 Mrads. However, in some cases a requirement for low power and high speed is imposed on the electronics. Generally, for these more restrictive cases the radiation tolerance is less than indicated in the figure. Consequently, even for the SSC fast radiation hardened electronics is a serious problem. For the ELIOSITRON the present state of hardened electronics falls short of need by at least an order of magnitude.

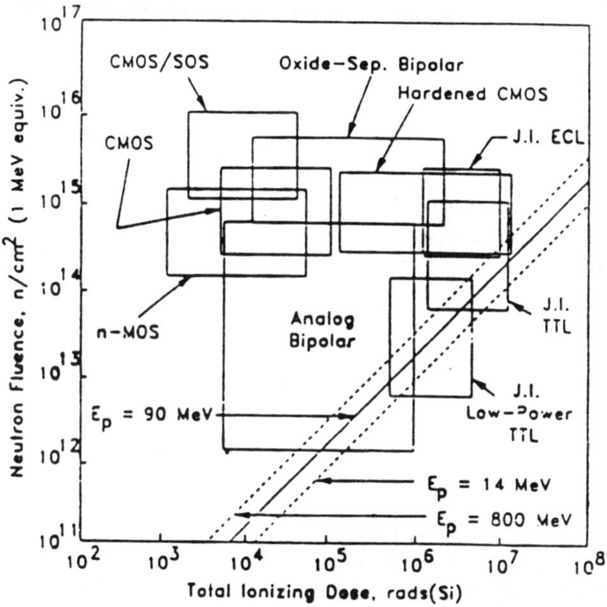


Figure 9. Radiation Tolerance of standard electronic technologies.[13]

## 4. Forward Accelerator Componants

The final focusing quad and collimater are subject to the radiation from the forward beam jets. Since most of the beam energy is in the forward beam jets it is easy to derive a simple formula for the power in the forward componants. The power in the forward direction is,

$$P = 8 \times 10^{-11} \cdot \sqrt{s} \cdot L \cdot \sigma_{inel} \tag{4}$$

For the SSC this is 320 watts/side and for the ELIOSITRON it is 18.5 kwatts/side. The final focussing quad must have adequate shielding in front to absorb as much of the power to prevent the quads from quenching the shielding and collimaters will have to be adequately cooled.

## 5. Conclusion

The radiation levels in high intensity colliders are over an order of magnitude larger than the radiation levels in the future SSC. To use present day technologies the active electronics will have to be located at a radius 4 times the present limit which is about 10 cm. In addition some of the detector technologies will also have to be located further away from the IP. Increasing the radius is very expensive since the channel count will scale as the square of the radius. Either less expensive electronics and detectors have to be developed or alternative technologies will have to be developed to meet the challange of the next generation of high luminosity and higher energy machines.

## 6. References

1. F. E. Paige et. al., *ISAJET 6.25* , CERN Pool W999..
2. Radiation Shielding Information Center, *HETC Monte Carlo High-Energy Nucleon-Meson Transport Code,* Report CCC-178, Oak Ridge National Laboratory (August 1981).
3. Group X-6, *MCNP-A General Monte CarloCode for Neutron and Photon Transport,* Los Alamos National Laboratory Report LA-7396-M Revised (April 1981).
4. H. U. Bengtsson, T. Sjostrand, *The Lund Monte Carlo for Hadronic Processes - Pythia version 5.4,* LU TP 87-3 (Jan 1987).
5. P. A. Aarnio, et. al., *Fluka86 User's Guide,* CERN TIS-RP/168 (1986).
6. M. B. Emmett, *The MORSE Monte Carlo Radiation Transport Code System,* Oak Ridge National Laboratory Report ORNL-4972 (February 1975).
7. J. Ranft, et. al., SSC Report SSC-149 (1987).
8. H. Fesefeldt, *The Physics of Hadronic Showers- Physics and Applications,* RWTH Aachen Pitha 85/02 (1985).
9. W. W. Engle, Jr.,*A Users Manual for ANISN, A One-Dimensional Discrete Ordinates Transport Code with Anisotropic Scattering,* K-1693, Union Carbide Corporation, Oak Ridge, Tn (1967).
10. R. E. Prael, et. al., *Users Guide to LCS: The LAHET Code System,* Los Alamos National Laboratory Report LA-UR-89-3014 ( September 1988).
11. D. M. Lee, et. al., *A Prediction of the Neutron and Charged Particle Backgrounds in the L* Detector",* Proceedings of the Symposium on Detector Research and Development for the Superconducting Super Collider, Fort Worth, Texas (October 1990).
12. D. E. Groom, *Radiation in the Interaction Region,* SSC Report SSC - SR- 1033 (1988).
13. Raymond and Peterson, IEEE- NS 34, 1622 (1987).
14. J. E. Gover,J. R. Srour, *Basic Radiation Effects in Nuclear Power Electronics Technology,* Sandia National Laboratory Report SAND85-0776 (May 1985).

# RESULTS OF IR WORKING GROUP

D. Ritson
*Stanford Linear Accelerator Center*
*Stanford University, Stanford, California 94309*
*and*
*Superconducting Super Collider Laboratory*
*Dallas, Texas 75237*

Working group on IR optics (D. Lee, LANL; R.Meinke, SSCL; D.Ritson, SLAC and SSCL; M.Syphers, SSCL)

ABSTRACT

The IP luminosity at the Eloisatron will direct very large fluxes of hadronic debris into the IR quads. For instance at $1.10^{35}$ $cm^2/sec$ the flux corresponds to 180 kilowatts. Already at the SSC fluxes in the neighborhood of 2 kilowatts are expected to require special handling. Scaling from SSC design experience we propose a configuration for the first IR quads at the Eloisatron capable of handling the heat load and radiation problems.

## Introduction

The best super magnet design at the SSC is estimated to handle ~200 Watts into a 15 m magnet and a maximum integrated dose of 1,000–5,000 Mrad.

The acceptable chromaticity from an IR is in the neighborhood of 100 units for present SSC chromaticity correction schemes. We assume that this will be unchanged at the Eloisatron. If correction schemes local to the IR are used this may underestimate acceptable chromaticities by a factor of two or so.

The required quadrupole aperture should be sufficient to accommodate beams separated by 10-15 $\sigma$. Allowing for a beam stay clear of $\sim 5\sigma$ a good field region of the order of $20 - 25\sigma$ is required. Typically actual quad bores needed are 2-3 times greater than the good field region and therefore at the Eloisatron IR quad bores should correspond to this or $\sim 40 - 50\sigma$.

Subject to these constraints the IR should be configured to maximize the luminosity at a fixed current corresponding to the $\beta^*_{IP}$ being minimized. We design below for a target luminosity of $\sim 10^{35}$ $cm^2$ /sec by scaling from SSC design configurations.

## SSC Designs

At 20 TeV and a nominal luminosity of $10^{33}$ cm$^2$ /sec there would be 320 watts of hadronic debris directed towards the IR quads a figure substantially less than the Eloisatron goal of 180 kilowatts.

We quote from design studies of Mokov et al for a standard IR triplet of 4 cm bore (design A) and a design B with the first member of the triplet being a 1 cm warm bore magnet followed by 6 cm bore quads with a 1 cm thick copper liner.

The triplet is split into a first lens Q1 and a second lens of two parts Q2 and Q3 and a third lens Q4. Each of the lens is approximately 12 m long and the face of Q1 is 20 metres from the IP. Mokov et al estimate that for design A the watts into Q1, Q2, Q3 and Q4 would be 50,15,26 and 15 respectively. The corresponding peak dosage is 320,192,220 and 96 Mrad/yr.

For design B the radiation and heat load for the warm bore Q1, is no problem and the watts into Q2,Q3 and Q4 are 3.4, 13.9, 2.1 with corresponding Mrads/yr of 20,84 and 12.

Scaled to the Eloisatron design A would miss by $\sim$ 100 and design B would miss by a factor $\sim$ 25.

## Eloisatron Designs

If quad bores are unchanged one can map an existing design by scaling all longitudinal lengths by the square root of the energy $E$. If the quad bore A is also changed (without changing the pole tip magnetic field) the scale factor S is $\sqrt{EA}$. This scaling results from the interplay of weaker gradients and longer quad lengths. To maintain constant chromaticity requires $\beta^*_{IP}$ to scale as S. Then $\beta_{max}$ in the IR quads scales $\propto S$. The maximum acceptable emittance $\epsilon_{max}$ scales as

$$\epsilon_{max} \propto \frac{A^2}{S^2 \beta_{max}}$$

or as

$$\epsilon_{max} \propto \frac{A^{1/2}}{E^{3/2}}$$

We consider first a low luminosity design where the constraints are simply lattice optics. Secondly we consider a full luminosity design that involves further compromises required to handle the heat and radiation loads.

*Low Luminosity IRs (Only Lattice Constraints)*

Leaving the quad bores unchanged and scaling by $\sqrt{E}$ the $\beta^*_{IP}$ will be increased from 0.5 m at the SSC to 1.2 m at the Eloisatron, the $\beta_{max}$ will go from 10 km to 22 km and the acceptable normalized emittance would be $\sim 1.4.10^{-6}$ m rad.

*Full Luminosity IRs Designed to Accommodate Heat and Radiation Loads*

To handle full luminosity we required a long region after the detector and prior to the first IR quad consisting of warm bore sweep field to remove and absorb the radiation from the IPs prior to the first quad. This region could either utilize dipole dogleg (a natural achromat) or straight dipole fields to provide separation and minimize the effects of long range beam beam interactions.

To achieve a satisfactory solution we tried successively larger scale up factors with associated larger bore magnets until we estimated the quads could safely absorb the radiation and heat loads.

From the previous discussion a simple $\sqrt{E}$ scale cannot handle radiation and heat loads. With a scale factor increased to three the rear quads still come too close to the beam-line and would be unlikely to meet the heat and radiation load requirements.

For a scale factor of four the bore of the rear quads becomes 12 cm and appears to lie outside the main cone of hadronic debris from the IP. This we believe could provide a feasible configuration for the IR quads.

Under these conditions the $\beta^*_{IP}$ is increased to 2 m, the $\beta_{max}$ to $\sim$ 40 km and the maximum normalized emittance would be $\sim 2.10^{-6}$.

Table 1 lists the required IP configuration corresponding to the above for a 100 TeV Eloisatron. Fig. 1 shows the associated beta values as a function of distance away from the IP.

The above are of at best only estimates. However in view of the speculative nature of the Eloisatron they are probably quite adequate at this point of time.

## Conclusions

We believe that even for fluxes of hadronic debris of 180 kilowatts into the IR quads that an adequate IR design is still possible. Our proposed design requires a quad configuration twice as long as that dictated solely by considerations of lattice optics. This permits the use of lower quad gradients and larger quad bores. The larger

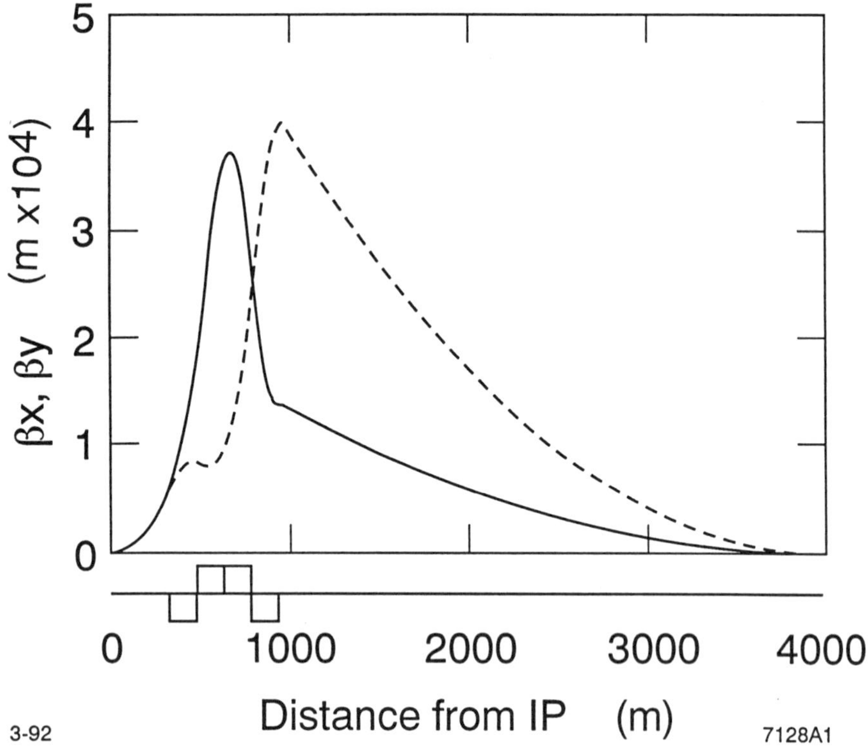
5
4
3
2
1
0
βx, βy (m x104)
0
1000
2000
3000
4000
Distance from IP (m)
3-92
7128A1

Fig. 1

quad bores and a long region of warm bore magnetic fields in the long drift from IP to IR quads should provide tolerable heat and radiation loads to the IR quads.

Thus heat and radiation load considerations will increase minimum achievable $\beta^*_{IP}$s and hence lower peak luminosities by a factor two relative to estimates derived solely from lattice optics.

| Eloisatron IR Quad Configuration | | | |
|---|---|---|---|
| Type | | Length (M) | Field at 6 cm (T) |
| | IP | | |
| DRFT | | 20.00000 | |
| DRFT | | 90.00000 | |
| QUAD | | 52.87838 | −4.0 |
| DRFT | | 1.00000 | |
| QUAD | | 46.65438 | 4.0 |
| DRFT | | 1.00000 | |
| QUAD | | 46.65438 | 4.0 |
| DRFT | | 1.00000 | |
| QUAD | | 52.87838 | 4.0 |
| DRFT | | 1000.00000 | |
| | FOCUS | | |

# Detector Part

# PARTICLE DETECTION AT ELOISATRON*

Heinrich Leutz
*CERN, Geneva, Switzerland*

## 1. Introduction

This workshop aimed at making the machine design compatible with the requirements of particle detectors to be installed at its beam crossing points. No specific signatures of new particles, which are expected to be produced at Eloisatron, are known yet. Therefore, the appropriate design should be an omni purpose detector surrounding the beam crossing point at nearly full solid angle with almost complete hermeticity. This omni purpose detector should contain inner tracking, followed by electro-magnetic and hadronic calorimetry and should finally be surrounded by muon detectors. These sub-detectors will suffer from radiation damage according to their distance from the vertex region and therefore inner tracking will be mostly affected by high luminosities and the subsequent production of minimum biased events.

Particle cross-sections and minimum biased background can at present only be approached by extrapolating corresponding data obtained from hadron colliders already in operation (Sp$\bar{p}$S, Fermilab) via the expectations for LHC and SSC finally to Eloisatron.

## 2. Scaling of particle cross-sections

Table 1 shows the rise of Z-pair production from 0.63 TeV (Sp$\bar{p}$S) to 100 TeV (half Eloisatron) centre of mass energy. The full cross-sections[1] for the DFLM[2] - structure functions span more than three orders of magnitude.

Table 1 : Z-Pair Production (pb)[1]

| Structure Functions | DFLM160 | DFLM260 |
|---|---|---|
| $p\bar{p}$ ; $\sqrt{S}$ = 0.63 TeV | $0.0693^{+0.005}_{-0.004}$ | $0.0610^{+0.005}_{-0.004}$ |
| $p\bar{p}$ ; $\sqrt{S}$ = 1.8 TeV | $1.17^{+0.04}_{-0.04}$ | $1.12^{+0.05}_{-0.04}$ |
| pp ; $\sqrt{S}$ = 6 TeV | $3.97^{+0.08}_{-0.06}$ | $4.06^{+0.1}_{-0.1}$ |
| pp ; $\sqrt{S}$ = 16 TeV | $13.7^{+0.04}_{-0.08}$ | $15.2^{+0.01}_{-0.02}$ |
| pp ; $\sqrt{S}$ = 40 TeV | $35.8^{+1}_{-1}$ | $43.6^{+1}_{-1}$ |
| pp ; $\sqrt{S}$ = 100 TeV | $77.4^{+3}_{-5}$ | $103^{+3}_{-5}$ |

---

* The work reported here is part of the LAA-Project

Expected top cross-sections are listed[3] in table 2 for assumed top-masses of 130 GeV and 200 GeV. The values are calculated from DFLM- and HMRSB-[4] structure funtions. The cross-sections for Eloisatron (200 TeV centre of mass) are estimated from graphical extrapolation by the author. The ratios between Eloisatron- and LHC-top-cross-sections vary between 390 and 80 depending on the top-mass.

Table 2: Total $\sigma_{t\bar{t}}$ (nb)[3]

| $m_{top}$ = 130 GeV | | |
|---|---|---|
| Structure Functions | DFLM260 | HMRSB |
| $p\bar{p}$ , 0.63 TeV | $2.5 \times 10^{-4}$ | $2.7 \times 10^{-4}$ |
| $p\bar{p}$ , 1.8 TeV | $2.23 \times 10^{-2}$ | $2.33 \times 10^{-2}$ |
| pp , 16 TeV | 4.35 | 3.96 |
| pp , 40 TeV | 21.85 | 18.34 |
| pp , 200 TeV | ~ 350 | |

| $m_{top}$ = 200 GeV | | |
|---|---|---|
| Structure Functions | DFLM260 | HMRSB |
| $p\bar{p}$ , 0.63 TeV | $7.3 \times 10^{-6}$ | $4.34 \times 10^{-6}$ |
| $p\bar{p}$ , 1.8 TeV | $1.9 \times 10^{-3}$ | $2.2 \times 10^{-3}$ |
| pp , 16 TeV | 0.636 | 0.621 |
| pp , 40 TeV | 3.89 | 3.50 |
| pp , 200 TeV | ~ 250 | |

Table 3 indicates cross-sections for Higgs production via the $H^0 \rightarrow Z^0Z^0 \rightarrow 4\mu$-decay-channel[5] for assumed Higgs-masses of 500 GeV and 1000 GeV. The centre of mass energies of the proton colliders are 16 TeV (LHC), 40 TeV (SSC) and 200 TeV (Eloisatron). The main production channels are gg-fusion and $W^+W^-$-fusion. The following inputs were used : $m_{top}$ = 130 GeV, BR $(H^0 \rightarrow Z^0Z^0)$ = 0.3, and BR $(Z^0 \rightarrow \mu\mu)$ = 0.03. The dominating irreducible background arises from all processes generating $Z^0$-pairs which subsequently decay into four muons. The signal to background ratios due to directly produced $Z^0$-pairs decaying into four muons is also listed in table 3. It improves by one order of magnitude from LHC- to Eloisatron centre of mass energy.

Table 3 : Cross-sections (in pb) for Higgs production[5] via $H^0 \to Z^0 Z^0 \to 4\mu$ and the corresponding background from direct $Z^0 Z^0$ pair production decaying into the $4\mu$-channel.

| $\sqrt{s}$ = 16 TeV | $M_H$=500 GeV | $M_H$=1000 GeV |
|---|---|---|
| $q\bar{q}$ | $6.6 \times 10^{-5}$ | $9.8 \times 10^{-6}$ |
| gg | $8.7 \times 10^{-4}$ | $3.2 \times 10^{-5}$ |
| $Z^0 Z^0$ | $8.5 \times 10^{-5}$ | $1.2 \times 10^{-5}$ |
| $W^+ W^-$ | $1.8 \times 10^{-4}$ | $2.6 \times 10^{-5}$ |
| Total | $12.0 \times 10^{-4}$ | $8.0 \times 10^{-5}$ |
| Background | $8 \times 10^{-2}$ | $5.3 \times 10^{-3}$ |
| Signal/Background | $1.5 \times 10^{-2}$ | $1.5 \times 10^{-2}$ |

| $\sqrt{s}$ = 40 TeV | $M_H$=500 GeV | $M_H$=1000 GeV |
|---|---|---|
| $q\bar{q}$ | $5.2 \times 10^{-4}$ | $1.2 \times 10^{-4}$ |
| gg | $5.0 \times 10^{-3}$ | $2.4 \times 10^{-4}$ |
| $Z^0 Z^0$ | $4.7 \times 10^{-4}$ | $1.1 \times 10^{-4}$ |
| $W^+ W^-$ | $1.0 \times 10^{-3}$ | $2.2 \times 10^{-4}$ |
| Total | $7.0 \times 10^{-3}$ | $6.9 \times 10^{-4}$ |
| Background | 0.18 | $1.8 \times 10^{-2}$ |
| Signal/Background | $3.9 \times 10^{-2}$ | $3.8 \times 10^{-2}$ |

| $\sqrt{s}$ = 200 TeV | $M_H$=500 GeV | $M_H$=1000 GeV |
|---|---|---|
| $q\bar{q}$ | $7.2 \times 10^{-3}$ | $2 \times 10^{-3}$ |
| gg | $5.6 \times 10^{-2}$ | $3.7 \times 10^{-3}$ |
| $Z^0 Z^0$ | $5.1 \times 10^{-3}$ | $1.5 \times 10^{-3}$ |
| $W^+ W^-$ | $1.0 \times 10^{-2}$ | $1.1 \times 10^{-3}$ |
| Total | $7.8 \times 10^{-2}$ | $8.3 \times 10^{-3}$ |
| Background | 0.54 | $6.1 \times 10^{-2}$ |
| Signal/Background | $1.4 \times 10^{-1}$ | $1.4 \times 10^{-1}$ |

### 3. Scaling of minimum biased events and dose rates

Minimum biased events scale at fixed centre of mass energy linearly with luminosity. For variable centre of mass energies but constant luminosity they scale with multiplicity and particle cross-section. They are by far the main source for the radiation dose around the crossing points of a hadron collider.

The program PYTHIA[6] predicts a large contribution of diffraction dissociation, whereas the Quark-Gluon String Model (QGSM)[7] enhances more the non-diffractive processes. These results in different multiplicity predictions[5], as displayed in fig.1, which shows in addition the measured values from Thomé et al (ISR)[8] and Alner et al (Sp$\bar{p}$S)[9]. The multiplicity distribution within a wide range of rapidity is flat according to PYTHIA and QGSM.

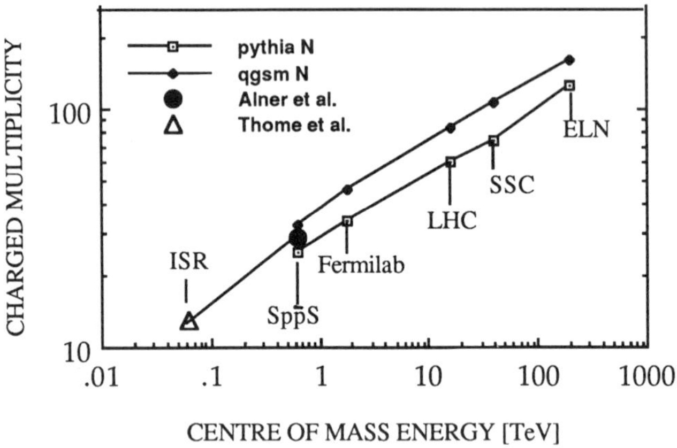


Fig. 1 Multiplicity predictions obtained from PYTHIA[6] and from QGSM[7] (Quark Gluon String Model).

The pp-(p$\bar{p}$)- inelastic cross-sections for different pp-(p$\bar{p}$)- colliders are plotted in fig.2 and extrapolated to 200 TeV centre of mass energy for Eloisatron.

This brings us finally to the dose rate Gy (Gray) per year ($Yr^{-1}$) expected for different pp-colliders. It can be expressed in the following way :

$$\text{Gy yr}^{-1} = \text{const}\frac{dE}{dX}\frac{1}{\sin\theta} L \times \sigma_{inel} \times H \times \left(2\pi D^2\right)^{-1} \sin\theta$$

with dE/dX $(\sin\theta)^{-1}$ meaning the stopping power [GeV $cm^2$ $Kg^{-1}$] for minimum ionizing particles emitted at the polar angle $\theta$, L the luminosity [$cm^{-2}s^{-1}$] of the collider, $\sigma_{inel}$ [$cm^2$] the inelastic cross-section for pp collisions, H the number of particles per collision and per unit of rapidity, D the distance from the collider axis and D/sin$\theta$ the length of the cylindrical detector arrangement. The term const contains the conversion from GeV into Joule (1.6 $10^{-10}$) and from seconds to year ($10^7$) if we assume a yearly operation of $10^7$ s which corresponds to about 115 days. Therefore, we finally end up with the dimension of Joule $Kg^{-1}$ $yr^{-1}$ which equals Gray (Gy) $yr^{-1}$ corresponding to the old unit of 100 RAD $yr^{-1}$.

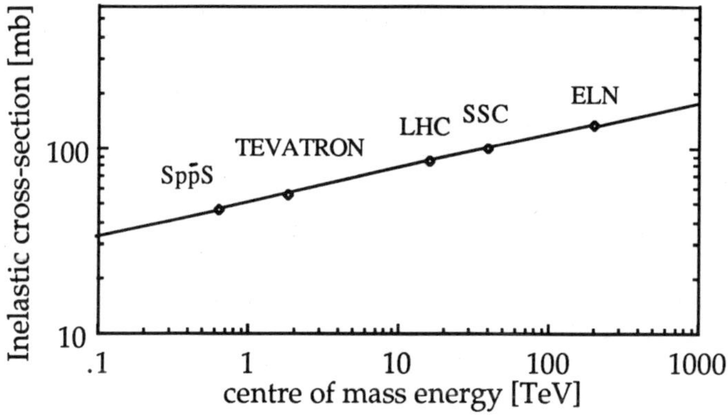


Fig. 2 Extrapolation of pp- ($p\bar{p}$) - inelastic cross-sections to Eloisatron- centre of mass energy (200 TeV). The inelastic values were obtained by subtraction of elastic cross-sections $\sigma_e$ from $\sigma_{total}$.

The relevant parameters for the dose calculations are listed in table 4. The annual dose rates are shown versus the distance from the collider axis in fig.3 for LHC, SSC and ELN with the decadic exponents for the luminosities indicated.

Table 4 : Parameters to calculate the annual dose rate

| PARAMETERS | | LHC | SSC | ELN |
|---|---|---|---|---|
| Stopping power | GeV $cm^2$ $Kg^{-1}$ | 2 | 2 | 2 |
| Luminosity | $cm^{-2}s^{-1}$ | $10^{34}$ | $10^{33}$ | $10^{34}$ |
| Inelastic cross-section | mb | 85* | 100* | 135 |
| Nb. of charged particles per unit of rapidity | | 6 | 7.5* | 13 |
| Annual dose rate at 1m dist. from collider axis | Gy $yr^{-1}$ | 240 | 38 | 830 |

* values are taken from D.E.Groom

A solenoidal magnetic field in the inner detector cavity changes these dose rates considerably. This case will be considered in the next contribution by C. D'Ambrosio.

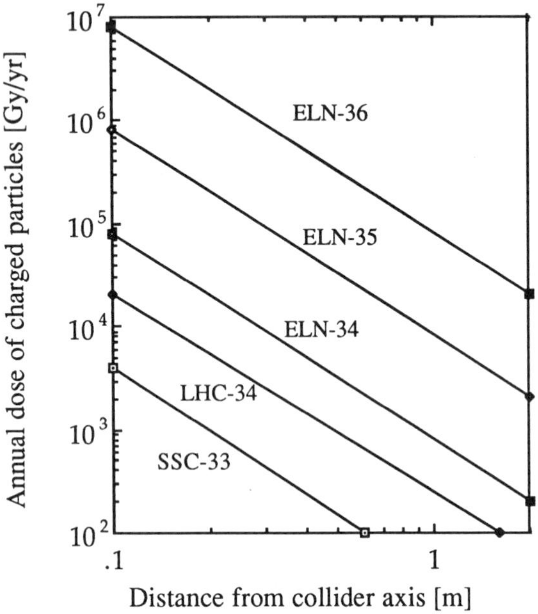


Fig. 3 Annual dose rates for different pp-colliders. The unit Gray (Gy) equals 1 Joule $Kg^{-1}$ or 100 RAD. The decadic exponents of the luminosity are indicated. These values hold for the inner cavity of a detector within a rapidity range of ±4. For calorimeters, starting at about 2 m distance, other dose rates must be considered[12].

## 3. Conclusions

As can be seen from table 5, the Higgs-cross-sections increase much faster with centre of mass energy than the minimum biased events, which are responsible for the radiation dose rate.This would be even more pronounced, for other new particles, where the production threshold is much closer to the ELN-centre of mass energy.

Table 5 : Ratios of Higgs-cross-sections compared with minimum biased background ratios between LHC- SSC- and ELN- centre of mass energies.

| Ratios | 40 : 16 TeV | 200 : 40 TeV | 200 : 16 TeV |
|---|---|---|---|
| Higgs cross-sections (mass 1000 GeV) | 8.6 | 12.1 | 104.0 |
| Min. biased background at same luminosity | 1.6 | 2.2 | 3.5 |

The radiation dose rate increases linearily with luminosity. To keep it within a tolerable level, it is much more efficient to increase the cross-section of rare particles. Therefore, the detector group recommends that the Eloisatron be built emphasizing the potential of an energy upgrade rather than a luminosity upgrade.

**Acknowledgments**

The author thanks Prof. Zichichi for his continuous support and encouragement and acknowledges the hospitality at the E. Majorana Centre in Erice.

**References:**

1) B.Mele, P.Nason and G. Ridolfi *Proceedings of the Large Hadron Collider Workshop, Aachen, Germany,*CERN 90-10, eds. G.Jarlskog and D.Rein, **Vol.II** (1990) 113

2) M.Diemoz, F.Ferroni, E.Longo, and G.Martinelli *Z.Phys.* **C39** (1988) 21

3) J.P. Guillet, P.Nason, and H.Plothow-Besch *Proceedings of the Large Hadron Collider Workshop, Aachen, Germany,*CERN 90-10, eds. G.Jarlskog and D.Rein, **Vol.II** (1990) 118

4) P.N.Harriman, A.D. Martin, R.G.Roberts and W.J. Stirling, *Preprint* DTP-90-04

5) F.Anselmo et al, Status report of the LAA-MSL group : CERN/DRDC 92-44

6) T.Sjostrand, *Comp. Phys. Comm.* **39** (1986) 347. T.Sjostrand and M. Bengtsson, *Comp. Phys. Comm.* **43** (1987) 367. H.U. Bengtsson and T.Sjostrand, *Comp. Phys. Comm.* **46** (1987) 43. T. Sjostrand and M. van Zijl, *Phys. Rev.* **D36** (197) 2019. T. Sjostrand, preprint CERN-TH.6488/92, May 1992

7) A.B.Kaidalov and K.A.Ter-Martirosyan, *Yad.Fiz.***39** (1984) 1545; **40** (1984) 211. A.B.Kaidalov and O.I.Piskunova, *Yad.Fiz.***41** (1985) 1278. Yu.M.Shabelski, *Yad.Fiz.***44** (1986) 186. Yu.M.Shabelski, *Z.Phys.***C38** (1988) 569

8) W.Thomé et al *Nucl.Phys.* **B 129** (1977) 365

9) G.J.Alner et al *Phys.Lett.* **138 B** (1984) 304

10) D.Denegri,*Proceedings of the Large Hadron Collider Workshop, Aachen, Germany,*CERN 90-10, eds. G.Jarlskog and D.Rein, **Vol.I** (1990) 56

11)D.E.Groom, Task Force Report : *Radiation Damage Testing at the SSC*, eds. W. Chinowsky and R. Thun SSCL-SR (1990) 1054

12) G.R. Stevenson, *Proceedings of the Large Hadron Collider Workshop, Aachen, Germany,*CERN 90-10, eds. G.Jarlskog and D.Rein, **Vol.III** (1990) 566

# CENTRAL TRACKING IN HIGH LUMINOSITY FUTURE COLLIDERS*

Carmelo D'Ambrosio
*CERN, Geneva, Switzerland*

Future colliders will achieve center of mass energies of the order of hundreds of TeV and luminosities in excess of $10^{34}$ cm$^{-2}$ s$^{-1}$. The aim of this paper is to furnish "simple tools" for a first quantitative approach for tracking in future supercolliders and to clarify the interconnections between collider- and tracker- parameters. Only central tracking will be considered.

## 1. Introduction

Future colliders will achieve center of mass energies of the order of hundreds of TeV and luminosities in excess of $10^{34}$ cm$^{-2}$ s$^{-1}$. In this scenario, it is important to establish simple criteria and relations to indicate the feasibility of tracking in the central detector region ($-1.4 < \eta < +1.4$) and to clarify the interconnections between collider- and tracker- parameters, aiming at an optimum for both.

Starting with the geometry of crossing beam and the related distribution of interaction vertices (Sect. 2), an inequality is given (Eq. 7) which relates accelerator parameters to detector performances: both must be optimized in order to eventually satisfy Eq. 7. Sect. 3 deals with the detector vertex resolution and shows the basic relations between tracker geometry and detecting element (cell) performance, projected on the beam axis.

The advantages of a solenoidal magnetic field in the tracking region are pointed out in Sect. 4 and Sect. 5: the first deals with momentum measurement and the second describes the effects of such a field on the radiation dose in the inner cavity. In addition, two examples are given (ELN-34 and ELN-35) in which all the developed relations are used to calculate Eq. 7 (Sect. 6).

The tracking region considered spans a pseudorapidity of $\pm$ 1.4 (central tracking). However, similar criteria can be employed for higher pseudorapidity regions, but in this case collision- and detection- processes are somewhat more difficult to describe.

* The work reported here is part of the LAA-Project.

Table 1 List of relevant parameters, their units and symbols used in the text

| Accelerator parameters | | Symbol |
|---|---|---|
| Centre of mass energy | TeV | $E_{cm}$ |
| Full beam crossing angle | μrad | $\alpha$ |
| Beam radius (68 % of particles) | μm | $\sigma_x$ |
| Bunch length (68 % of particles) | m | $\sigma_b$ |
| Bunch crossing time | ns | $t_b$ |

| Interaction parameters | | |
|---|---|---|
| Standard luminosity | $cm^{-2}s^{-1}$ | $L_0$ |
| Length of luminous region | m | $l^*$ |
| Inelastic cross-section | $cm^2$ | $\sigma_{inel}$ |
| Avr. Nr of charged part. per unit of rap. and per event | | H |
| Distribution of vertices per mm | | $n_v(z)$ |
| Avr. Nr of vertices per bunch crossing | | $\langle n_v \rangle$ |
| Avr. Nr of vertices per mm (for 68 % of interaction) | | $\langle n_v^* \rangle$ |
| Momentum, trans. Mom. and min. bias avr. trans. Mom. | GeV/c | $p$, $p_t$, $\langle p_t \rangle$ |

| Detector parameters | | |
|---|---|---|
| Det. precision in (r,z)-plane | μm | $\sigma_z$ |
| Det. precision in (r,φ)-plane | μm | $\sigma_\varphi$ |
| Distance from the beam axis | m | r |
| Distance between outer- and inner det. shell | m | $\Delta r_t$ |
| Number of hits | | N |
| Multiple scattering angle | rad | $\vartheta_{ms}$ |
| Thickness of det. in unit of radiation length | | t |
| Thickness of det. to achieve one hit in unit of rad. length | | $t_0$ |
| Det. precision on the beam axis (z-axis) | mm | $\delta z$ |
| Total and transverse Mom. resolution | | $\delta p/p$, $\delta p_t/p_t$ |

## 2. Geometry of crossing beams and distribution of interaction vertices

In a collider, the particle bunches interact in the crossing region, where the two beams hit each other. Since the bunches are of gaussian shape both in length and in transverse dimension[a], the related crossing region will also be of gaussian shape (fig. 1).

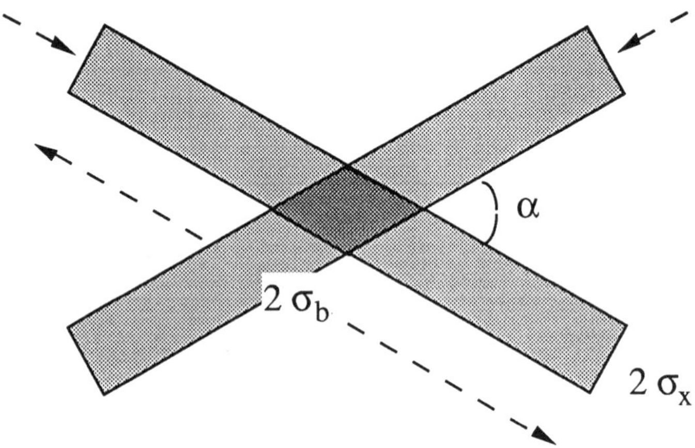


Fig. 1 Geometry of crossing beams.

The standard luminosity is defined as

$$L_0 = \frac{f\mathcal{N}^2}{4\pi\sigma_x^2}\,\frac{F^2}{q} \qquad [\mathrm{cm}^{-2}\mathrm{s}^{-1}] \tag{1}$$

where

$$q = \sqrt{1+\left(\frac{\alpha}{2}\cdot\frac{\sigma_b}{\sigma_x}\right)^2} \tag{2}$$

F is the filling factor, f the bunch collision frequency, $\mathcal{N}$ the number of particles per bunch, $\alpha$ the full beam crossing angle and $\sigma_{b,x}$ the bunch -length and -radius respectively (see fig. 1). However, due to the gaussian longitudinal dimension[b], the luminosity will depend on z (z is the value of the z-axis and its origin is at the centre of the crossing region). In fact, we can write[1]

---

[a] The full bunch length is defined as $4\sigma_b$ and the full beam radius as $2\sigma_x$, for circular beams and for 95% of the particles.

[b] Here, we do not consider the transverse distribution, since in general $\sigma_x \ll \sigma_b$ .

$$L(z) = L_0 \frac{q}{\sqrt{\pi}\sigma_b} e^{-\left(\frac{q}{\sigma_b} z\right)^2} \tag{3}$$

which integrated between $-\infty$ and $+\infty$ gives correctly $L_0$.

The distribution of interaction vertices is related to the luminosity via the relation

$$n_v(z)dz = \sigma_{inel}\, t_b\, L(z)dz \tag{4}$$

and the average number of produced vertices[c] per bunch crossing is

$$< n_v >= \sigma_{inel}\, t_b\, L_0 \tag{5}$$

where $\sigma_{incl}$ is the inelastic cross-section and $t_b$ is the bunch crossing time. $n_v(z)$ is shown in fig. 2 for different colliders.

From fig. 2, we learn that the main difference between Eloisatron, LHC and SSC is in the widths of the distributions. The consequences can be fully recognized by calculating the average number of vertices in the centre of the respective distributions. We define it as

$$< n_v^* >= \frac{0.68 < n_v >}{l^*} = \frac{0.68 < n_v >}{\sqrt{2}\sigma_b} \sqrt{1 + \left(\frac{\alpha}{2}\frac{\sigma_b}{\sigma_x}\right)^2} \tag{6}$$

where $l^* = 2\sigma_b/\sqrt{2}q$ is called the luminous region and includes 68% of the interactions and $<n_v{*}>$ is in $mm^{-1}$. Note the term containing the beam crossing angle $\alpha$, which influences the length of the vertex distribution.

$<n_v{*}>$ sets the minimum required vertex resolution, in order to resolve each of the vertices[d]. This can be written in terms of an inequality between $<n_v{*}>$ and the precision $\delta z$ (in mm) required to the tracker to measure a vertex position on the beam axis:

$$< n_v^* > 2.36\delta z < 0.5 \tag{7}$$

---

[c] The actual total number of vertices in one bunch crossing follows Poisson statistics with an average of $<n_v>$ .

[d] For a discussion on the need of resolving each vertex, see the contribution of U. Gensch in these same proceedings .

The factor 0.5 or better is dictated by the requirement of having a valley to peak ratio of at least ~10% between two vertex maxima and 2.36 $\delta z$ represents the FWHM of the tracker resolution curve.

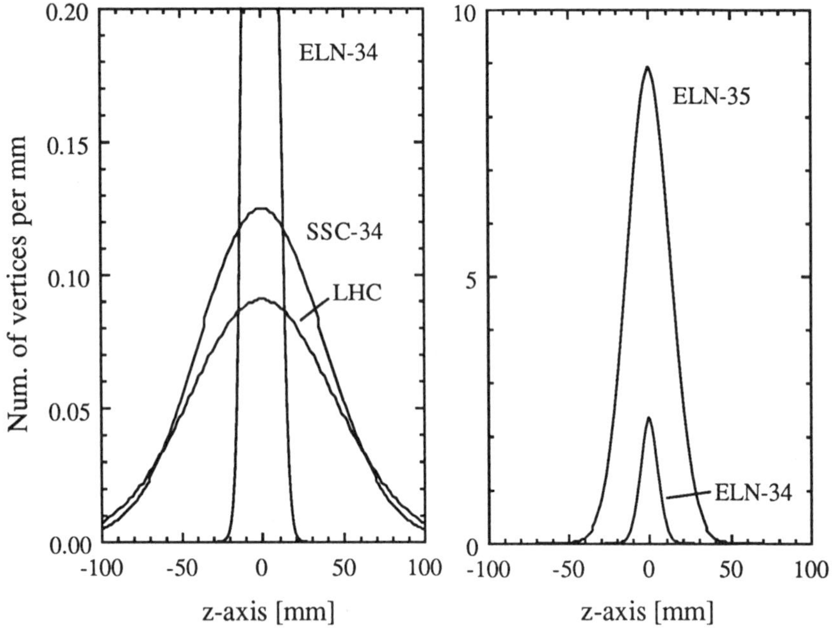


Fig. 2 Distribution of interaction vertices along the beam axis for different accelerators. SSC luminosity was set to $10^{34}$ only for comparison: its number of vertices corresponds to the ordinate reading divided by 10 for the proposed luminosity of $10^{33}$ cm$^{-2}$ s$^{-1}$.

In table 3, the values obtained from Eqs 5 and 6 are summarized: between LHC and ELN-35 there is a factor ~21 for $<n_v>$ and a factor 76 for $<n_v^*>$ ! The first essentially originates from an increase of luminosity (fact. 10) and an increase in inelastic cross section (fact. 1.6), but the second is mainly due to the different beam parameters involved in the calculation of the luminous region (fact. 3.5).

Inequality 7 tells us that the product between two quantities must be smaller than 0.5. In fact, both depend on specific and different parameters: $<n_v^*>$ depends only on accelerator parameters, while $\delta z$ only on the tracker performance. Therefore, to satisfy Eq. 7 means: first, an optimization of $\delta z$ and second, changing some accelerator parameters – in particular the full beam crossing angle $\alpha$ – in order to minimize $<n_v^*>$ (see the contribution of H. Leutz and the recommendations on these same Proceedings).

Table 2 Values of <nv*> for various beam crossing angles.

| α [μrad] | ELN-34 | | ELN-35 | |
|---|---|---|---|---|
| | $<n_v{}^*>$ | l* | $<n_v{}^*>$ | l* |
| 250 | 1.82 | 11.2 | 11.03 | 16.6 |
| 150 | 1.11 | 18.4 | 6.84 | 26.8 |
| 100 | 0.76 | 26.8 | 4.84 | 38 |
| 50 | 0.43 | 47.1 | 3.06 | 60 |

Table 3 Values of Eqs 5 and 6 for different colliders

| BEAM PARAMETERS | | LHC | SSC | ELOISATRON | |
|---|---|---|---|---|---|
| Luminosity | $cm^{-2}s^{-1}$ | $10^{34}$ | $10^{33}$ | $10^{34}$ | $10^{35}$ |
| Average number of vertices $<n_v>$ | | 13 | 1.7 | 30 | 270* |
| Avr. Nr of vert. per mm (68% of interactions) $<n_v{}^*>$ | $mm^{-1}$ | 0.09 | 0.014 | 1.8 | 6.8 |

* for a beam crossing time of 20 ns (see Detector group recommendations)

## 3. Vertex resolution

The main task of a central tracker in a general purpose detector will be to track high momentum particles from the outer layers to the respective production vertices. For example: – four high $p_t$-muons detected in the outer muon chambers need to be linked to their (eventually) common vertex and to measure their momenta; – high $p_t$-electrons giving a signal in the electromagnetic calorimetre have to be identified and then tracked back to their vertex.

In future colliders, the position of the beam is well known in the (r,φ)-plane ($\sigma_x$), while the vertex distribution along the z-axis spans a length of ±l* for each bunch crossing (fig. 3). Therefore, the tracker has to provide the vertex positions along the beam axis with a precision δz such to satisfy the detector part of Eq. 7.

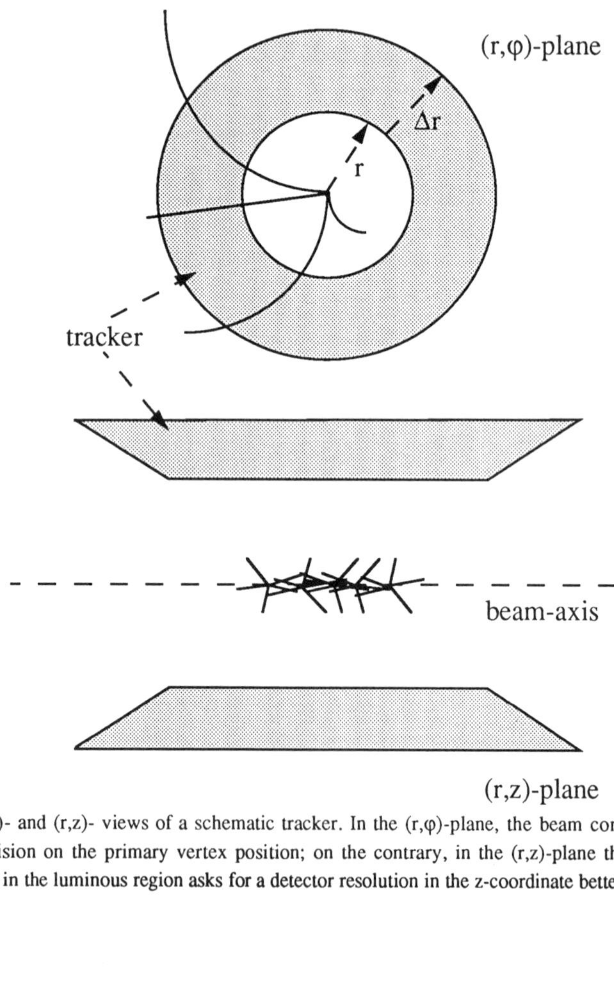


Fig. 3 (r,φ)- and (r,z)- views of a schematic tracker. In the (r,φ)-plane, the beam constraint provides enough precision on the primary vertex position; on the contrary, in the (r,z)-plane the production of multivertices in the luminous region asks for a detector resolution in the z-coordinate better than 1 mm.

For simplicity we suppose to have a cylindrical tracker with respect to the z-axis, composed of concentric shells containing detector elements with z-precision $\sigma_z$. Then the vertex precision can be described with the following equation:

$$\delta z^2 \approx \frac{\sigma_z^2}{N} \left[ g(\frac{r_{in}}{\Delta r_t}) \right]^2 + N\ (r_{in}\, \vartheta_{ms})^2 \tag{8}$$

where N means the number of hits, $r_{in}$ the distance between beam axis and inner shell, $\Delta r_t$ the "tracker thickness" and $\vartheta_{ms}$ the multiple scattering angle per hit

$$\vartheta_{ms} \approx \frac{0.0136}{p} \sqrt{t_0} \tag{9}$$

with $t_0$ in units of radiation length being the minimum material thickness to achieve one signal hit and p the momentum in [GeV/c]. $g(r_{in}/\Delta r_t)$ is a function of the ratio between tracker inner shell radius $r_{in}$ and "tracker thickness" $\Delta r_t$. The first term of Eq. 8 is due to the measuring error, while the second term represents the multiple scattering error. For a "three shell tracker", the measuring error amounts to

$$\frac{\sigma_z^2}{N} \left[ g(\frac{r_{in}}{\Delta r_t}) \right]^2 = \frac{\sigma_z^2}{N} \left[ \frac{5}{2} + 6 \frac{r_{in}}{\Delta r_t} \left( 1 + \frac{r_{in}}{\Delta r_t} \right) \right] \tag{10}$$

where $N \geq 3$ and N/3 is the number of hits per shell. In fig. 4, $g(r_{in}/\Delta r_t)$ is shown versus $r_{in}/\Delta r_t$ .

The second term in Eq. 8 arises from the material present into the inner cavity. This can be active (detector elements) or passive ( support structure, beam pipe, etc...)[e] material and care must be paid that it be as small as possible, since it will also increase the production of secondaries and gamma conversions. As the multiple scattering error is inversely proportional to the momentum, high $p_t$-tracks will be considerably less affected by it.

The vertex resolution improves increasing the tracker thickness, i.e. the number of signal hits. But the number of hits has to be balanced between the measuring term of Eq. 8 which improves with the hit number and the second term representing the multiple scattering error which increases with the material layer. The vertex resolution is also linearly dependent on the inner shell position in both terms of Eq. 8 and therefore it will be essential to go as close as possible to the beam axis. However, in Sect. 5 we shall see that there is a major constraint in reducing $r_{in}$: the radiation field present in the tracking cavity, which decreases with $r^{-2}$.

---

[e] For simplicity, in Eq. 8 only the active part is taken in account.

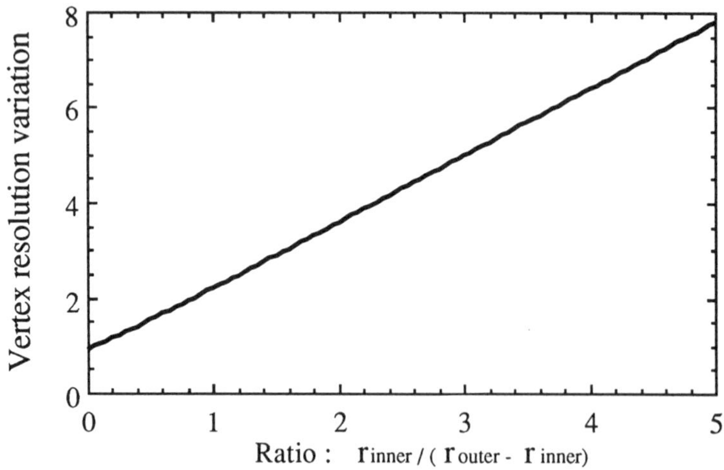


Fig. 4 Variation of the vertex resolution $\delta z$ ($g(r/\Delta r_t)$) versus the ratio inner shell radius over "tracker thickness".

## 4. Momentum resolution

If a solenoidal magnetic field is present in the inner cavity, then the tracker will provide sign and momentum of the interesting particles. Since these have high transverse momenta, their track curvature in the $(r,\varphi)$-plane will measure the transverse component of $\bar{p}$, while in the $(r,z)$-plane they continue to be straight tracks. This means that $p_t = 0.3B/c$, where B is the magnetic induction in Tesla and c is the curvature in $m^{-1}$, while the parallel component is simply $p_z = p_t/tg\vartheta$, where $\vartheta$ is the polar angle and is calculated from the track projected in the $(r,z)$-plane. The consequent momentum resolution will be[2]

$$\left(\frac{\delta p}{p}\right)^2 \approx \left\{\left(\frac{Gc}{N}\right)^{1/2} \times \left(\frac{\sigma_\varphi}{r_{out}^2}\right) \times \left(\frac{p\sin\vartheta}{0.3B}\right)\right\}^2_{curv.} + \left\{t^{1/2} \times \left(\frac{0.016\sin\vartheta}{0.3B\,r_{out}}\right)\right\}^2_{ms} +$$

$$+\left\{\left(\frac{Gsl}{N}\right)^{1/2} \times \left(\frac{\sigma_z\cos\vartheta}{r_{out}}\right)\right\}^2_{sl} + \left\{\left(\frac{t}{\sin\vartheta}\right)^{1/2} \times \left(\frac{0.014}{p\,tg\vartheta}\right)\right\}^2_{ms} \qquad (11)$$

$\sigma_{\varphi,z}[m]$ means respectively spatial precision in $\varphi$ and z, $r_{out}[m]$ the distance in the $(r,\varphi)$-plane between the beam axis and the outermost measured point (i.e., the radius of the outer shell), t the material layer in units of radiation lengths, N the number of hits along the track, $G_C$ a number, which varies between 256 for optimum hit spacing and 720 for continuous hit distribution along the track[f], and finally $G_{sl}$ which varies in the same way between 2 and 10.

An example of the advantage of a central tracker with solenoidal magnetic field for momentum resolution is shown in fig. 5. In fact, the measurement of muon momenta with only the muon chambers (which are mounted on the outest layers of the whole detector) is degraded by multiple scattering for momenta between 1 GeV/c to 1000 GeV/c and by the muon energy loss in the calorimeter and filter material for higher momenta. On the contrary to measure momenta with only the inner tracking worsens the resolution for momenta larger than 1000 GeV/c. Connecting both systems, a gain in momentum resolution over the whole range of momenta is clearly obtained.

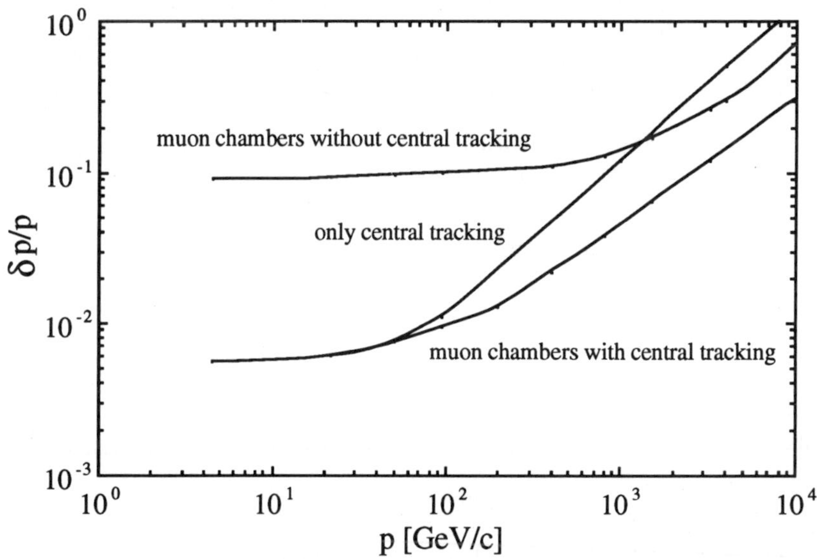


Fig. 5 Muon momentum resolution versus muon momentum, using only the muon chambers, the inner tracker or both.

## 5. Radiation dose with a solenoidal magnetic field

Apart from the advantages on momentum resolution and the possibility of determining the charge sign, a solenoidal magnetic field can be in addition useful in containing charged minimum bias particles within a region close to the beam pipe.Consequently most of the central tracker elements remain free from this unpleasant background and their radiation dose as well as their occupancy are strongly reduced.

---

[f] In fact, the brackets in relation (1) should read $(720/N+4)^{1/2}$ for this case.

The radiation dose $D_0$ without magnetic field amounts to

$$D_0(r) = \frac{C}{2\pi r^2} \tag{12}$$

where r is the distance of a point from the beam axis and the value of the constant C depends on the accelerator parameters (see fig. 3 in the contribution of H. Leutz).

However, in presence of a solenoidal magnetic field along the z-axis, the radiation dose will depend also on the coordinate z and on "boundary conditions", like calorimeter or magnet positions. This is due to low-$p_t$ tracks (mostly charged pions) which first curl in the tracking volume and then decay, leaving their products curling too. The net effect is an increase in the radiation dose for small distances and a strong decrease for large distances. In practice, the magnetic field acts like a filter for low $p_t$-particles, allowing only the high-pt particles to cross the tracking volume.

To have an analytical estimation of this effect, we perform a complete calculation, which is shown in the appendix and its result is

$$D(r;B,\eta_{max}) = D_0(r)\left[2 < n_c > \int_{x_r}^{x_c} f(x')dx' + \int_{x_c}^{\infty} f(x')dx'\right] \tag{13}$$

where $<n_c>$ is the average number of curls in the pseudorapidity range $0 < \eta < \eta_{max}$, $f(x) = x\, e^{-x}$ is the transverse momentum distribution of min. bias particles approximated from Monte Carlo data and $x = 2p_t/<p_t> = 0.3Br/<p_t>$ where $<p_t>$ is the average transverse momentum of the min. bias particles. $<p_t>$ (in GeV/c) depends logarithmically on the centre of mass energy $E_{cm}$ (in TeV) and can be approximated by the equation[3]

$$<p_t> \approx 0.44 + 0.071 \log E_{cm} \tag{14}$$

therefore the distribution f(x) will scale accordingly for different accelerators. To be more precise, D(r) represents the average radiation dose on a cylindrical surface of radius r, which extends between $\pm\eta_{max}$[g].

Eq. (13) is composed of two terms: the radiation dose without magn. field $D_0$ and a term in parenthesis, which shows:

– the magnetic field makes particles curl and the average number of curls is $<n_c>$;

– as the curling radii depend on $p_t$, only a fraction of particles will reach a certain distance r, and this is represented by the two integrals. In particular:

- particles curling between 0 and r do not contribute to the radiation dose, i.e. the integral starts at $x_r$;

[g] The reason we calculate an average is that for small pseudorapidities, as it is for the central region, the dose will not change much at different pseudorapidities.

- for those curling between r and $r_c$ (where $r_c$ indicates the position of the solenoid or of the calorimeter), the contribution is twice their avr. number of curls per pseudorapidity (at each curl a particle passes twice through the shell at radius r);
- finally, those who would reach $r_c$ or more contribute only once, since they are absorbed by the outer detectors.

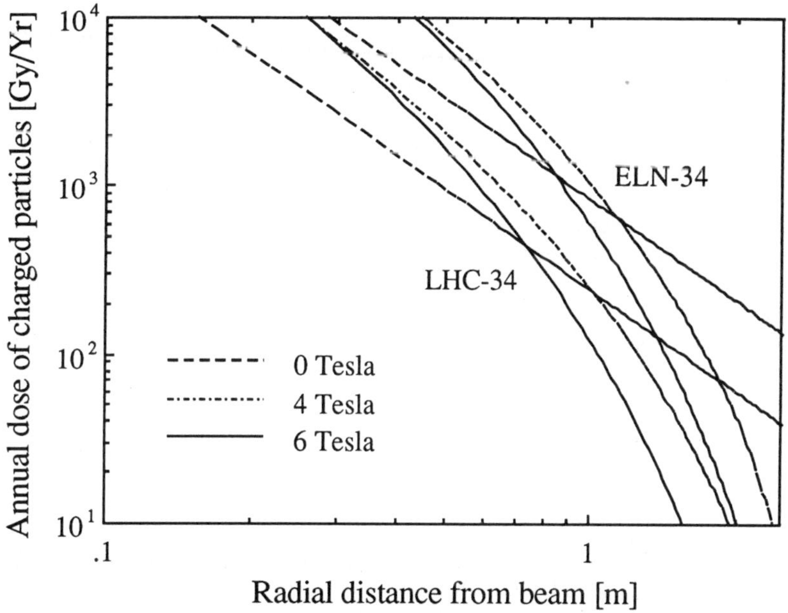


Fig. 6 Radiation dose in Gray per Year of operation. The curves for LHC-34 and ELN-34 at different magnetic fields are shown. The curves stop at r = 2.5 m, where the calorimeter or the solenoid are placed for convenience.

In fig. 6 this situation is shown for LHC and ELN-34 and for three different values of the magnetic induction. Up to a certain radius the radiation dose with magnetic field is higher than the one without magnetic field; for larger radii, however, the radiation dose falls rapidly to much smaller values than the no-B case: the region in which this transition takes place is called "cross-over region" (for LHC it is around 0.6 ÷ 0.8 m and for ELN 0.8 ÷ 1.2 m). Although only two cases are shown, the "cross-over region" always exists and it is important to establish its value for any tracker design.

## 6. Examples

We describe now two cases in which detector- and accelerator- parameters are optimized by using the relations of the previous paragraphs and then Eq. 7 is checked. These examples concern ELN-34 and ELN-35. The parameters of these two ELN versions are listed in the table of the recommendations.

- We choose the inner detector solenoidal magnetic induction, B = 6 Tesla (4 Tesla would not change much the results).
- Referring to present detector technologies, we consider the case of three concentric shells, each composed of four superlayers of scintillating fibres. In two superlayers, the fibres are stretched parallel to the collider axis and indicate the ($\varphi$, r) coordinates, in the other two, the fibres run at angles of a few degrees with respect to the parallel ones and provide via their corresponding stereo view the z coordinate of a particle track[2]. We choose[4] $\sigma_\varphi = 30\mu m$, $\sigma_z = 250\ \mu m$, N = 60. It has good z-precision and many hits (for a comparison between detector techniques, see table 1 of the contribution of T. Gys).
- Then we define the maximum radiation dose per year that the detector can stand: let us say $D_{max}$ = 10 KGy/year for 10 years, which means 100 KGy (or 10 MRad) total dose.
- The outer shell radius is fixed: $r_{out} = 2$ m.
- The minimum distance $r_{in}$ between beam axis and inner shell of the tracker is calculated using Eq. 13,

$$D_{max} = \frac{C}{2\pi r_{in}^2}\left[2 < n_c > \int_{x_{r_{in}}}^{x_c} f(x')dx' + \int_{x_c}^{\infty} f(x')dx'\right] \tag{15}$$

and the "tracker thickness" $\Delta r_t = 2 - r_{in}$.
- Finally, z-vertex precision and momentum resolution can be calculated from Eq. 8 and Eq. 11, and the inequality Eq. 7 is checked.
- If needed, $<n_v{}^*>$ has to be decreased by modifying beam crossing angle, beam radius and bunch length. The new accelerator parameters are shown in the table of the recommendations.

### *6.1 ELN-35*

From relation 15 we obtain for ELN-35 a minimum distance $r_{in} \approx 1$ m, which also means a distance between inner and outer tracker shell $\Delta r_t = 1$ m. Then, the vertex resolution (2.36 $\delta z$) can be calculated from Eq. 8 and is shown for our case versus the momentum in Fig. 7: below 100 GeV/c momenta, multiple scattering dominates. In Fig. 8, the vertex resolution is plotted versus the number of hits for three different momenta. For low momenta increasing the number of hits does not improve the resolution, while at high momenta there is a considerable improvement.

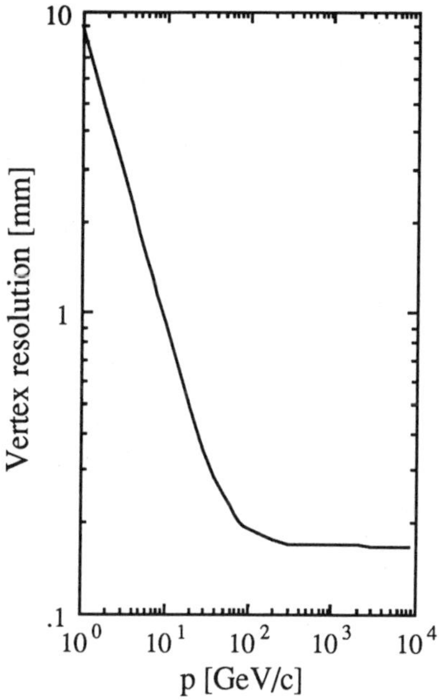


Fig. 7 Vertex resolution for scintillating fibres versus momentum (N = 60).

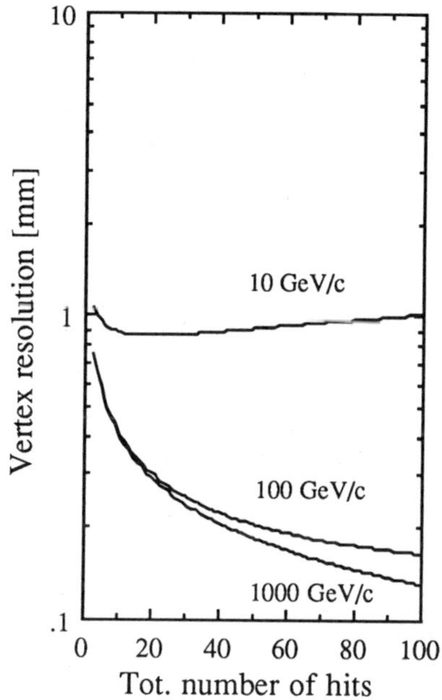


Fig. 8 Vertex resolution versus total number of hits for three different momenta.

From Figs 7 and 8 we learn that at 10 GeV/c the resolution is only 1 mm and from Table 2 that $<n_v{}^*>$ is 6.8 [$mm^{-1}$], which is far from satisfying Eq. 7 as shown in Table 4. Even for momenta higher than 100 GeV/c and reducing sensibly the beam crossing angle $\alpha$, Eq. 7 cannot be satisfyied (Table 4).

Table 4 Comparison of values obtained from left-hand side of Eq. 7 for ELN-35, for various momenta and beam crossing angles. Resolution of each of the interaction vertices would require values < 0.5.

| p [GeV/c] | $<n_v{}^*>$ 2.36 $\delta z$ | | |
|---|---|---|---|
| | $\alpha$ = 150 μrad | $\alpha$ = 100 μrad | $\alpha$ = 50 μrad |
| 10 | 6.84 | 4.84 | 3.06 |
| 100 | 1.37 | 0.97 | 0.61 |
| 1000 | 1.16 | 0.82 | 0.52 |

### *6.2 ELN-34*

For ELN-34, the radiation dose is everywhere a factor 10 smaller: as a consequence the minimum distance $r_{in}$ is reduced to 0.5 m and the tracker thickness $\Delta r_t$ becomes 1.5 m. The vertex precision from Eq. 8 for scintillating fibres is now

$$\delta z^2 \approx \frac{(1.4\sigma_z)^2}{N} + \frac{(1+0.01N)}{p^2} 0.5^2 10 \ \text{[mm]} \tag{16}$$

For 10 GeV/c the vertex resolution is 0.48 mm, but at 100 GeV/c is already 0.12 mm and stays constant for higher momenta.

Table 5 Comparison of values obtained from left-hand side of Eq. 7 for ELN-34, for various momenta and beam crossing angles. Resolution of each of the interaction vertices would require values < 0.5.

| p [GeV/c] | $<n_v{}^*>$ 2.36 $\delta z$ | | |
|---|---|---|---|
| | $\alpha$ = 250 μrad | $\alpha$ = 100 μrad | $\alpha$ = 50 μrad |
| 10 | 0.87 | 0.37 | 0.21 |
| 100 | 0.22 | 0.09 | 0.05 |
| 1000 | 0.20 | 0.08 | 0.05 |

According to Table 5, we learn that all values satisfy the condition of relation 7 with the exception of the 10 GeV/c case at 250 μrad crossing angle.

From last two examples, it is clear that one of the main task of a central tracking system will be to distinguish and to measure the position of each of the vertices produced in one bunch crossing. In particular, for ELN-34 it is possible to find values of the different parameters (accelerator and detector) which consent Eq. 7 to be satisfied (Table 5), while for ELN-35 this is not possible (Table 4), even using the mostly advanced detector technologies. Although progress may be made in this field, it is pointed out in the Detector Group Recommendations that an increase of center of mass energy would be much more remunerative in term of physics discoveries than an increase of luminosity (as for ELN-34 and ELN-35).

## 7. Conclusions

For future supercolliders, accelerator experts and detector designers will have to collaborate very closely. As this workshop was aimed at improving such a collaboration, I have shown one aspect of the interconnection between accelerator– and detector– design related to the central tracking of particles.

In particular, one of the toughest issues is represented by the multivertices production per beam crossing and by separating and detecting them along the beam axis via the central tracker. They will be distributed over the crossing region according to the equations developed in Sect.2. We have shown with simple analytical relations how the resolution of this measurement (which depend totally on detector parameters) can be related to the basic beam parameters in the crossing region and, through an inequality (Eq. 7), we described how to optimize vertex resolution and beam crossing angle.

The advantages of a solenoidal magnetic field have been pointed out. Together with the possibility of determining the charge sign and of measuring the momentum of charged particles, it improves the momentum resolution for muons if we link the central tracking to the muon chambers outside the calorimeter. Over a certain distance from the beam axis (cross-over region), the solenoidal field reduces drastically the radiation dose. This results in a looser constraint on the radiation hardness of the detector components.

Finally, two examples are given (ELN-34 and ELN-35) in which it is shown that, in order to satisfy Eq. 7, the full beam crossing angle of the colliders in question needs to be changed.

## Acknowledgments

The author thanks Prof. A. Zichichi for his continuous support and encouragement and acknowledges the warm hospitality at the E. Maiorana Centre in Erice. He wishes to thank Dr H. Leutz for the useful discussions and the careful reading of the manuscript.

## Appendix. Tracks curling in a solenoidal field

Let us suppose solenoidal field lines parallel to the collider axis, and that a charged particle is produced at z = 0 (fig...). The number of turns $n_t$ within the coverage $r \cot\vartheta_{max}$ of the tracker becomes for this particle:

$$n_t = \frac{1}{\Delta z} r \cot\vartheta_{max} \tag{1A}$$

with r being the maximum distance of the curling particle from the collider axis, corresponding at the same time to the helix diameter and to a tracker shell position. $\Delta z$ indicates the increment of particle displacement along the collider (z) axis after each helix turn:

$$\Delta z = v_z T = \pi d \cot\vartheta \tag{2A}$$

where T means the time needed for a particle starting with velocity v ($v_z$ is its z projection) and polar angle $\vartheta$ to accomplish one revolution. Combining relations (3) and (4) and replacing $\cot\vartheta$ by $\sinh\eta$, where $\eta$ is the pseudorapidity, yields:

$$n_t = \frac{\sinh\eta_{max}}{\pi \sinh\eta} \tag{3A}$$

Eq. 3A holds in a general way for particles without restriction on their curling diameters, which are related to the magnetic flux density B and to the transverse particle momentum $p_t$:

$$r = \frac{p_t}{0.15\,B} \tag{4A}$$

According to Eq. 3A, the number of helix turns would become infinite, if the particle is emitted at zero rapidity ($\vartheta$ = 90°). This discontinuity is avoided by decays or interactions of these particles. The number of turns $n_d$ within the decay length $l_d$ of an unstable particle becomes $n_d = l_d/s$ where s means the length of one helix turn:

$$s = \frac{\pi r}{\sin\vartheta} = \frac{\pi p_t}{0.15\,B \sin\vartheta} \tag{5A}$$

if we assume again, that the particle is emitted at z = 0 with polar angle $\vartheta$. The decay length can be written:

$$l_d = \frac{p_t c \tau}{m_0 \sin\vartheta} \tag{6A}$$

with $\tau$ meaning the particle lifetime, $m_0$ its rest mass and c the velocity of light. Combining Eqs 5A and 6A, we obtain for the number of helix turns within a decay length:

$$n_d = \frac{0.15\,B\,c\,\tau}{\pi\,m_0} \tag{7A}$$

Now, if the ratio of Eq. 3A and Eq. 7A becomes unity, we obtain the polar angle $\vartheta_0$ above which the curling unstable particles decay inside the tracker:

$$\cot\vartheta_0 = \sinh\eta_0 = \frac{m_0}{c\tau}\frac{\sinh\eta_{max}}{0.15\,B} \tag{8A}$$

For charged pions ($m_0 = 0.14$ GeV/$c^2$; $c\tau$-length = 7.8 m), which constitute the vast majority of particles in minimum bias events, we obtain polar angles $\vartheta_0$ of 77° or 88° for a rapidity coverage of $\eta_{max} = 1.4$ and for magnetic flux densities of 1 T and 6 T respectively. Pions starting with larger polar angles decay within the tracker after 2.7 (at 1 T) or 16 (at 6 T) turns. For the few longer lived or stable particles the decay length can be replaced by the interaction length.

Finally, to obtain the average number of curls per particle, we have to integrate the number of curls over the whole pseudorapidity coverage, weighting with the relative number of charged particle per unit of pseudorapidity per interaction H:

$$< n_c >= \frac{\int_0^{\eta_0} n_d\,H\,d\eta + \int_{\eta_0}^{\eta_{max}} n_t\,H\,d\eta}{\int_0^{\eta_{max}} H\,d\eta} \tag{9A}$$

where $\eta_0$ is defined in Eq. 8A. The integrals can be solved analitically, giving as result

$$< n_c >= \frac{1}{\pi}\frac{\sinh\eta_{max}}{\eta_{max}}\left\{\frac{\eta_0}{\sinh\eta_0} + \ln\left[\frac{\tanh\left(\frac{\eta_{max}}{2}\right)}{\tanh\left(\frac{\eta_0}{2}\right)}\right]\right\} \tag{10A}$$

Eq. 10A represents the average number of curls for a particle regardless of its helix diameter. However for each minimum bias event the produced particles will have a certain distribution in transverse momentum and only those with a $p_t$ bigger than 0.15 B $r_{in}$ will cross the tracker shells. Moreover, particles with $p_t$ bigger than 0.15 B $r_c$, where $r_c$ is the position of the calorimeter or of the solenoid, will be absorbed outside the tracker without curling and only those having a $p_t$ in between these two values will really curl in the tracking volume.

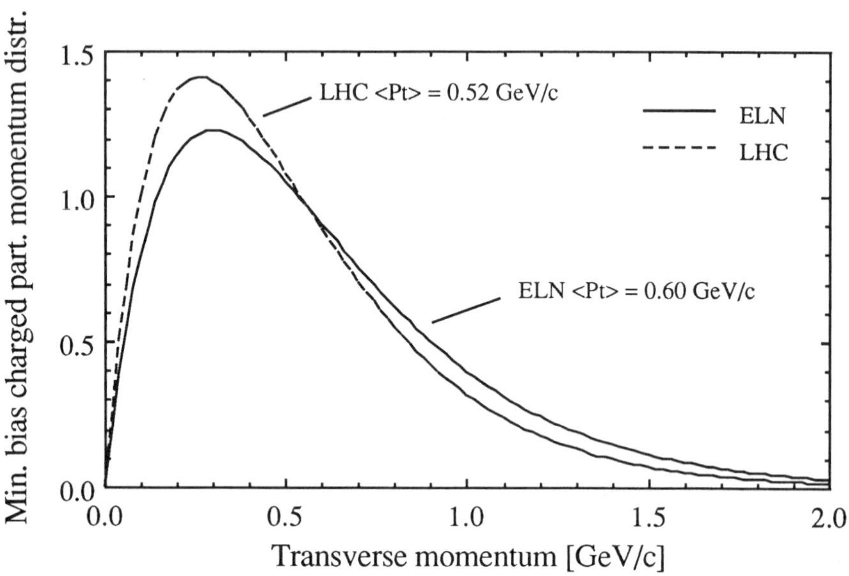


Fig. 1A Transverse momentum distribution of min. bias particles versus transverse momentum for LHC (broken line) and ELN (full line) as calculated from Eqs 11A, 12A and 14. The respective average transverse momenta are also shown. To obtain the positions of the distribution maxima just divide $<p_t>$ by two.

The $p_t$-distribution can be well approximated by the function f(x)

$$f(x) = x e^{-x}, \quad \text{and} \quad \int_0^{\infty} f(x)\,dx = 1 \tag{11A}$$

where

$$x = \frac{2 p_t}{< p_t >} = \frac{0.3 B r}{< p_t >} \tag{12A}$$

and $<p_t>$ is the average transverse momentum of the min. bias particles (fig. 1A). Then, the average number of curls through a tracker shell at a distance r from the beam axis and covering from 0 to $\eta_{max}$ of pseudorapidity is

$$(< n_v > H \eta_{max}) < n_c > \int_{x_r}^{x_c} f(x')dx' \tag{13A}$$

where the parenthesis represents the number of particle generated in the $\eta_{max}$ tracker coverage.

Finally, the total number of crossings per unit of shell surface per year of operation, multiplied by ΔE gives the radiation dose per year,

$$D(r;B,\eta_{max}) = \frac{C}{2\pi r^2}\left\{2 < n_c > \int_{x_r}^{x_c} f(x')dx' + \int_{x_c}^{\infty} f(x')dx'\right\} =$$

$$= \frac{C}{2\pi r^2}\left\{2 < n_c > \left[(1+x_r)e^{-x_r} - (1+x_c)e^{-x_c}\right] + (1+x_c)e^{-x_c}\right\} \qquad (14A)$$

where $$C = (3.2\ 10^{-3} < n_v > H\ t_b^{-1}) .$$

## References


1 C. D'Ambrosio and H. Leutz, in *Perspective for new detectors in future supercolliders*, eds L. Cifarelli, R. Wigmans and T. Ypsilantis (World Scientific Publishing Co., Singapore, 1991) 18.

2 C. D'Ambrosio, U. Gensch, T. Gys, H. Leutz and S. Schlenstedt, *Nuclear Instruments and Methods* **A322** (1992) 20.

3 D. Denegri, in *Large Hadron Collider Workshop* , CERN 90–10 **Vol.1** (1990) 56.

4 C. D'Ambrosio, H. Leutz, T. Shimizu and O. Shinji, *Nuclear Instruments and Methods* **A325** (1993) 161.

# READ-OUT AND DATA ACQUISITION FOR TRACKING DEVICES AT FUTURE COLLIDERS*

Thierry Gys
*CERN, CH-1211 Geneva 23, Switzerland*

This paper discusses some aspects of readout and data acquisition of three tracking detector candidates – semi-conductor microstrips, microstrip gas counters and scintillating fibres – for use in experiments at future high-luminosity and high-energy hadron colliders, in paricular Eloisatron, LHC or SSC. From the limits of basic detector parameters (e.g. time response and space resolution) and readout architectures, recommendations are made for the design values of the luminosity, bunch crossing frequency and energy of such colliders.

## 1. Introduction

As physicists seek very rare events with production cross-sections in and below the nanobarn range (see contribution of H. Leutz, these proceedings), they are faced with the task of building colliders of ever higher particle energy and increased luminosity. For this purpose, different hadron collider designs are developed such as Eloisatron[1] (ELN) operating at 100 TeV per beam and at a luminosity of $10^{34}$ $cm^{-2}$ $s^{-1}$, LHC[2] (7.7 TeV per beam at $10^{34}$ $cm^{-2}s^{-1}$ luminosity) and SSC (20 TeV per beam at $10^{33}$ $cm^{-2}s^{-1}$ luminosity). In parallel, intensive R&D work on detectors is carried out, initiated in 1987 by the LAA Project[3]. Indeed, it is crucial to develop detector techniques able to fulfil the requirements of the above colliders. In particular, tracking detectors must be very fast and very precise in terms of space resolution. Their readout scheme must be designed according to the collider bunch collision frequency and the high multiplicity events resulting from the collisions of very energetic protons at a high luminosity. Candidates meeting these demands are semi-conductor microstrips, microstrip gas chambers (MSGC) and small diameter scintillating fibres. The purpose of this paper is to evoke some aspects of readout and data acquisition of these detectors, and, from their relevant characteristics, to set limits on important design parameters of future hadron colliders. In particular, attention will be paid on the bunch separation time and the luminosity.

Tracking detectors will have to sustain large radiation levels due to the high flux of high-energy particles from the proton-proton interactions and to low energy neutrons from cascades and showers in the surrounding calorimeter. Consequently, these detectors will have to be designed in a radiation-hard technology. This important problem will not be developed in the present paper, since it already forms the object of other contributions in these proceedings (see C. D'Ambrosio, D. Groom, H. Schoenbacher and F. Wulf).

---

* The work reported here is part of the LAA Project.

This paper is organized as follows : after a brief description of the principle, time response and space resolution of the detectors, a multi-level trigger system is introduced. The detector readout schemes are then detailed, with a particular focus on the different approach between semi-conductor strips and MSGCs on one side, and scintillating fibres on the other side. A discussion compares the advantages and drawbacks of the various techniques, and upper limits on related collider design parameters (particularly their luminosity and bunch crossing rate) are finally given.

## 2. Tracking detectors, their response time and their space resolution

There exist mainly three different media for particle tracking : semi-conductors, gases and scintillators[4]. They all provide signals or hits along the particle path. With subdivisions into adequate detector elements – narrow-spaced wires or strips for gases; strips, pixels or pads for semi-conductors; fibres or capillaries for scintillators – these hits indicate space coordinates. In this paper, we will refer to detector types which present time and space resolutions adequate for future hadron colliders. Consequently, we will limit our discussion to semi-conductor microstrips, microstrip gas counters and small diameter scintillating fibres.

### *2.1. Semi-conductor detectors*

Among semi-conductor detectors, single-sided silicon strips are the most widely used ones[5,6]. Their structure is represented in fig. 1 :

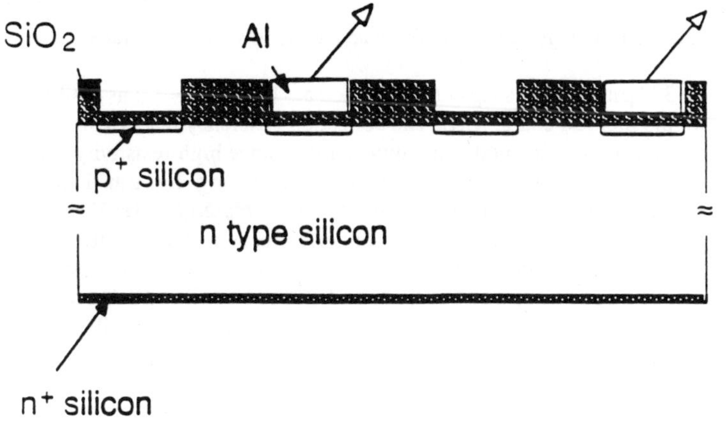


Fig. 1 Cross-section of a semi-conductor strip detector (from ref. 6).

One side of an n-type, high resistivity (5 to 10 kΩ cm) silicon wafer is implanted with highly-doped $p^+$ strips that form diodes. A layer of silicon dioxide, a fraction of a micron thick, is grown on top of the diodes, forming the coupling capacitor to the readout lines coated with aluminum. Those metal lines are connected to the electronics. The other side of the wafer is highly doped with $n^+$ implants to assure good ohmic contact. This back contact is earthed while the biasing voltage is applied to the diodes through individual polysilicon resistors (typically 1MΩ). The voltage (-100V) is negative such that the pn junction is reversed bias, resulting in a full depletion of the bulk silicon. The strips are typically 10 μm wide and 100 mm long. The strip pitch ranges from 25 to 50 μm, in order to match the diffusion width of charge carriers. In most cases, only one strip every 2, 3 or 4 is read out.

When a minimum ionizing particle crosses such a detector, electron-hole pairs are generated homogeneously along its trajectory (at a rate of 80 pairs per micrometer for silicon). The charge collection time is a determining factor for the semi-conductor detector speed. In a simple model[7], and assuming a highly purified material, the induced charge Q(t) at the readout is equal to :

$$Q(t) = \frac{1}{2} q \frac{dE}{dx} \frac{w}{\varepsilon} \left[ \left( 1 - \exp\left(-\frac{t}{\tau_e}\right) \right) + \left( 1 - \exp\left(-\frac{t}{\tau_h}\right) \right) \right]$$

In this equation, q is the electron charge, dE/dx is the ionization loss in the semi-conductor, w is the detector thickness, $\varepsilon$ is the mean energy required to create one electron-hole pair. $\tau_{e(h)}$ is the electron (hole) transit time constant given by :

$$\tau_{e(h)} = \frac{w^2}{2\, \mu_{e(h)}\, V}$$

with $\mu_{e(h)}$ the electron (hole) mobility and V the reverse bias voltage.

For a 300 μm thick Si strip detector biased at -100 V, $\tau_e \approx 3.0$ ns while $\tau_h \approx 9.4$ ns. Fig. 2 shows the induced charge Q(t). The collection efficiency reaches 90 % after 15 ns. The values obtained experimentally are quite similar, since high-resistivity silicon wafers can be manufactured. GaAs is in theory much faster : for a 180 μm thick detector (a thickness creating as much signal as 300 μm of silicon) biased at -100 V, $\tau_e \approx 0.2$ ns and $\tau_h \approx 4.1$ ns (fig. 3). In practice, however, imperfections in the crystal lattice produce traps for charge carriers, which decrease the charge pulse amplitude compared to that obtained from the same amount of ionization deposited in silicon[8]. On the other hand, GaAs presents the advantage of being more radiation hard than existing silicon detectors.

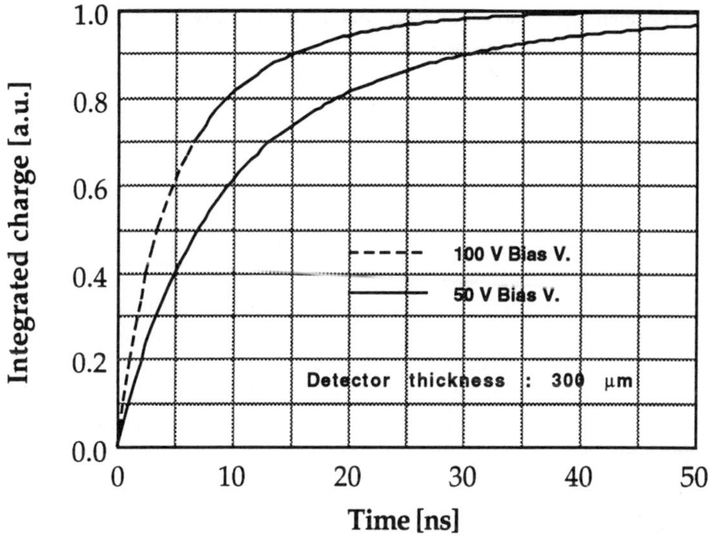


Fig. 2 Theoretical induced charge Q(t) as a function of time for a Si strip detector.

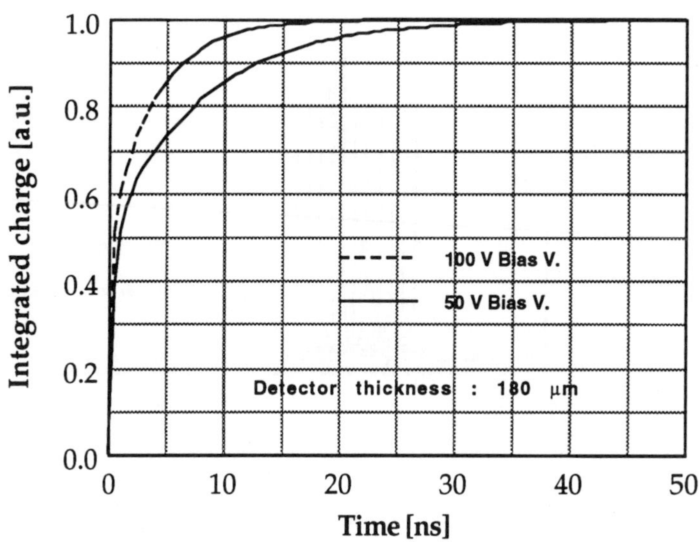


Fig. 3 Theoretical induced charge Q(t) as a function of time for a GaAs strip detector.

The spatial precision σ of semi-conductor detectors is directly related to the pitch p *of the readout strips* by the usual relation :

$$\sigma = \frac{p}{\sqrt{12}}$$

Similarly, the two-track resolution is equal to twice the readout pitch.

*2.2. Micro-strip gas counters*

The Micro-Strip Gas Counter (MSGC) detector is based on the idea of reproducing intensity and structure of the electric field typical of a multi-wire proportional chamber, but to a reduced geometrical scale[9–11]. The MSGC consists essentially of a sequence of alternating thin conductive anode and cathode strips placed on a 0.5 mm thick insulating support (fig. 4). A drift electrode defines a region of charge collection (about 2 mm thick) and application of appropriate potentials on anodes and cathodes creates a proportional gas multiplication field. Primary electrons, created in the drift region by the incoming ionizing radiation, drift toward the anode and are multiplied in the amplification region around the anode strips. Typically, the anode strips have a pitch of 200 μm, are 5 μm wide while the cathodes are 50 μm wide.

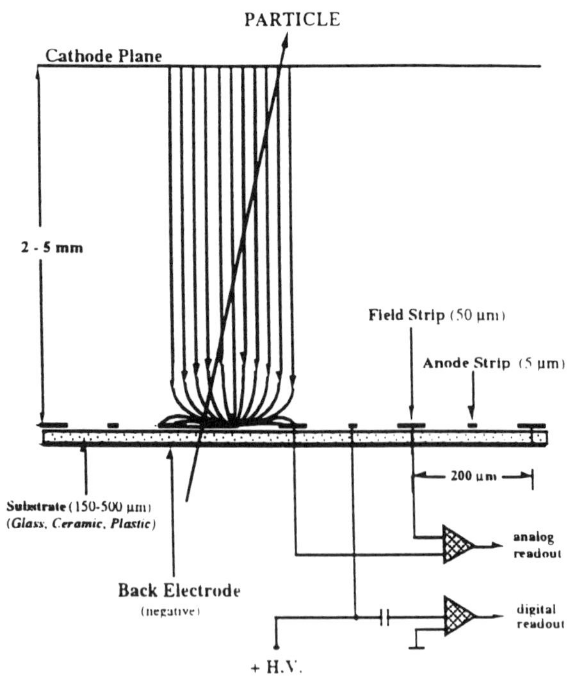


Fig. 4 Cross-section of a Micro Strip Gas Counter detector.

The drift gas gap dimension of the MSGC determines important characteristics of the detector. With a thick gap, the induced charge signal is larger and the gain fluctuations are reduced. However, a larger transversal diffusion occurs and the final electron cloud is broader, inducing a degradation of the space resolution. Similarly, the drift time of the electrons is longer and the overall detector time response is affected. In parallel, the choice of the gas mixture is crucial. A promising candidate is DME/$CO_2$ 60/40. It features a high primary ionization density of 46.6 $cm^{-1}$ and a low transversal diffusion coefficient of 50-60 μm/√mm. The electron drift velocity reaches 70 μm/ns under operating fields close to 10 kV/cm[12].

To achieve the required electron multiplication, the gas gap should be about 2 mm thick. This reflects on the electronic pulse duration via the drift time of the electrons within this gas gap. Therefore, the pulse duration amounts to between 30 and 50 ns, which means two and three bunch crossings of the collider, if we assume bunch separation times of 15 to 20 ns. The ionization cluster number is around 10, which is sufficient for a good detection efficiency. The space resolution depends on whether the readout is binary or analog, and is applied on the cathode or the anode strips. For comparison with the other tracking media, the case of binary readout of the cathode strips will be considered. The spatial precision $\sigma$ and the two-track resolution are then given by relations similar to the ones used for semiconductor microstrips.

### *2.3. Scintillating fibres*

A scintillating fibre[13] consists of a core (in most cases polystyrene) doped with a scintillator, and surrounded by a cladding with a lower refractive index. Ionization losses of a charged particle crossing the fibre induce scintillation light. The fibre traps a small fraction (≈ 10 %) of this light and guides it, via many subsequent total reflections at the core-cladding interface, to the fibre end. There, it is in most cases detected by the photocathode of an image intensifier. For central tracking, realistic fibre diameters range from 30 μm to 100 μm; they represent a good compromise between the required spatial and 2-track resolutions (asking for small diameters), and the reflection losses of light, increasing linearly with decreasing diameters. Such thin fibres must be produced as fused coherent bundles with edge dimensions between 1 and 3 mm. The detected photon, or "hit", density along the particle path ranges typically between 1 and 10 per mm of fibre thickness, depending on the position of the crossing point of the particle with respect to the readout.

The measured time distribution P(t) of a 1 mm fibre (PS core doped with 1 % wt PMP) is displayed in fig. 5. It results from the convolution of the fibre emission time distribution E(t) and the measuring system resolution function R(t)[14] :

$$P(t) = \int E(t') \, R(t-t') \, dt'$$

The resolution function may be approximated by a Gaussian distribution with standard deviation $\sigma_t$. The emission time distribution can be described by a fast ($\tau_{fast}$) and a slow ($\tau_{slow}$) decay time process of the form :

$$E(t) = \frac{\exp(-\frac{t}{\tau_{fast}}) + R\exp(-\frac{t}{\tau_{slow}})}{1 + R}$$

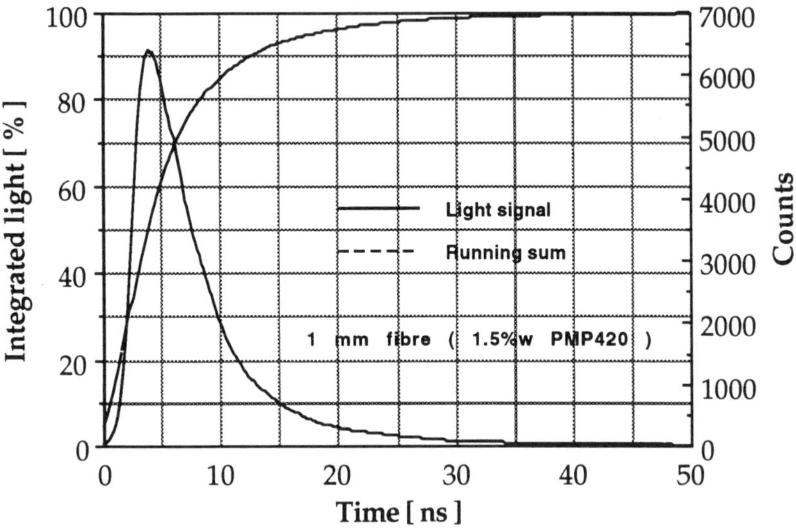


Fig. 5 Measured light output as a function of time for a small diameter scintillating fibre. The running sum of the light signal is also displayed.

In the case of fig. 5, $\sigma_t$ is equal to 0.9 ns, $\tau_{fast}$ amounts to 3.2 ns and $\tau_{slow}$ to 10 ns, while the ratio R of the slow to the fast components is 0.4. Taking away the time response of the system, one deduces that after 12.5 ns, 95 % of the light have been emitted.

The intrinsic optical properties of 60 μm hexagonal fibres have been studied[15] and yield a spatial precision (sigma) of 20 μm and a two-track resolution of 50 μm. These figures are slightly degraded by the optical resolution of the readout system, consisting mainly of an electromagnetically-focussed image intensifier (see paragraph 5).

### 3. Principle of a multi-level trigger system

Following the experience of previous and current experiments (e.g. at LEP or HERA), a trigger system based on several levels is envisaged for the data acquisition of experiments at future colliders[16]. The principle of such a system is that crude but fast decisions at a first level can be refined at subsequent levels using more detailed information and more complicated processing algorithms. These data reduction levels are needed to ultimately meet the data storage and off-line analysis capabilities (fig. 6).

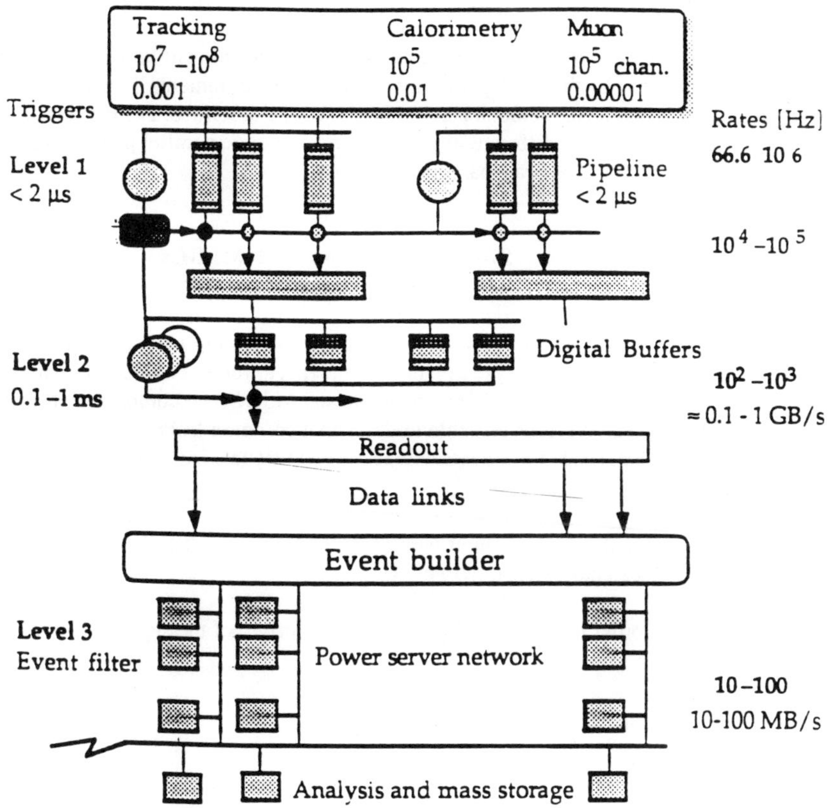


Fig. 6 Principle of a multi-level trigger system (from ref. 16).

Prompt triggers will be based on signatures of electrons, muons, jets and missing transverse energy, the basic selection parameter being a threshold on the transverse momentum (for electrons and muons) or energy (for jets and missing energy). It is commonly agreed that the first-level trigger (for LHC) will take about 1 to 2 μs to provide its decision. However, a bunch crossing occurs every 15 to 25 ns. Consequently, the information must be stored *for each channel* during the first-level processing : this will be done by storage devices called pipelines. These analogue or digital memories must be able to accept new information *every bunch crossing*, and store it until the first-level decision is available. After, the information is either discarded or transferred to the second-level acquisition system. The delicate interface between the first-level pipeline and the second-level buffer must preferably be dead-time free. In the particular case of tracking detectors where the occupancy of each cell is very low, this interface should include a zero suppression facility, in order to compress the data before further processing.

With no doubt, the most challenging part for data acquisition at future colliders will be the front-end area. In the next paragraphs, current developments for the front-end electronics of semi-conductor detectors and MSGCs will be described. For the scintillating fibre front-end, a completely different approach using analogue parallel processing of the data will be the subject of a separate paragraph.

## 4. Front-end electronics for Si-GaAs detectors and MSGCs

Substantial R&D work has already been done to develop new architectures and high-speed, high-density readout systems for experiments at future colliders. In particular, since the charge signals of semi-conductor detectors and MSGCs are comparable and of the order of 2 $10^4$ electrons, it seems possible to develop a common front-end electronics for both types of detectors. The requirements to read out such signals are rather demanding and are as follows :

- low noise (typically an Equivalent Noise Charge ENC less than 2 $10^3$ electrons), in order to reach a signal-to noise ratio of 10.
- high speed, in order to be able to localize the time origin of the signal within a single bunch crossing.
- minimum pile-up.
- minimum dead-time.
- analogue or digital read-out, depending on the spatial precision needed.
- data availability for level 2 trigger processing.
- minimum power consumption (of the order of 1 to 2 mW per channel).

Among the above requirements, the high speed is contradictory with low power consumption and low noise, and a compromise is necessary. Of course, it is essential to identify the time origin of a signal with the precision of a single beam crossing interval. In this context, the RD-20 Collaboration has developed an original readout scheme[17] based on a charge sensitive preamplifier and shaper, providing low-noise amplification with sufficiently low power, an analogue delay and buffer followed by an analogue signal processor in order to recover the initial signal pulse from the shaped amplifier pulse (fig. 7). This system is described below.

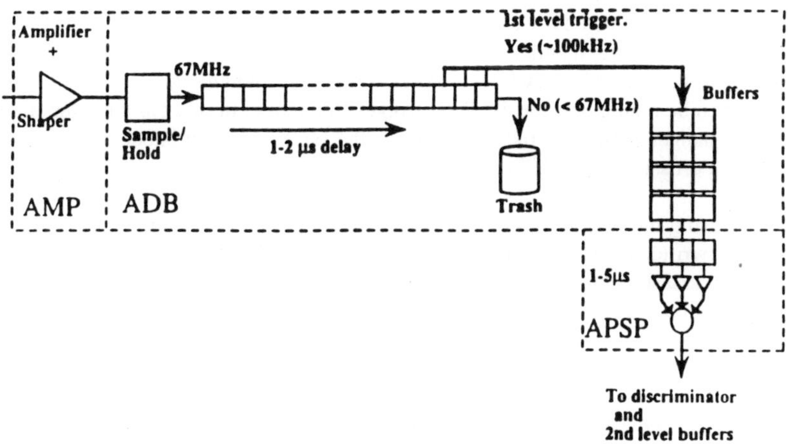


Fig. 7 Schematic diagram of the front–end electronics developed for the readout of silicon microstrips and MSGCs (from ref. 19).

### *4.1. Signal preamplification and shaping*

The preamplifier is a conventional charge-sensitive amplifier, and the shaper is a CR-RC filter. To achieve low power, the shaping amplifier uses a long shaping time of 45 ns. Clearly, the choice of a peaking time that produces a pulse shape extending across several beam-crossing intervals increases the detector occupancy, and thus the risks of pile-up. Consequently, this shaping technique only applies for tracking detectors with very low intrinsic occupancies (i.e. fast response and small detector cell size).

For a first prototype, the noise performance has been measured[17] and is equal to :

$$ENC = (400 + (65 \times C_d))\ \text{electrons}$$

where $C_d$ is the detector capacitance in pF. The power consumption is 1.6 mW. New prototypes are in progress and their power consumption is expected to be below 1 mW.

### *4.2. Analogue delay and buffer (ADB)*

The output of the shaper is sampled at the bunch crossing frequency. Each sample is stored in an ADB cell during the 1st level trigger processing time. Each cell consists of a capacitor, a write switch and a read switch[18]. While the write switch register enables one capacitor to be charged every bunch crossing, the read switch register enables a *different* capacitor to be read out at the 1st level trigger frequency. In this pipeline, the number of cells per channel is given by the ratio between the 1st level trigger latency and the sampling (i.e. bunch crossing) period.

During the period in which no 1st level trigger occurs, samples are discarded by being overwritten by new samples. Every time a 1st level trigger occurs, three samples are transferred to a signal processor. The average time between each positive 1st level trigger is expected to be ≈ 10 μs. Consequently, if the subsequent processing time is shorter, the system is deadtime free, providing additional buffers in the ADB can store data from 1st level triggers close in time.

### *4.3. Analogue pulse shape processor (APSP)*

The APSP[19] is simply the inverse filter of the CR-RC shaper. The original signal pulse can be recovered by performing the deconvolution of the shaped signal. It can be shown that for the CR-RC filter in question, the impulse response of the deconvolution filter is finite and uses only three consecutive samples in time : by forming a weighted sum of the three samples which have been selected from the ADB, a new pulse with a much reduced tail is obtained. The time identification of this signal is then possible.

Because the APSP only operates on data selected by the first level trigger, the filter operation consumes very little power (< 1 mW). A prototype has been tested and performed satisfactorily the required functions of the filter.

## 5. Optoelectronic readout of scintillating fibres

### *5.1. Optoelectronic delay*

Within the framework of the LAA Project and the RD7 Collaboration, a novel optoelectronic device has been developed for the readout of small diameter scintillating fibres[20]. It is able to delay in a controlled way optical images produced by a fibre tracker, and to select and amplify a small fraction of them. The proposed delay system is a 600 or 900 mm long (for a 1 or 2 μs delay), electromagnetically focussed image intensifier. The tube is divided in 6 sections separated by 5 grids labelled G1 to G5 (fig. 8).

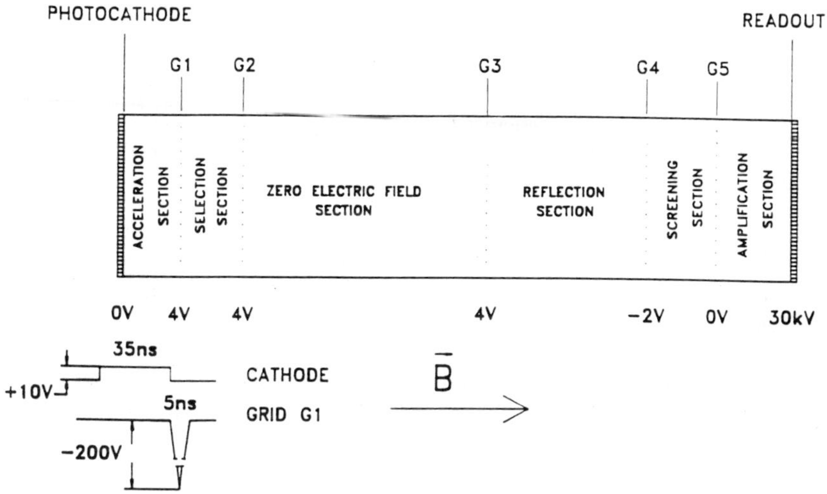


Fig. 8 Basic scheme of the optoelectronic delay tube.

The readout requirements are fulfilled in the following way :

- track points coming from the scintillating fibres are detected by a photocathode and converted into photoelectrons.

- the delay is achieved by *to and fro* drifts of very low-energy photoelectrons in the first four sections of the tube.

- the space resolution of the track points is ensured by the detector magnetic field, parallel to the photoelectron trajectories.

- the required time resolution is obtained by compensating the inevitable velocity spread of the photoelectrons with different drift lengths.

- the gating is realized by blocking the photocathode before applying a 5 ns negative pulse on grid G1.

- the selected track points are accelerated (at 30 kV) before hitting a silicon pixel array.

– a fast track reconstruction is provided by random access readout of the hit pixels (see next paragraph).

Two scale model prototypes (300 mm long) have already been tested[21]. They include all the elements schematically depicted in fig. 8, except the readout, which is not yet a pixel array as finally foreseen, but a fast phosphor screen. Experimental results show encouraging performances in terms of optical signal delay, selection and amplification capabilities. The time resolution has been investigated in detail. It reaches 7 ns for a 400 ns delay. The space resolution has been measured in two ways : with an optical test pattern, it amounts to 25 μm in a 0.7 T magnetic flux density. With cosmic rays detected by a scintillating fibre target and a hodoscope, it is 110 μm in a 0.03 T magnetic flux density. This value will inevitably improve at higher magnetic fields.

*5.2. Readout scenario*

The choice of the delay tube solution for an analog pipeline has the following advantages :

– it is an imaging device by itself, and can perform parallel processing of up to $4\ 10^5$ microfibres. As a consequence, it yields a low cost per channel.

– it preserves both excellent time resolution (within the bunch crossing period) and space resolution (within the microfibre diameter).

– it has a low power consumption (less than 1 W per tube, i.e. less than 10 μW per fibre) and thus minimizes heat production.

The readout is a silicon pixel device directly bombarded with the photoelectrons of the images selected in average every 10 μs by the 1st level trigger. Since the mean energy to create an electron-hole pair in silicon is 3.6 eV, and with an acceleration voltage of 30 kV, the expected pixel signal is $8.3\ 10^3$ electrons. This asks for a very low noise electronics similar to the one developed by the RD-19 Collaboration[22].

*5.3. Data compression*

Fibre tracking detectors have an occupancy of the order of 0.1 % which leads to read out about 10 hits per square centimeter of silicon. Clearly, the use of a fast zero-suppression technique is needed in order to compress the data to be sent to the second level trigger processor. One method is to transfer the addresses and eventually the analog pulseheight values of those detector channels which fulfill given constraints (such as exceeding a certain threshold). A solution to this problem has been proposed by the Collège de France as part of the RD19 Collaboration developping silicon micropattern detectors[21]. The zero suppression circuit, called Sparse Data Scan[23], realizes a selective readout of the hit pixels, and solves the ambiguity problem of the vertical and horizontal projections. Each hit pixel is used as a position memory and activates two switches connected to it : the first one indicates that there is at least one pixel hit on the line. The second switch establishes a logical connection between line and column, during the scanning of each line. Lines and columns are thus scanned and encoded *if and only if they are hit*, according to their priority. All empty lines and columns are skipped. If needed, the readout of the analog information can be added. A 16 × 8 pixels prototype has been realized. The chip supplies an external bus with the 7 bits addresses of the hit pixels at a maximum clock rate of 10 MHz.

## 6. Discussion

Table 1 summarizes the important properties of semi-conductor, gas and scintillating fibre tracking detectors.

**Table 1** Comparison of relevant parameters between different tracking detector media. $X_0$ means radiation length and H the material layer required to achieve 1 hit.

| Parameters | | Si strips | MSGCs | Fibres |
|---|---|---|---|---|
| Number of hits | $(0.1\ X_0)^{-1}$ | 15–30[(a)] | ~10-15[(b)] | 60-200[(c)] |
| Cell busy time | ns | < 15 | ~50 | < 15 |
| Intrinsic spatial precision ($\sigma$) | μm | 14[(d)] | 58[(e)] | 20[(f)] |
| 2-track resolution | μm | 100 | 400 | 50 |
| Occupancy ($10^{34}cm^{-2}s^{-1}$ luminosity) | % | < 1 | 2-3 | < 1 |
| Number of channels | $m^{-2}\ H^{-1}$ | $2{\times}10^5$[(g)] | $5{\times}10^4$[(h)] | $8{\times}10^3$[(i)] |
| Heat production | $W\ m^{-2}\ H^{-1}$ | 200[(j)] | 50[(j)] | < 1[(k)] |

(a) 0.3 mm detector + 0.3 mm electronic layer per hit, higher number without electronics.

(b) Arrangement of F. Angelini et al., CERN-PPE 91-122.

(c) 1.5 (5) hit(s) per mm at 2 m (0.1 m).

(d) Silicon strips with 50 μm pitch, divided by √12.

(e) Micro strip gas counter with 200 μm pitch, divided by √12.

(f) Measurements of C. D'Ambrosio et al, CERN-PPE 92-108.

(g) Silicon strip of 50 μm pitch and 10 cm length.

(h) Micro strip gas counter of 200 μm pitch and 10 cm length.

(i) 10 000 pixels of 60 μm × 600 μm each, at the readout end of one delay tube. They cover 3.6 $cm^2$ of fibre endsections. For a fibre layer thickness H of 600 μm (i.e. one hit in average), this corresponds to a subtended detector surface of 1.2 $m^2$ (= 10000 × 60 μm × 2 m). See C. D'Ambrosio et al., NIM A322 (1992) 20-33.

(j) Assuming 1 mW per channel.

(k) Delay tube pixels will only be activated by the first-level trigger.

From table 1, it appears clearly that semi-conductor detectors and scintillating fibres are intrinsically very fast. They can thus afford bunch crossing separation of the order of 20 ns, provided, for semi-conductors, the readout does not degrade their time resolution. Semi-conductors and fibres also have excellent spatial precision and two-track resolution needed for good vertex reconstruction and momentum resolution (see contribution of C. D'Ambrosio, these proceedings). A consequence of these properties is that their occupancy is low. They could thus be confronted with high multiplicity events produced in collisions of very high energy protons, like in Eloisatron.

The properties of MSGCs are not so favourable : their intrinsic response time integrates two to three bunch crossings, and their two-track resolution, in particular, is too large. Occupancy problems are bigger and have consequences on track reconstruction (see contribution of U. Gensch, these proceedings). However, MSGCs have the advantage to provide intrinsic amplification of the signal, which is not the case for fibres and semi-conductors.

Due to their original readout scheme, the fibres have no problem of power consumption. On the contrary, MSGCs, and even more semi-conductors produce so much heat that a cooling system is necessary to keep the temperature low, and consequently the noise of the electronics at an acceptable level. This increases the complexity of a full tracker detector and the amount of material inside the detecting cavity.

The occupancy of a detector being by definition the ratio of hit cells to the total number of cells, it determines the number of detector elements to be accessed after the 1st level trigger and also the total time needed to read their address and eventually their analog content. A too high occupancy would obviously increase this read out time to a value close to the average 1st level trigger period. This is of course unacceptable : a solution would be a longer trigger period, but it would ask for more stringent event selection condition, for a longer trigger latency and thus for an even more complex readout architecture. Using a detector design keeping the occupancy low is in consequence quite crucial.

## 7. Conclusion

From the above discussion, some recommendations can be made about several collider parameters. The bunch crossing time should stay within the time range 15-25 ns for the following reasons : it is similar to the intrinsic time response of Si, GaAs and scintillating fibre detectors, while MSGCs already integrate three bunch crossings. It is adapted to the fibre readout, but limits the readout of Si, GaAs and gas detectors where a trade-off is necessary between speed and power consumption.

Luminosities above $10^{34}$ $cm^{-2}s^{-1}$ provoke intolerable radiation levels and too high occupancies for vertex detectors. Inner tracking detectors could eventually stand this situation if they are implemented in a high enough solenoidal field (see contribution of C. D'Ambrosio, these proceedings). However, problems arise because track reconstruction efficiencies decrease and the readout time after the 1st level trigger reaches the average repetition time of this trigger. For reasons of further processing, this rate is limited to 100 kHz or less. An increased trigger reduction factor would have as a consequence a loss of physics, which is not acceptable.

The key parameter in future colliders seems to be the center-of-mass energy : a high value like in ELN will increase the cross sections as well as the signal-to-noise ratios (see contribution of H. Leutz, these proceedings). The higher multiplicity induced by a higher beam energy is acceptable by tracking detectors.

## 8. Acknowledgments

The author thanks Professor Antonino Zichichi for his encouragement and support and for his kind hospitality at the Ettore Majorana Centre for Scientific Culture. He also gratefully acknowledges discussions with Dr H. Leutz and all the members of the LAA group at CERN.

## 9. References

1. Zichichi, A., The ELOISATRON Project (August 1988).
2. Design Study of the Large Hadron Collider, CERN 91-03.
3. Zichichi, A., The LAA Project, ICFA Instrum. Bull. 3 (1987) 17.
4. C. D'Ambrosio et al., NIM A322 (1992) 20-33.
5. G. Lutz and A.S. Schwarz, CERN 89-10 p. 272.
6. N. Bingefors et al., CERN-PPE 92-173.
7. P.A. Tove and K. Falk, NIM 29 (1964) 66-68.
8. S.P. Beaumont et al., NIM A321 (1992) 172-179.
9. A. Oed et al., NIM A263 (1988) 351.
10. F. Angelini et al., CERN 89-10 p. 465.
11. F. Angelini et al., CERN 90-10 p. 222.
12. J. Schmitz, NIM A323 (1992) 638.
13. H. Leutz, CERN 89-10 p. 176.
14. C.M. Hawkes et al., CALT-68-1598 (Nov. 1989).
15. C. D'Ambrosio et al., CERN-PPE 92-108.
16. N. Ellis et al., CERN-ECP 90-17.
17. CERN-DRDC 92-28 RD20 Status Report.
18. F. Anghinolfi et al., CERN-ECP 90-13.
19. N. Bingefors et al., CERN-PPE 92-70.
20. T. Gys et al., CERN 90-10 p. 261.
21. T. Gys et al., CERN-DRDC 92-42 LAA Status Report.
22. F. Anghinolfi et al., CERN-ECP 91-26.
23. J.-J. Jaeger et al., LPC 92-11.

# TRACKING AND PATTERN RECOGNITION
# AT HIGH LUMINOSITY ACCELERATORS

U.GENSCH

*Deutsches Elektronen Synchrotron-IfH Zeuthen*
*Platanenallee 6, 15738 Zeuthen*
*Germany*

## ABSTRACT

Tracking will play a vital role in experiments at future hadron colliders. The task of pattern recognition in tracking detectors is outlined. As an example track finding is being discussed for a central tracker composed of scintillating fibres. The performance of such detector fulfils the stringent requirements for the experimentation at supercolliders.

## 1. Introduction

Future hadron colliders are designed for high energies and luminosities, but the rich physics potential opened up will be only accessible if the experimental technique can master the new challenges. The goal of experiments foreseen at high luminosity accelerators as LHC, SSC or ELOISATRON is to study very rare "gold plated" events. The events are hidden in a tremendous background. In general, it will be not sufficient just to tag interesting final states. Instead, it will be necessary to perform an almost complete reconstruction of heavy objects or specific topologies like Higgs-Bosons, SUSY-particles, three Gauge-Boson vertices etc. .

In section 2 a glance is given at the overwhelming experimental problems. The arguments for tracking are also summarised and its importance for extracting physics is being stressed. The main part of this contribution is section 3. The possible layout of a central tracking detector based on scintillating fibres is sketched. The performance of such a tracker is analysed and the interplay is discussed between detector design parameters and ambiguities, occupancy, resolution and track reconstruction. Finally, a simple and fast track finding algorithm is presented which could be used to decide on stiff tracks at the second trigger level.

## 2. Experimental problems, arguments for tracking

In Fig. 1 is shown a simulated Higgs event produced in pp collisions at $\sqrt{s}$=40 TeV in the "LAA test detector " [1]. The experimental difficulties become at least partially clear from the figure. The μ's originating from the gold plated Higgs decay, $H \rightarrow ZZ \rightarrow 4\ \mu$, are separated but the central region of the detector is extremely busy due to the large number of secondaries produced.

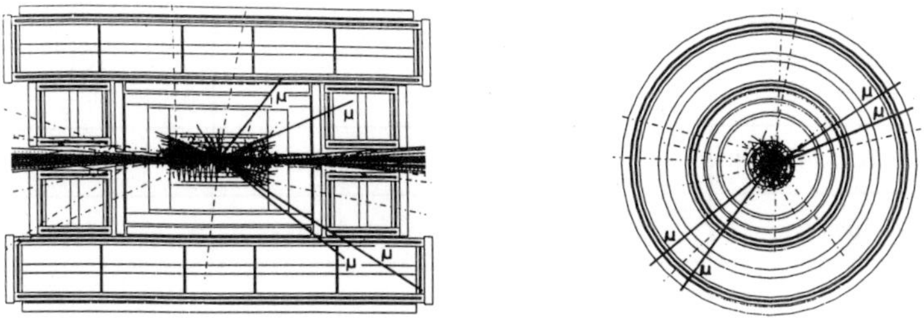


Figure 1 Display of a simulated Higgs event produced in pp collisions at $\sqrt{s}$=40 TeV in the "LAA test detector". The Higgs mass is set to 750 GeV and the Higgs decays via $H \rightarrow ZZ \rightarrow \mu\mu\mu\mu$. Only charged tracks with p > 0.1 GeV are shown. Both Z's have a transverse momentum of at least 80 GeV. The total event multiplicity is about 550.

It is evident that any proposed detector must have high granularity. Other obvious but important constraints are its ability to stand high rates and a fast and pipelined readout. To identify an interesting bunch crossing demands a very efficient and flexible triggering system. A detector working at the LHC at a luminosity of $\mathcal{L} = 10^{34}$ cm$^{-2}$ sec$^{-1}$ will see about $10^9$ "normal" pp collisions per second whereas a Higgs Boson with a mass of 500 GeV is expected about every 100 seconds [2]. This raises the question if an optimal strategy exists to perform experiments searching for such rare events : is it the best way to opt for an ultimate increase of the luminosity or is it more preferable to go towards higher energies ? An obvious price to be paid for working at extreme luminosities is the linearly increasing number of pp collisions per crossing causing severe problems for the experimentation. The step towards higher energies seems to be the more promising way. This is underlined by the energy dependent ratios of cross sections collected in table 1.

Ratios are given for Higgs ($H \to ZZ \to 4\mu$) and for Z-pair production ($ZZ \to 4\mu$) and for minimum bias events. Input are the cross sections ($\sigma$*BR) expected at LHC ($\sqrt{s}$ = 16 TeV), SSC ($\sqrt{s}$ = 40 TeV) and ELOISATRON ($\sqrt{s}$ = 200 TeV) [1].

Table 1 Ratios of cross sections for $H \to ZZ \to \mu\mu\mu\mu$ and $ZZ \to \mu\mu\mu\mu$ production in pp interactions. The values for minimum bias events are also included.

| Ratios | 40 : 16 TeV | 200 : 40 TeV | 200 : 16 TeV |
|---|---|---|---|
| Higgs production (mass = 750 GeV) | 7.4 | 13.5 | 101 |
| Min. bias background | 1.6 | 2.2 | 3.6 |
| ZZ - pairs | 3 | 4.8 | 14.3 |

Experiments performed at the ELOISATRON will profit from the very different energy dependencies expected for signal processes and background channels. It is expected that the value of $\sigma$*BR(Higgs $\to ZZ \to 4\mu$) grows by two orders of magnitude between LHC and ELOISATRON whereas the most dangerous background channel, namely the production of a Z pair leading to 4 $\mu$'s in the final state will increase only by a factor of 14. For minimum bias events delivering the overwhelming noise in any detector, only a rather weak increase of the cross section will be observed. These simple considerations underline that it will be more promising to perform experiments at higher energies rather than at extreme luminosities.

The task of tracking in future collider experiments is the subject of many reviews. A comprehensive discussion can be found e.g. in ref. 3. The obvious benefits from tracking as isolation cuts , vertex position, sec. vertex tagging etc. are widely acknowledged. Our interest is especially directed on central particle tracking. Central particle tracking sub detectors are naturally located inside the free barrel space of the detector surrounded by hadron and electromagnetic calorimeters as well as large area muon chambers. It may be surprising to note that inner tracking is vital even for experiments intended to look for gold plated events, like $pp \to H \to ZZ \to 4\mu$. Monte Carlo simulations performed for the ATLAS and CMS set-ups show that the completion of a dedicated $\mu$ detection system by a precise inner tracking improves the mass resolution expected for the Higgs Boson considerably [4,5]. The trigger performance will also benefit from a more precise measurement of the muon tracks.

After glancing the problems there is still need for some more detailed remarks. Should the central tracking be inside a magnetic field or should the tracker be located in a field free space? The experiments foreseen at future colliders will include a muon detector. A strong magnetic field has to be supplied to allow the measurement of the muon momentum, but there remain the two options : $B_{cent}$ on or off. The central tracker will be located in a field free space if a toroid is chosen, but it is $B_{cent} \neq 0$ if a large solenoid magnet delivers the magnetic field. The pro's and contra's of the two options are summarised :

**$B_{cent} = 0$**

The tracking detector will be simpler and cheaper, therefore the overall cost will be reduced. Pattern recognition might be easier since there will be no loopers on the inner tracking areas (shells). A restriction of the physics outcome will be the price to be paid, as becomes obvious if one compares with the $B_{cent} \neq 0$ option.

**$B_{cent} \neq 0$**

Central tracking will provide the sign of the electric charge of particles and their momenta. It will improve the momentum resolution even for very high energetic tracks. On the first sight pattern recognition might become more complicated, but a strong solenoidal field will facilitate track finding of stiff tracks, since the occupancy will become very small at large radii and high $p_t$ tracks are easy to find and to trace in the r,φ plane. This opens the possibility to use tracking information in a second level trigger.

In any case, independent of the presence of a magnetic field central particle tracking will improve the performance of the experiment. The z-coordinate of the primary vertex will be measured. Tagging of secondary vertices will be possible if the experiment is also equipped with a high resolution vertex detector. Isolation cuts can be applied using the measured directions of charged secondaries, improving the photon identification and the performance of the calorimeter.

The last comment deals with the required vertex resolution. The knowledge of the interaction vertex in the presence of multiple events per bunch crossing producing each a large number of charged particles will be important for the performance of experiments at colliders. A measurement of the z-coordinate of the primary vertex will improve the interplay of different detectors, the rejection of background will be more efficient and the decision on rare/exotic phenomena will be safer. C.d'Ambrosio discussed in his contribution to these proceedings the interconnections between collider- and trackerparameters [6] and especially the connection of beam geometry and vertex distribution. The required vertex resolution, $\delta_z$ , is estimated using the inequality

$$< n_v^* > 2.36\ \delta_z < 0.5$$

$n_V^*$ denotes the number of interactions vertices per mm. In table 2 are summarised the average number of vertices and the required resolution in the z-direction for different colliders.

Table 2 average nr. of vertices and required vertex resolution for different colliders

| BEAM PARAMETERS | | LHC | SSC | ELOISATRON | |
|---|---|---|---|---|---|
| Luminosity | $cm^{-2}s^{-1}$ | $10^{34}$ | $10^{33}$ | $10^{34}$ | $10^{35}$ |
| av. nr. of vertices $\langle n_V \rangle$ | | 13 | 1.7 | 30 | 270 |
| av. nr. of vert. per mm (68% of interactions) $\langle n_V^* \rangle$ | $mm^{-1}$ | 0.09 | 0.014 | 1.8 | 6.8 |
| required vertex resolution, $\delta z$ | mm | 13 | 84 | 0.65 | 0.17 |

The constraints on a tracking system are very stringent for both ELOISATRON set-ups. At the LHC a rather modest resolution is required, whereas the determination of the vertex position will be easier at the SSC. Nevertheless a warning is necessary, as these numbers scale up with the number of interactions per bunch crossing. Therefore, asking for higher $\mathcal{L}$ or longer bunch separation will complicate vertex finding also at LHC or SSC. The physics goal to observe heavy quarks via tagging of secondary vertices sets additional constraints on the resolution and performance of the central particle tracker. A z-resolution of about 1 mm is required in order to enable an efficient linking of charged tracks from a central outer tracker downwards to a high resolution vertex detector.

## 3. Central tracking with a fibre detector

### 3.1. Detector layout

In section 2 it was stressed that there is a need to equip future experiments with tracking facilities otherwise the performance of these experiments will suffer substantially from a loss of good physics. In this section the properties of a central particle tracking detector are discussed. There exist mainly three different media for particle tracking : gases, semiconductors and scintillators. A comprehensive discussion of the advantages and problems applying the different media can be found in ref. 7. We opt for scintillating fibres as detector elements since their application offers good spatial resolution, reasonable two-track resolution, sufficient hit densities and a manageable number of readout channels. Occupancy will be not at all a problem for a fibre tracker. In fig.2 is shown a schematic view of a fibre tracker.

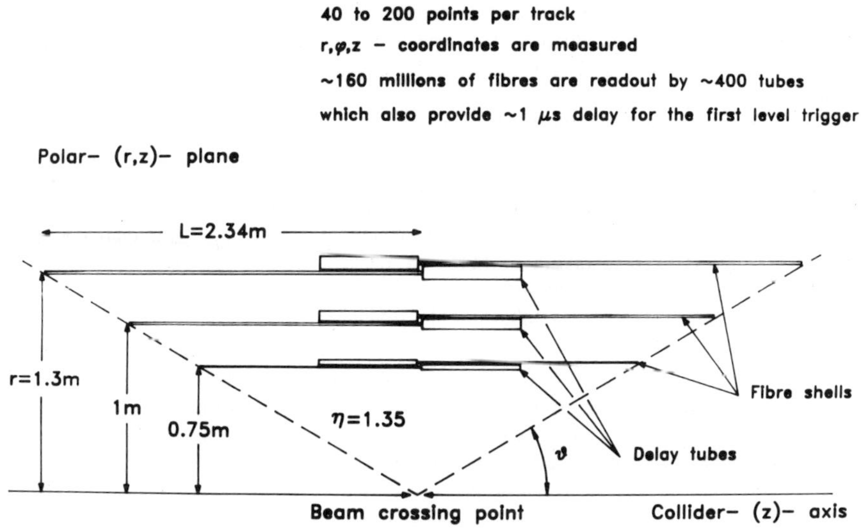


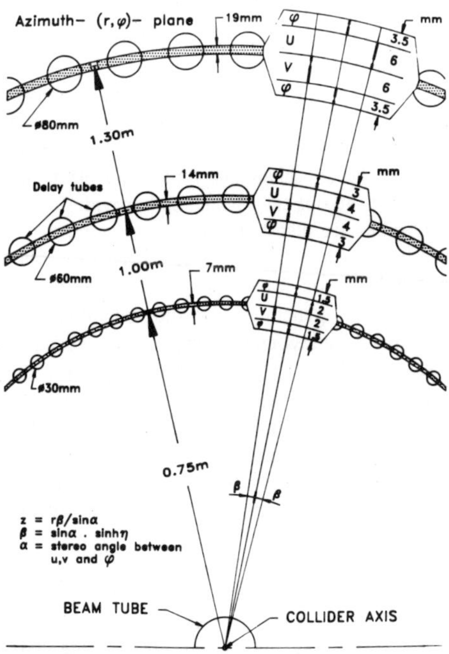


Fig.2 Schematic view of the central particle tracker composed of scintillating fibres. The upper part shows the r,z - projection of the detector, the lower part displays the r,φ - view.

The pseudo rapidity acceptance of the tracker is $|\eta| < 1.7$ at $\sqrt{s} = 16$ TeV, covering the central region of particle production. The tracker is located inside a rather strong solenoidal magnetic field of B = 3 - 4 T. The three cylindrical tracker shells surrounding the beam pipe are placed at rather large radii. This lowers the occupancy and guarantees that the radiation dose is tolerable. Due to the large distances from the collider axis low momentum tracks ("curling tracks") will not spoil substantially the performance of the tracker [7], on the contrary the strong field enables fast track finding at the outermost shell. Each tracker shell is composed of four super layers with different angular orientations of the fibres. The plastic fibres consist of a polystyrene core surrounded by a polymethylmetacrylate cladding. The diameter of the hexagonal fibres is fixed to 60 μ. The fibres are packed into bundles with rectangular cross section and edge dimensions between 1 and 3 mm. These bundles are arranged into the super layers. The magnified parts of fig.2 illustrate the subdivision of the shells into the layers. The u and v layers composed of oblique fibres are placed in between $\varphi$ layers with fibres aligned parallel to the collider axis. The (u,v) fibres are oriented at an angle $\alpha$ with respect to the parallel ones, providing via the stereo view the z coordinate of a particle track, whereas the $\varphi$ fibres directly measure its azimuth angle. The r coordinate of the charged particle passage is given by the location of the light emitting fibre within the layer. The layers (and shells) have different thickness in order to optimise the distribution of hits along the track improving the momentum resolution and enabling pattern recognition.

The fast readout of the 160 millions fibres of the tracker is a challenge by itself. It has not only to master the huge number of channels but also to meet the stringent timing constraints set by the bunch crossing. These constraints demand ambitious solutions. In addition, it might be essential to organise the readout in such way that trigger signals can be derived from the information delivered by the tracker. This of course requires efficient pattern recognition algorithms combined with appropriate hardware realisations. T.Gys [8] gives a detailed discussion of all aspects important for the design of the fibre tracker readout. Here we only sketch the layout. The key element of the tracker readout is the so called delay tube, an electromagnetically focused image intensifier. The tube represents an optical pipeline and provides a gating mechanism for the first trigger level. About 400 tubes are distributed on the circumference of the tracker as it is indicated in fig.2. The fibre bundles are directly coupled onto the photocathode fibre optics window where the emitted light is converted into photo-electrons. Each tube covers $(1 \text{ to } 4) * 10^5$ fibre channels. The images generated for the individual bunch crossings are delayed and stored by to and fro drifts of the low energy photo-electrons through the first sections of the tube. This analog pipeline provides a 1 to 2 μs delay of the optical images. Tests of prototype tubes showed, that the performance will be excellent, the achievable time resolution will be smaller than the bunch crossing time and the space resolution will be well within the microfibre diameter [9]. After applying the gating pulse from the first level trigger the selected image is accelerated via a 30 kV voltage towards the readout plane which is a silicon pixel detector directly bombarded with the electrons corresponding to the recorded track points.

The pixel size of the actual design is $60*600\ \mu^2$, i.e. each pixel will collect the photoelectrons from 10 fibres. This summation over the fibres is performed along the radius but not over $\varphi$, therefore the resolution for high momentum tracks will be not affected. The lateral displacement of a "straight" track with a $p_t$ of 10 GeV/c traversing 10 fibres is within the fibre diameter. The loss of individual track points (hits) will be compensated at least partially due to an increased efficiency. The very fast readout will be realised by a special zero suppression circuit (so called sparse data scan). At present, a prototype chip is under preparation working at a clock rate of 10 MHz which will readout a 16*8 pixel matrix [10], i.e. covering an area of about 0.5 $cm^2$ of the tracker readout plane. The address of a single fired pixel will be obtained within about 100 ns. A great advantage of the fibre tracker is the very low occupancy. 30 min. bias events will cause an averaged hit density of about 0.1 hit per $cm^2$. It seems feasible, that a highly parallel readout of the complete tracker can provide the address matrix of the hit pixels within a few μs, leaving time to form a second level trigger decision using the tracker information.

Additional work and refinement is needed to come from this sketched scenario to the final readout scheme, one could imagine of an optimisation of the pixel sizes, matching of fibre layers with the delay tubes, parallel processing and built in logic decisions to ease pattern recognition etc..

### 3.2. Tracker performance

Several issues important for the performance of a tracker are discussed. A derivation of tools allowing a quantitative discussion of tracking at super colliders is given by C.d'Ambrosio in his contribution [6] and a comparison of different tracker media can be found in ref. [7]. This approach is confined to the fibre tracker introduced in the preceding section.

The occupancy O denotes the fraction of busy detector cells with respect to the total number of tracker cells. O is depending on collider parameters and properties of the final states of the pp interactions and of course on the detector layout. It is

$$O = (\mathcal{L}\sigma n t_b) x \left( \frac{S \eta_{\max}}{2\pi r} \right) ,$$

where $\mathcal{L}$ is the collider luminosity and σ the cross section of the hadronic interaction. n stands for the number of minimum bias particles per unit of rapidity and per interaction and to the time $t_b$ between two bunch crossings. The detector specific quantities are : $\eta_{max}$ the rapidity interval covered, r the distance of the detector cells from the collider axis and the fibre diameter S (transverse cell dimension). Such calculation provides only a first approximation, since several effects as the influence of the magnetic field, particle decays, secondary interactions etc. are not considered, more realistic numbers are obtained from Monte Carlo simulations. Minimum bias events were generated with Pythia 5.5 and passed through GEANT to calculate the occupancies for nearly experimental conditions.

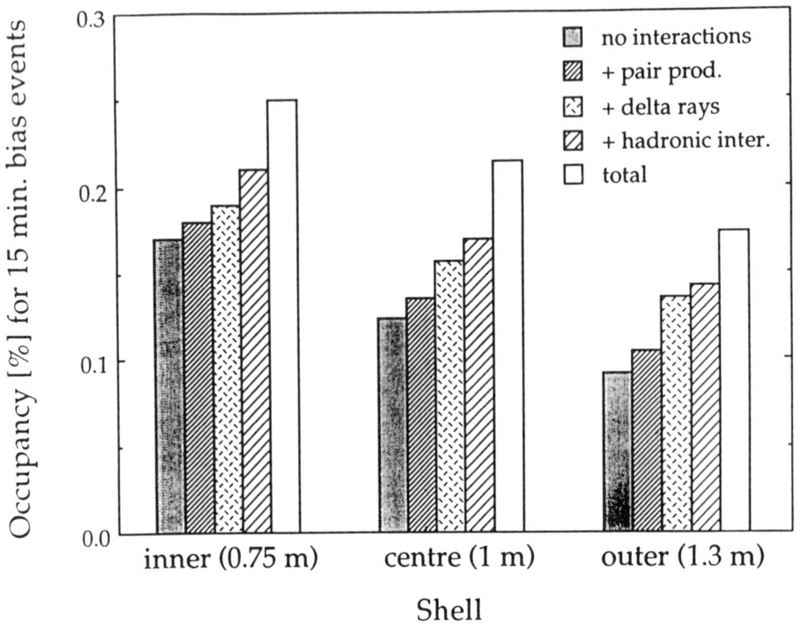


Fig. 3 Occupancies of the inner, centre and outer shells of the fibre tracker. The fibre diameter is 60 μ.

In fig. 3 the simulation results are shown displaying the occupancies expected for the three individual shells of the fibre tracker. For the simulations it was assumed that the tracker will be located at the LHC running at a luminosity of about $\mathcal{L} = 10^{34}\ \mathrm{cm}^{-2}\ \mathrm{sec}^{-1}$. The piling up of the different contributions - charged secondaries from the interaction vertex, pair production, delta rays and secondary hadronic interactions - to the total values is indicated. It was shown, that the agreement between values calculated for the occupancy and simulation results is quite reasonable for the special case of no mag. field and absence of any secondary processes [7]. It is seen in fig.3, that the occupancies will be very small for all three shells. The "highest" value will be observed at the inner shell with about 0.25 %, whereas the total occupancy of the outermost shell is well below 0.2 %. The presence of a strong solenoidal magnetic field of about 3-4 T causes charged particles with low pt to curl at a smaller radii, therefore the bulk of charged tracks is not reaching the outer shell The high granularity of the fibre tracker and its small busy time will assure these extreme small occupancies, which will be of great importance for any track finding procedure.

The performance of a tracker is also deteriorated due to ambiguities. The design of the fibre tracker presented in fig. 2 allows to limit their frequency remarkably. In fig. 4 is shown an enlarged view of the readout plane of a super layer with the hit pattern produced by the passage of a charged particle.

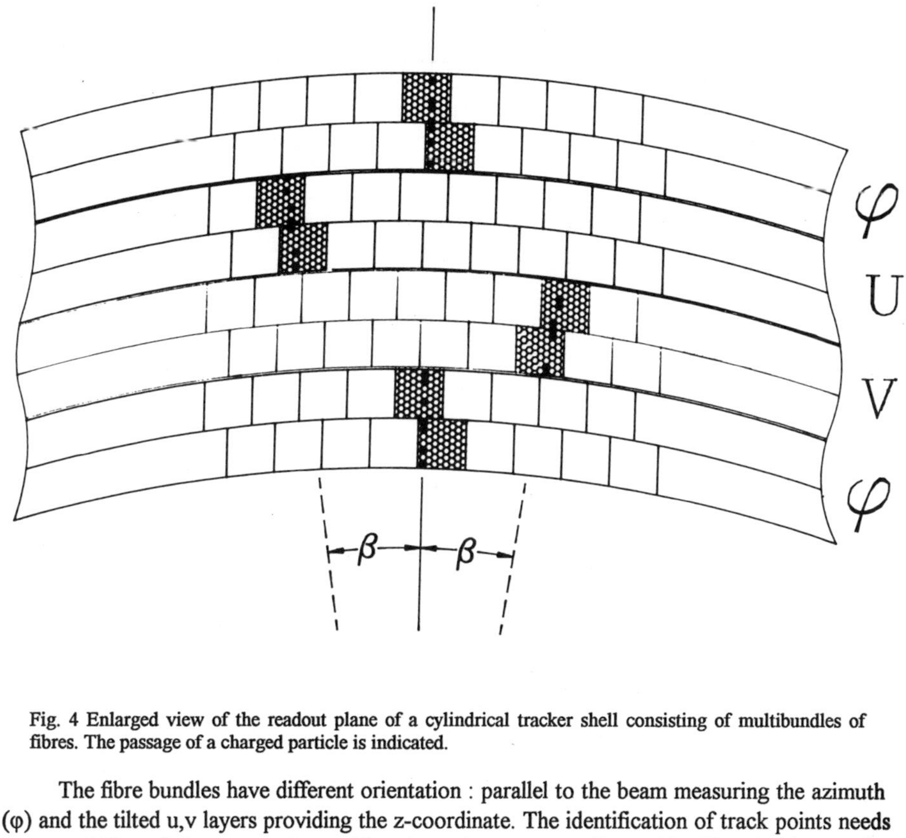


Fig. 4 Enlarged view of the readout plane of a cylindrical tracker shell consisting of multibundles of fibres. The passage of a charged particle is indicated.

The fibre bundles have different orientation : parallel to the beam measuring the azimuth ($\varphi$) and the tilted u,v layers providing the z-coordinate. The identification of track points needs the association of the symmetric u,v hit pair to the corresponding $\varphi$-hit. The u,v hits are displaced by an angle $\beta$ with respect to $\varphi$, as indicated in fig. 4. A compromise has to be chosen [7] : the z-precision is improved with increasing the stereo angle $\alpha$, whereas the confusion for track finding is growing with sin$\alpha$, since the angular section to be scanned for a symmetric u,v hit pair is $\beta_{max} \sim \sin\alpha$. Obviously, larger values of $\beta_{max}$ enhance the chance to collect faked symmetric u,v pairs. The number of (u,v)-doublets symmetric within $|\beta_{max}|$ was calculated for several values of $\alpha$ and different fibre diameters. It turned out, that for occupancies below 1% which are easily reached with 60 $\mu$m diameter fibres and stereo angles of 4-8 ° the fraction of ambiguous pairs is sufficiently small (< 0.1 for 0.5 m shell radius and < 0.001 for 1.4 m shell radius and 2 T magnetic flux density). A further reduction can be obtained if the tracker shells will be constructed with slightly different stereo angles [7].

The vertex resolution was also analysed in ref. [7]. The estimations were done for the tracker shown in fig. 2 with the stereo angle $\alpha$ varying between 4 and 16 °. It was demonstrated, that the achievable z-resolution (2.3 mm and 0.6 mm, resp.) will be sufficient to discriminate between the primary vertices produced per one bunch crossing in a LHC or SSC experiment at $\mathcal{L} = 10^{34}$ cm$^{-2}$ sec$^{-1}$.

The momentum resolution ($\delta p/p$) expected for charged tracks reconstructed with the fibre tracker at B=3.3 T is shown in fig. 5 [7]. The number of hits per track was set to 15 being a rather modest assumption and the beam position was used as a constraint (|30 μm|).

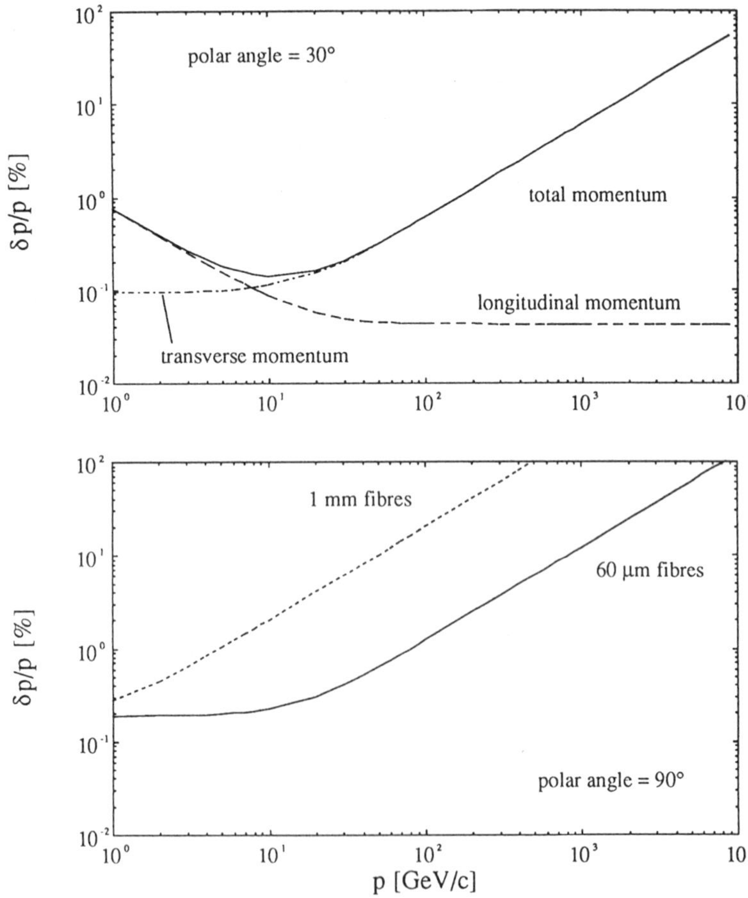


Fig. 5 Momentum resolution ($\delta p/p$) versus p for different polar angles.

The upper part of fig.5 shows (δp/p) for tracks with 30° . Below 10 GeV/c the resolution is dominated by multiple scattering. The lower part of fig.5 shows the resolution for tracks at 90°. It is demonstrated that the tracker composed of 60 μm diameter fibres allows reasonable momentum measurements even for tracks with a $p_t$ of more than 1 TeV. The benefit from a good spatial resolution is demonstrated by the second curve, the use of fibres of 1 mm diameter will limit the track reconstruction at considerably smaller transverse momenta. Finally, fig.6 presents the momentum resolution for tracks under 90° polar angle reconstructed only within the outermost shell. The result is encouraging, high pt tracks can be tagged even at 100 GeV/c using the hits from the outer shell only, enabling track finding with a single shell as described in the next section.

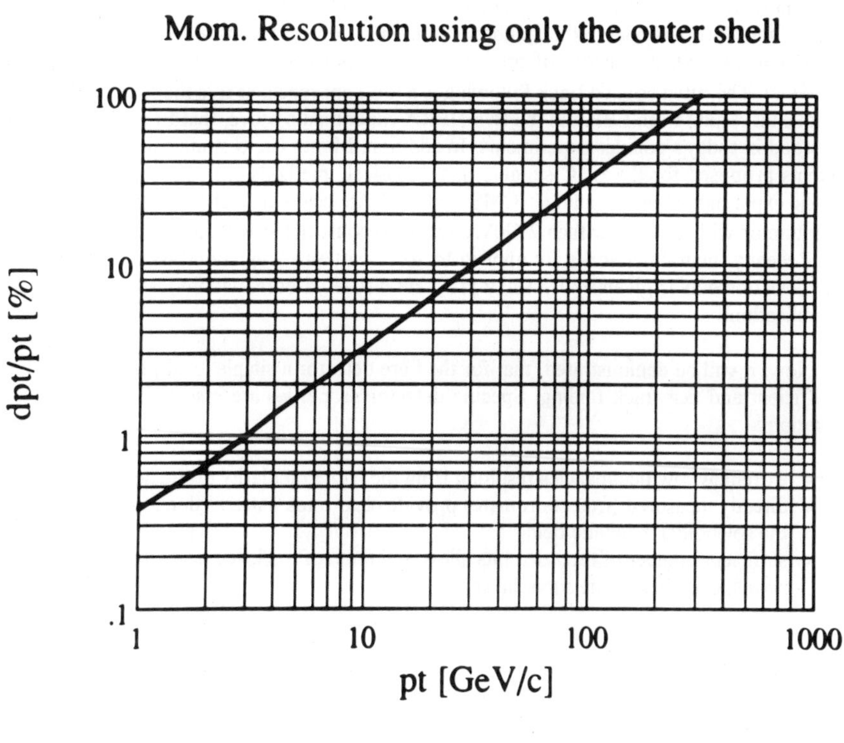


Fig. 6 Momentum resolution for tracks at 90° using only the outermost shell.

### 3.3. Track finding

The Higgs event shown in fig.1 demonstrates how interesting physics final states will appear in experiments at future pp colliders. Track finding will be challenging and furthermore complicated due to the required high luminosity. The tracking detector will not only see charged secondaries from a gold plated event but will be also exposed by the debris of charged particles produced in non interesting minimum bias events. The numbers collected in table 2 show, that LHC-(ELOISATRON-) experiments running at a modest luminosity of $\mathcal{L} = 10^{34}$ cm$^{-2}$ sec$^{-1}$ have to cope with 13 (30) background events per bunch crossing.

The different track finding algorithms are in general classified as global or local methods [12]. The solution chosen for a specific tracker depends of course on the hardware and will be different if one asks for complete track reconstruction and momentum measurement or if only a subset of data is being used to form a trigger signal. Global methods, as histogramming, inverse mapping, template matching or the minimum spanning tree, are treating all measured space points simultaneously, in the same way and independent on their order. The time t consumed by the algorithm is proportional to the number N of points. Local methods use only sub samples of points (local groups), the history of track finding is important and the time needed goes like $t \sim N^{\alpha(1..3)}$. The often applied track following is a local approach, as it is the road method. The experimental results from LEP and the two HERA experiments prove, that track finding is well understood in today's collider experiments. In addition, Monte Carlo studies performed for experiments proposed for the LHC suggest that "classical" methods will work also in this very complex experimental surrounding : CMS and ATLAS claim that charged tracks with high momentum or with a $p_t$ larger than 2 GeV/c will be reconstructed with an efficiency of ≥ 98 % [4,5]. Recently, neural networks are applied to solve pattern recognition tasks. A summary of actual developments is given by B.Denby [13]. He discussed both, Monte-Carlo exercises and real applications.

Now, it will be demonstrated, that for the fibre detector a simple local procedure will enable efficient and fast track finding. Specific detector properties are essential for a successful application :

- extremely low occupancies resulting from the high granularity of the detector
- directly measured (r,φ) coordinates provide clear track patterns due to the solenoidal magnetic field
- sufficient number of recorded hits allow definition of track segments, rather than single points, even in a single super layer

The first task will be the recognition of tracks with high transverse momentum. In the presence of a strong solenoidal magnetic field the patterns of such stiff tracks are very simple in the r,φ-plane : one expects circles or (within a single super layer) even straight lines. Fast track finding benefits -1- from these clear patterns and -2- from the large hit densities measured [7] : tracks can be located using data of the φ-layers from only one tracker shell. Tests performed with 60 μm diameter fibres have shown, that 2-3 photoelectrons are registered on the photo cathode of the optoelectronic readout per mm of crossed scintillator. Track finding will be discussed using the tracker layout shown in fig. 2. It will be sufficient to look only at a small area of the readout plane, the region of interest can be confined to one readout-tube or its built-in pixel detector. The search will be started at the outermost shell which is located for the example of fig.2 at a radius of 1.30 m. The bulk of low $p_t$ tracks curls inside this shell, due to the magnetic flux density of about 3 T and the occupancy will be even at a luminosity of $\mathcal{L} = 10^{34}$ cm$^{-2}$ sec$^{-1}$ below 0.2 %, as illustrated in fig. 3. Fig. 7 shows an enlarged view of the readout plane.

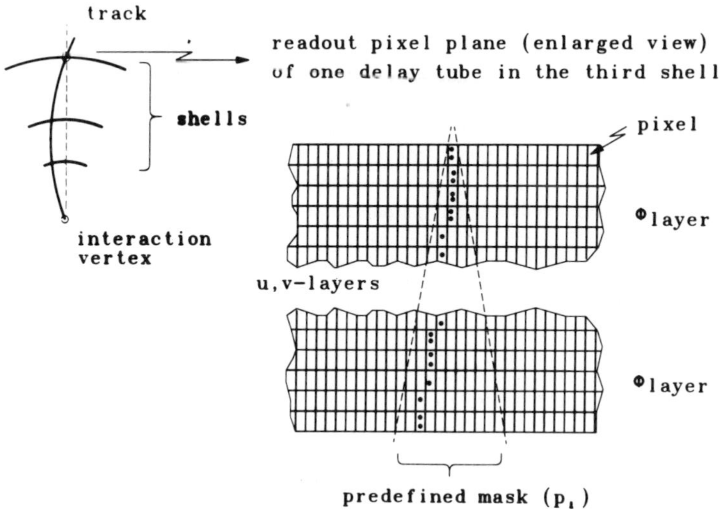


Fig.7 Search for tracks with a large transverse momentum. Shown is an enlarged view of the pixel detector at the end section of a fibre shell with a track traversing both φ-layers. The cone corresponding to the transverse momentum is indicated. The dots indicate the individual hits as recorded by the fibres.

The drawing illustrates schematically the passage of a charged track with a transverse momentum of about 50 GeV/c through the two φ-layers of the outermost shell. Indicated are the generated hits in the fibres. As already stated in sect. 3.1 the readout via the pixel detector will integrate radially over 10 fibres, this will cause a loss of individual hits but will increase the efficiency. The hits recorded have a displacement of about 0.25 mm or 4 fibre(-pixel)widths. The distance between the φ-layers is 12 mm. Assuming to have obtained the address matrix of hit pixels resulting from the "sparse data scan", stiff tracks can be found via the following procedure :

- start the search for hit pixels at largest r (first row of the matrix)
- take first hit pixel found as starting point of track, skip if row is empty
- confine scan to region within predefined mask, corresponding to the chosen $p_t$ cut
- count hit pixels within the mask, in φ only neighboured ones are accepted (columns should not differ by more than 3)
- accept track if #counts > set value, e.g. 2/3*12 pixels (thickness of 2 φ-layers)

The timing behaviour of this procedure was tested. The search algorithm, written in C, and the input pixel set (matrix of pixel addresses) were downloaded to a C40-processor (TMS320C40,30 MHz). The area scanned is 100 ($\varphi$) * 12 (r) pixels ~ 0.4 $cm^2$. First results of our tests are encouraging. A single track combined with 10 % background hits was found within 30 $\mu$s. This estimation seems to be rather realistic considering the expected values for occupancy and frequency of charged tracks with a $p_t$ large enough to reach the outer shell. A possible realisation in a second level trigger scheme will need cheaper solutions, therefore we will also study the timing behaviour of the algorithm using appropriately designed XILINX chips [14]. A useful by-product of the procedure described will be the number of tracks with a transverse momentum larger than a chosen $p_t$-cut. In addition, a rough and fast determination of the z-position of the primary vertex will be possible if the hits generated in the u,v layers are considered, this will be important for any experiment at a future super collider [15]. A further optimisation of the tracker layout, e.g. concerning the arrangement and the number of shells and layers and the thickness of layers, will also improve the track finding. As stated in section 3.1 efficient track finding needs a highly parallel organised readout to cope with the number of output channels.

## 4. Conclusions

From our discussion it seems to be indispensable to equip future collider experiments with sophisticated tracking facilities located inside a solenoidal magnetic field. We opted for a fibre detector as central particle tracker. It was demonstrated that such detector can fulfil the requirements of an experiment at a pp collider.

## Acknowledgements

The author wishes to thank Dr.'s H.Leutz, H.Leich, C.d'Ambrosio and P.Wegner for many useful discussions. He acknowledges the warm hospitality at the E.Majorana Centre in ERICE.

**References**

1. F.Anselmo et al., The Monte Carlo Simulation Laboratory of LAA,CERN/DRDC/92-4, LAA status report.

2. D.Denegri, Proceedings of the Large Hadron Collider Workshop, Aachen 1990, CERN 90-10, vol. I, p. 56.

3. D.Saxon, Proceedings of the Large Hadron Collider Workshop, Aachen 1990, CERN 90-10, vol. I, p. 339.

4. CMS, letter of intent, CERN/LHCC/92-3.

5. ATLAS, letter of intent, CERN/LHCC/92-4.

6. C.d'Ambrosio, contribution to these proceedings.

7. C.d'Ambrosio, T.Gys, H.Leutz, U.Gensch and S.Schlenstedt, Nucl. Instr. and Meth., A322(1992)20.

8. T.Gys, contribution to these proceedings.

9. T.Gys et al.,CERN-DRDC 92-41.

10. J.-J. Jaeger et al., CERN-ECP 91-26.

11. H.Leutz, contribution to these proceedings.

12. H.Eichinger, Nucl. Instr. and Methods, 176(1980),417.

13. B.Denby, The use of neural networks in High Energy Physics, FERMILAB-Pub-92/215-E.

14. U.Gensch, H.Leich, U.Schwendicke, P.Wegner, internal note in preparation.

15. U.Gensch, S.Schlenstedt, Proceedings of the Large Hadron Collider Workshop, Aachen 1990, CERN 90-10, vol. III, p. 248.

# CALORIMETRY FOR HIGH ENERGY AND HIGH LUMINOSITY HADRON COLLIDERS

A. Contin
*European Organization for Nuclear Research (CERN)*
*CH-1211 Genève 23, Switzerland*

ABSTRACT

An overview is given of the requirements to calorimetric detectors in future very high energy ($\geq$ 200 TeV) and very high luminosity ($\geq 10^{34}\,\mathrm{cm}^{-2}\mathrm{s}^{-1}$) hadron colliders. Some specific detector options are examined and an example is given of a calorimeter which can operate even in the very harsh forward region of the apparatus.

## 1. Physics

The calorimeter will be one of the most important components of a detector for future supercolliders. It will have the primary task of precisely measuring and localizing the energy flow of each event, thus characterizing, independently from the subsequent hadronization processes, the dynamics of the fundamental interactions. It will have to play an important and perhaps predominant role in particle identification to select the events containing leptons. Finally, it will be the main element in the trigger.

The energy resolution of the calorimeter should be as good as possible but, above all, the calorimeter must have a fast, uniform, and linear response with a gaussian lineshape. Moreover, the $e/\pi$ discrimination signal must be easy to extract. These last requirements make it possible to use the calorimeter signal to build an efficient first level trigger.

A calorimeter adequate for a supercollider experiment must also be hermetic, radiation resistant, and manageable.

The physics signals to be measured by the calorimeter system in multi-TeV hadronic interations range from several TeV, to measure the high tail of the transverse momentum spectrum of jets, to some GeV, to tag isolated leptons or $\gamma$'s and to measure the energy loss of muons traversing the detector, down to some MeV, to have a good energy resolution on hadrons.

The capability to meet some of these demands will be affected by the pile-up of energy which occours as a result of the large number of interactions per beam crossing expected in a very high luminosity collider. In the following, the resuts of a simulation of proton-proton interactions at $\sqrt{s} = 200\,\mathrm{TeV}$ generated using the PYTHIA 5.6 / JETSET 7.3 MonteCarlo programs [1] with the defaults options ("minimum bias") are presented.

*1.1 Total cross-section*

The total cross sections estimated by PYTHIA5.6 for the different inelastic processes are the following:

| | |
|---|---|
| *single diffractive:* | 30 mb |
| *double diffractive:* | 3 mb |
| *all other subprocesses:* | 126 mb |
| *TOTAL:* | 159 mb |

With a bunch separation of 15 ns, the number of events per bunch crossing can be approximated to:

$$\sigma \times L \times \Delta t = 24 \times L,$$

with $L$=luminosity in units of $10^{34}\,\mathrm{cm}^{-2}\mathrm{s}^{-1}$.

*1.2 Relevant particle distributions*

Each event will produce on average 220 stable particles, mainly charged hadrons and $\gamma$'s from neutral hadron decay. As the effects of these two kind of particles are different both for what concerns the pile-up and for the radiation damage, they will be treated separately in the following.

Fig. 1a and 1b show the multiplicity distribution for $\gamma$'s and stable hadrons, respectively.

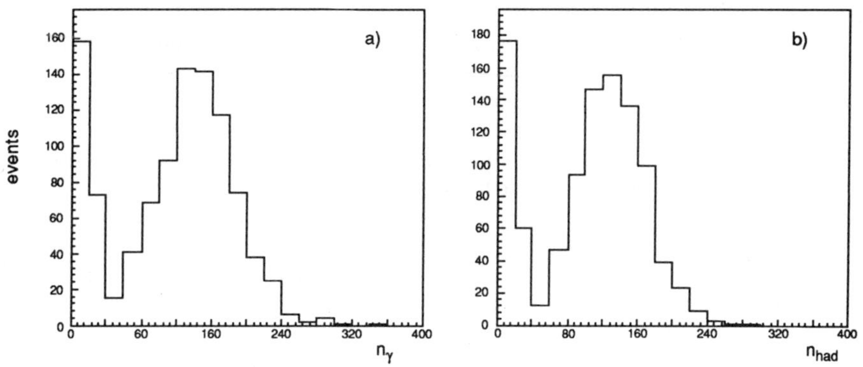


*Fig. 1. Multiplicity distributions for $\gamma$'s (a) and stable hadrons (b) in minimum bias proton-proton interactions at $\sqrt{s} = 200$ TeV.*

The peak at low multiplicities is mainly due to diffractive processes. The angular distributions are shown in Figs. 2a and 2b, in terms of the pseudorapidity $\eta = -\log_e[\tan(\frac{\theta}{2})]$, where $\theta$ is the angle at which the particle is produced with respect to the incident beams. A large fraction of particles are produced in a very narrow cone around the beam direction. The "central" and "forward" regions, arbitrarily defined here as $|\eta| \leq 3$ and $3 \leq |\eta| \leq 5$ (corresponding to $5.7° \leq \theta \leq 174.3°$ and $0.8° \leq \theta \leq 5.7°$ $174.3° \leq \theta \leq 179.2°$, respectively) are also indicated in the figure.

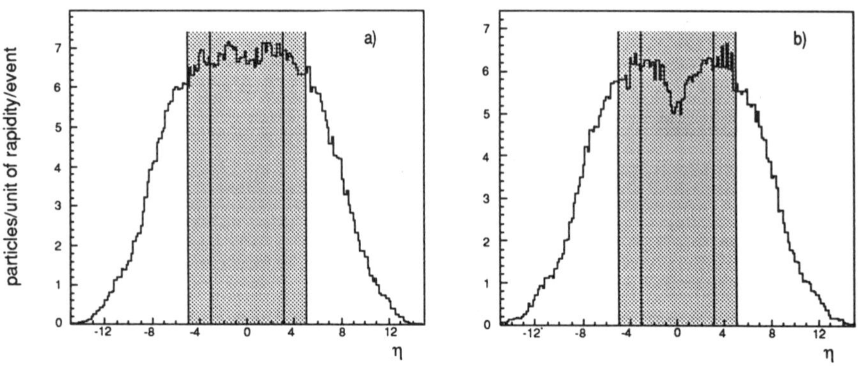


*Fig. 2. Pseudorapidity distributions for $\gamma$'s (a) and stable hadrons (b) in minimum bias proton-proton interactions at $\sqrt{s} = 200$ TeV.*

It is interesting to note the difference between the pseudorapidity distributions expected at ELOISATRON ($\sqrt{s} = 200\,\text{TeV}$), and at LHC($\sqrt{s} = 16\,\text{TeV}$). These are shown in Fig. 3 for stable hadrons.

The main effect of the increase in energy is to enhance the production of particles with very small angle w.r.t. the beams, at production angles which are very difficult to be covered by the detector. We will see in the following that this is not a very serious drawback for what concernes hermeticity, if one limits to the study of transverse momentum imbalance. However, a major loss in acceptance is expected for exclusive channels like Higgs particle decay.

The transverse momentum distributions for $\gamma$'s and stable hadrons are shown in Figs. 4a and 4b. The average $p_T$ is 300 MeV for $\gamma$'s and 500 MeV for hadrons.

The cumulated energy in two interval: $|\eta| \leq 3$ and $3 \leq |\eta| \leq 5$ are shown in Figs. 5a, 5b, 5c and 5d. Even if most of the particles group into one or more high energy jets, the bulk of particle production is still at quite low values of energy.

Finally, Fig. 6 shows the missing transverse energy expected in minimum bias events, if a coverage down to $|\eta| \leq 5$ and perfect resolution is assumed.

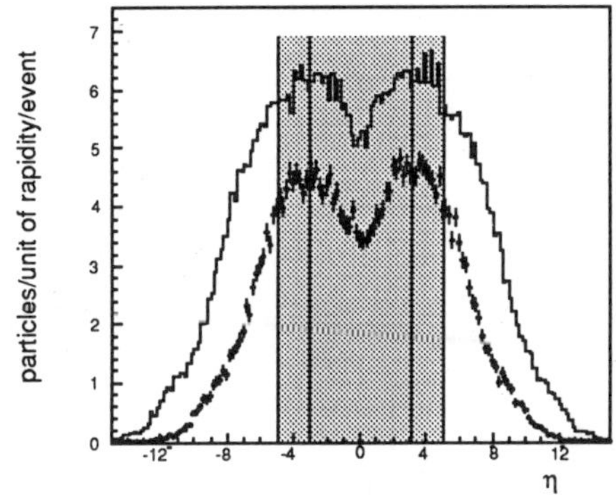


*Fig. 3. Pseudorapidity distributions for stable hadrons in minimum bias proton-proton interactions at $\sqrt{s} = 200$ TeV (solid line), and at $\sqrt{s} = 16$ TeV (points).*

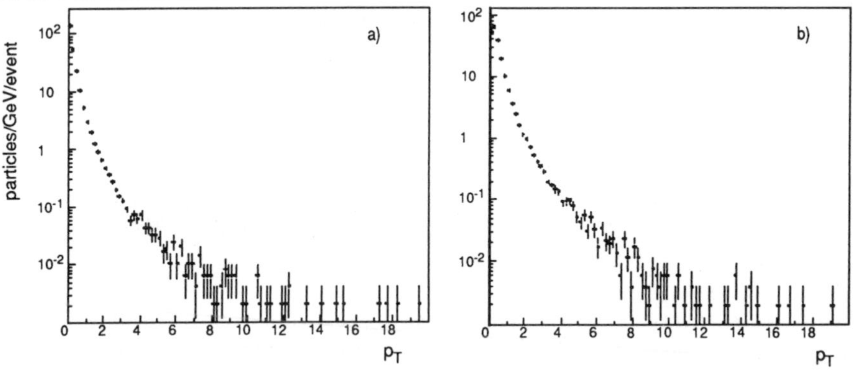


*Fig. 4. Transverse momentum distributions for $\gamma$'s (a) and stable hadrons (b) in minimum bias proton-proton interactions at $\sqrt{s} = 200$ TeV.*

## 2. Detector simulation

In order to compute the background expected from minimum bias events in a quasi-real calorimeter system, the following assumptions were made:

a) The region $|\eta| \leq 3$ is covered by a detector with an azimuthal symmetry and an eliptical shape in the beam plane (inner radius 150 cm, length 5.6 m). Each calorimeter cell has a $3 \times 3\,\text{cm}^2$ front face cross-section.

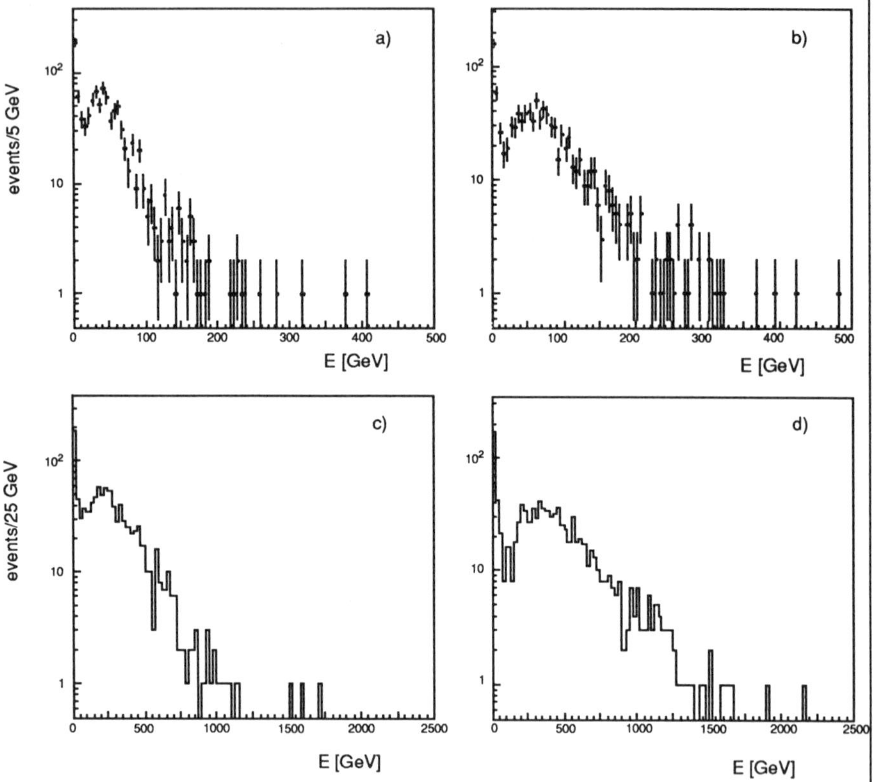


*Fig. 5. Total energy distributions for γ's [a): $|\eta| \leq 3$, b) $3 \leq |\eta| \leq 5$] and stable hadrons [c): $|\eta| \leq 3$, d) $3 \leq |\eta| \leq 5$] in minimum bias proton-proton interactions at $\sqrt{s} = 200\,\text{TeV}$.*

b) In the region $3 \leq |\eta| \leq 5$, the detector is a truncated cone placed at a distance of 16 m from the interaction point with an external radius of 160 cm at the front face, projective towards the interaction point. A 20 cm hole around the beam line determines the limiting acceptance in $\theta$ ($\theta \geq 0.8°$). The cell size is $5 \times 5\,\text{cm}^2$.

Fig. 7 shows a schematic view in the beam plane of the calorimeter system. The granularity of the calorimeter varies between $\Delta\eta \times \Delta\phi = 0.02 \times 0.02$ at $\eta = 0$ and $\Delta\eta \times \Delta\phi = 0.2 \times 0.1$ at $\eta = 5$ (see Fig. 9a).

This simplified geometry has been used to determine the energy pile-up due

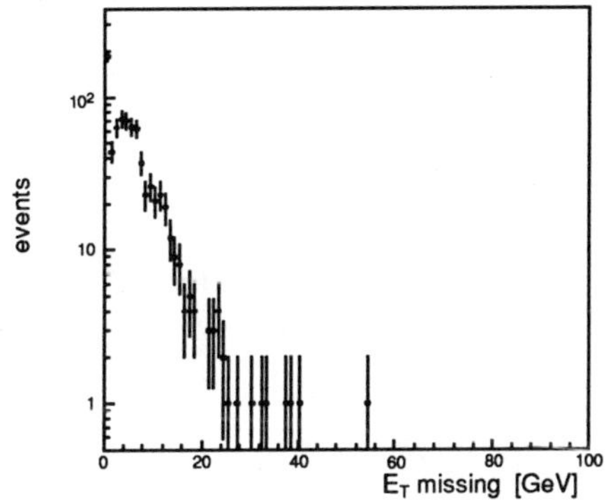


*Fig. 6. Missing transverse energy distributions for minimum bias proton-proton interactions at $\sqrt{s} = 200\,\text{TeV}$. A coverage down to $|\eta| \leq 5$ and perfect resolution is assumed.*

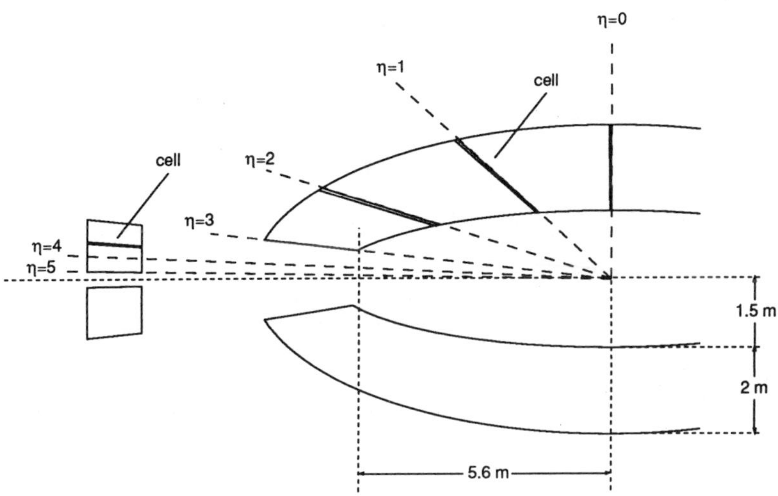


*Fig. 7. Schematic view in the beam plane of the calorimeter system used in the simulation.*

to multiple interactions in the same bunch crossing. The result is shown in Fig. 9b as a function of $\eta$ and for a luminosity $L = 10^{34}\,\text{cm}^{-2}\text{s}^{-1}$ (24 minimum bias events per bunch crossing). The conclusion is that energy pile-up for single-particle (electrons) detection is a problem only in the very forward region of the detector.

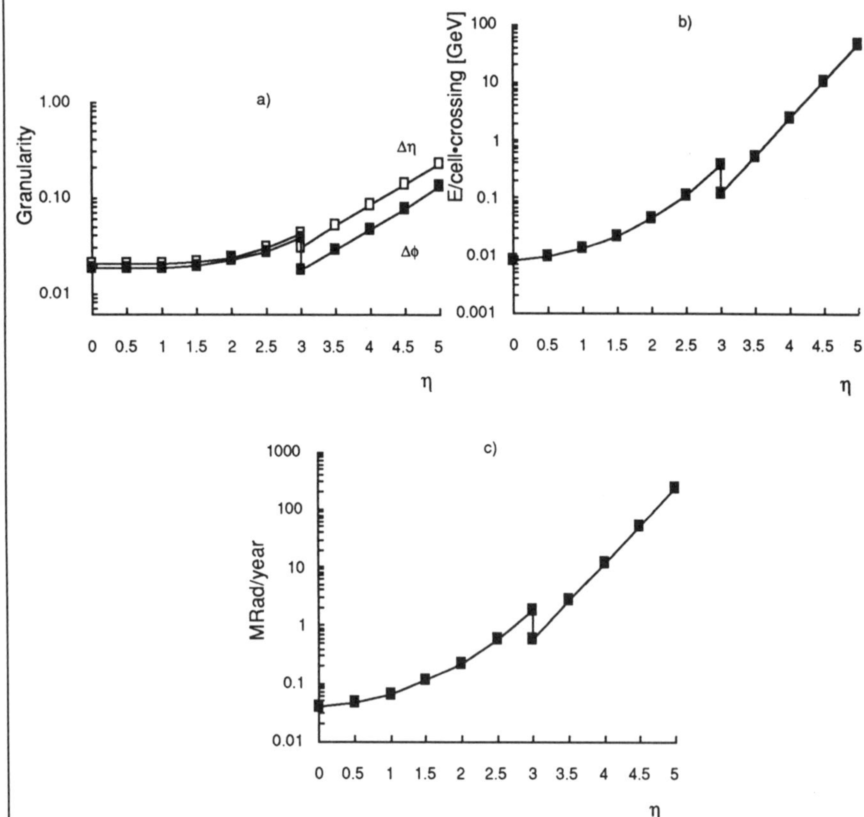


*Fig. 9. Results of the simplified detector simulation described in the text.*
*a) Detector granularity as a function of $\eta$.*
*b) Energy pile-up per cell per crossing due to minimum bias events at $L = 10^{34}\,\mathrm{cm}^{-2}\mathrm{s}^{-1}$, as a function of $\eta$.*
*c) Radiation dose due to minimum bias events at $L = 10^{34}\,\mathrm{cm}^{-2}\mathrm{s}^{-1}$, as a function of $\eta$.*

However, most of the interesting physics at ELOISATRON will produce jets of particles. Due to the angular spread of the particles inside the jet, several cells must be added together. Defining as $R = \sqrt{\Delta\eta^2 + \Delta\phi^2} = 0.4$ the limiting radius of the jet cone, the resulting pile-up is shown in Fig. 10. Even more important is the contribution of the event-by-event variation of the pile-up which adds to the energy resolution of the calorimeter. An added resolution of the order of some percent is expected on TeV forward jets.

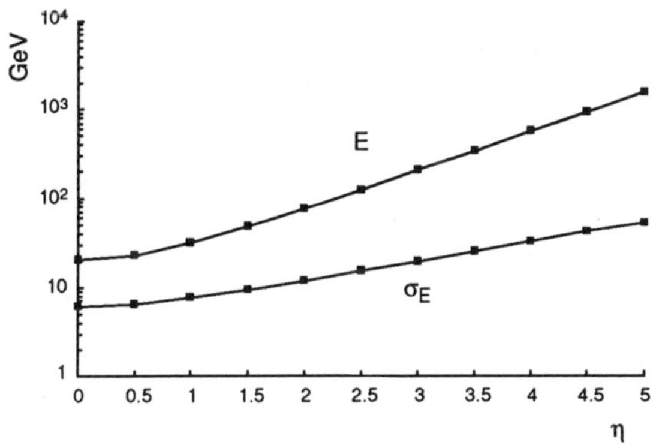


*Fig. 10. Energy pile-up on jets and event-by-event energy fluctuation as a function of $\eta$.*

Depending on how the calorimeter is built (see section 3), the energy deposited might cause radiation damage on the sensitive material or on the electronics. As an example, the radiation dose deposited in a fibre/lead calorimeter at the peak of the electromagnetic shower is shown in Fig. 9c. Radiation levels between one and several hundreds MRad/year are expected in the forward calorimeter at a luminosity $L = 10^{34}\,\mathrm{cm}^{-2}\mathrm{s}^{-1}$.

The results quoted above are not very sensitive to the total centre-of-mass energy of the interaction, but scale linearly with the luminosity. Therefore, extremely radiation-hard materials must be used if a good calorimetric coverage is wanted at very high luminosities.

## 3. An overview of possible calorimeters for ELOISATRON

The requirements on the calorimeter system for an ELOISATRON detector are the following:

a) Fast response time ($\leq 15\,\mathrm{ns}$).
b) High radiation resistance (up to the GRad level).

c) Good energy resolution (constant term less than 1% at least for the central calorimeter).
d) Small shower size.
e) Good space resolution ($\sim 1\,\text{mm}$).
f) Compact construction.

In the following sections, a brief review of the calorimeter techniques which are presently under development is given.

*3.1 Electromagnetic calorimeters*

A very good electromagnetic energy resolution (stochastic term of the order of 1%) can be achieved only by homogeneous calorimeters. Two chioces for the active material are presently under consideration: noble liquids (Xenon or Krypton) and inorganic crystals.

Noble liquids have a potentially very good energy resolution ($\sigma/E \sim 0.4\%/\sqrt{E}$) due to the possibility of a simultaneus measurement of the scintillation light and of the ionization produced by the shower [2]. However, with the constraint of high signal speed the read-out must be limited to the scintillation light only, resulting in a stochastic term in the energy resolution of about $1 \div 2\%$. The other relevant characteristics of noble liquids are an intrinsic speeds below 10 ns and radiation resistance due to recirculation. Limitations on the radiation resistance could come from the front-end electronics, if directly attached to the detector.

Energy resolutions of the order of or better than $1\%/\sqrt{E}$ can be also obtained, in principle, by inorganic crystals [3]. Data on radiation damage are not yet available on big crystals, but one can expect a radiation resistance up to several megarads if the crystal is sufficiently pure. Crystals can also be very fast ($\leq 10\,\text{ns}$) if the read-out is limited to the fast component of the scintillation light.

Finally, both solutions allows for good segmentation and good space resolution.

*3.2 Integrated electromagnetic and hadronic calorimeters*

The need for an extremely precise electromagnetic calorimetry decreases with increasing interaction energy. Sampling calorimeters may be an acceptable alternative if the constant term can be kept low ($\leq 1\%$). Moreover, they can measure both the electromagnetic and the hadronic showers with the same basic design, making the detector simpler and less expensive. Under certain conditions, sampling calorimeters can also achieve compensation, i.e. equal response to electrons and hadrons, thus providing gaussian signals for hadronic showers (an important condition for the trigger, as mentioned in section 1). Absorber materials with small radiation and interaction lengths (lead, iron or copper) allows for very compact detectors.

Several techniques are under study in view of their possible use in SSC/LHC detectors. Some of them could also be considered for ELOISATRON.

*Scintillator-based* calorimeters use either scintillating fibres running (almost) parallel to the incident particle direction or scintillator tiles transverse to it as active material.

The lead/scintillating fibre technique has been developed during the last five years in the framework of the LAA Project at CERN [4]. The basic idea was to build a compensating calorimeter by tuning the lead:scintillator ratio to about 4:1. Then, a good energy resolution could only be achieved by using small (1 mm diameter) scintillating fibres in order to increase the sampling frequency. Measurements on several prototypes have demonstrated the very good characteristics of this calorimeter for what concerns:

- the signal speed (rise time $\sim$ 1 ns, FWHM $\sim$ 4 ns for electrons, $\sim$ 16 ns for hadrons);
- the hadronic energy resolution ($(\sigma/E)_{\mathbf{hadrons}} = 30\%/\sqrt{E} + 1\%$);
- the constant term in electromagnetic energy resolution ($0.28 \pm 0.10\%$ with a stochastic term of about 15%) [5];
- the fine segmentation ($\sigma_{\mathbf{position}} = 5\,\mathrm{mm}/\sqrt{E}$ for electrons and $30\,\mathrm{mm}/\sqrt{E}$ for hadrons with 10 and 50 $\mathrm{cm}^2$ cells, respectively);
- the $e/\pi$ discrimination ($\leq 10^{-3}$ for 50 GeV particles at the first trigger level) [5].

In addition, the high density of the calorimeter ($\lambda = 9.8\,\mathrm{cm}$, $X_0 = 0.75\,\mathrm{cm}$, $r_M = 2.0\,\mathrm{cm}$) makes it very compact, and its structure provides full hermeticity.

Extensive radiation damage tests have been performed on different samples of fibres. The best results have been obtained for 3-HF fibres at 500 ppm doped with Y7 with fourinate cladding‡, which show very small damage after 5 Mrad.

The "tile calorimeter" have been proposed for the SDC detector at SSC [6]. Scintillator tiles, placed perpendicularly to the incident particle direction, are interleaved with absorber (lead or iron) plates. The light produced in each tile is collected by a scintillating fibre doped with a wavelength shifter imbedded in the tile and transmitted to the light detectors through clear fibres. This technology allows to keep the very good electron/hadron discrimination capabilities of a longitudinally segmented calorimeter while improving on hermeticity and granularity with respect to the classical tile/wavelength shifter bar calorimeters. As already mentioned, in order to keep a good electromagnetic energy resolution thin tiles are required. Light yield and collection efficiency are therefore the main problems of this type of calorimeter. Given a sufficiently high photostatistics, similar performances as the lead/scintillating fibre calorimeter are expected.

The advantages of *liquid-based* calorimeters, using either liquid argon or TMS / TMP as active material, derive from the direct collection of charge which lead to stable and well calibrated response. Moreover, they are flexible and easly segmented (both laterally and longitudinally), with a read-out insensitive to magnetic field. Radiation problems can in principle be solved by recirculating the liquid.

A large R&D programme is being pusued to increase the speed of charge collection in order to be compatible with the requirements of supercolliders. In particular, the possibility of fast shaping of the signal has been tested [7]. Results indicate that a pulse peak at about 40 ns intruduce an acceptable electronic noise

‡ Manifactured by Kuraray Ltd., Tokio, Japan.

of 300 MeV in the so-called "accordeon" liquid Argon calorimeter. This new geometry [8] foreses accordion-shaped electrodes running parallel to the particle direction, and produced very good results in terms of electromagnetic resolution ($\sigma/E = 10\%/\sqrt{E} \oplus 0.3\%/E \oplus 0.2\%$), position resolution ($\sigma_{\text{position}} = 3.6\,\text{mm}/\sqrt{E}$), and angular resolution ($\sigma_{\text{direction}} = 75\,\text{mrad}/\sqrt{E}$, with two longitudinal samplings $12\,X_0$ each). Data on hadronic shower measurements with this calorimeter design are not yet available.

Ionizing liquid calorimeters working at room temperatures ("warm liquids") have been under development since many years. After the pioneering work carried out by the UA1 Collaboration, a large R&D effort has been undertaken by the WALIC Collaboration [9]. Results indicate that sufficiently long ionization electron lifetimes can be achieved with either TMP or TMS. Under certain conditions, the high hydrogen content of these liquids may also allow for compensation, even if at the expenses of a rather poor electromagnetic energy resolution. Several problems have still to be solved in a consistently way, in order to make this technology a mature one.

*Gas-based* calorimeters have been used in several large experiments. Their biggest advantages are low cost, high segmentation flexibility, and ease of operation. However, poor energy resolutions are generally obtained from these devices, which are also intrinsically slow, due to the drift time of the ionization electrons to the collecting wires.

Recently, single-gap, gaseous detectors working in the avalanche mode (Parallel Plate Chambers) have been tested for possible use in multi-TeV detectors [10]. Advantages are a very fast signal (few ns resolution with 1 mm gap) and high radiation resistance.

A new calorimeter concept, based on the detection of the Cherenkov light emitted by the shower particles in quartz fibres, is described in the following section. This calorimeter, which is now being considered as a solution for the very forward ($3 \leq |\eta| \leq 5$) region of LHC, may represent the ideal calorimeter for a high-luminosity ELOISATRON.

## 4. The lead/quartz fibres calorimeter

In the course of the NA38 experiment ("Research of a Plasma of Quarks and Gluons produced by the collision of 200 GeV/nucleon sulfur ions on uranium, looking at muon pairs and especially at the suppression of the $J/\psi$ resonance"), it appeared that for the future NA50 experiment (same as NA38, but using Pb-Pb interactions and due for mid 1994), the use of a hadronic calorimeter placed at 0° after the target and measuring the energy of the projectile spectators, was mandatory. This calorimeter, called Zero Degree Calorimeter (ZDC), is described in the NA50 proposal [11] which was accepted by the SPSC early 1992. The requirements for this apparatus are:

a) Radiation hardness: it can be computed that the deposited dose for every month of use will be of the order of 1 GRad.

b) Time response: the beam intensity will be around $2 \times 10^7$ Pb/sec and the data acquisition being done event per event, the pulses have to be back to zero in less than 15 ns to avoid pile-up.

It is apparent that these requirements are very similar to the ones imposed on ELOISATRON calorimeters.

The idea was then to use the lead/scintillating fibre calorimetry technique with the replacement of the plastic scintillating fibers by quartz fibers which have been measured to be highly resistent to radiation. Light would be provided by the Cerenkov effect produced when the shower particles cross the fibers. This effect is extremely fast.

Irradiation of quartz fibers was performed with 2 MeV electrons striking the fibers at 45°. The Cerenkov light output as a function of radiation dose for different kind of fibres is shown in fig. 11. The doped quartz clad fiber shows no sign of damage up to 2.5 GRad while the polymer clad response decrease after 1 GRad due probably to plastic reticulation.

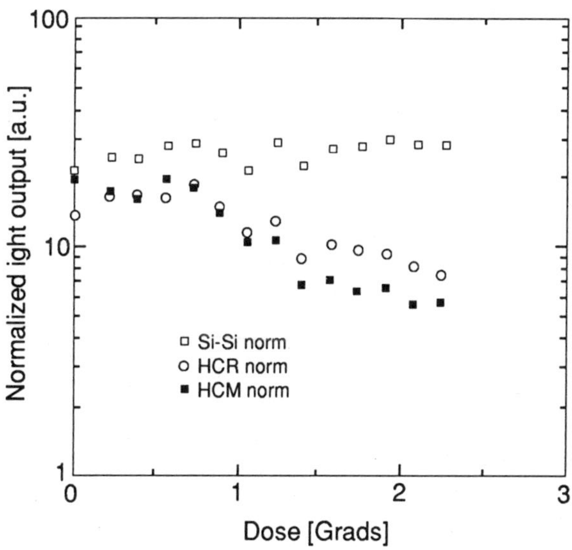


*Fig. 11. Cherenkov light induced by 2 MeV electrons in 3 different quartz fibers. Si-Si stands for quartz cladding. The two others have PMMA cladding. Light normalization was done through a non-irradiated LED-lit fiber.*

*4.1 Simulations*

B. Grosdidier from Strasbourg, has just finished his thesis [12] on the subject of Monte-Carlo simulation of heavy ion induced showers in the ZDC by associating the Lund model program called Fritiof with the well known Geant [5]. Geant has been modified to include a library of heavy ions and a model to create heavy ion fragments (the projectile spectators). The Cherenkov light response in the fibers has been parametrized and computed as a function of the angle and of the minimum distance between the particle trajectory and the fiber axis. The tables (see fig. 12 as an example) obtained that way for different $\beta$ of the particles are incorporated in Geant and called during the tracking of the shower particles. They show that the best angle between fibers and particles is 45°.

This program has been used to simulate the response of the ZDC in different configurations. The results obtained are:

a) The optical pulse is back to zero in about 3 ns.
b) The linearity is exremely good between 0.2 and 33 TeV.
c) The resolution needed for the ZDC can be easily reached for a lead to fiber volume ratio of 40 ($\sigma/E = 250\%/\sqrt{E}$), minimizing the price for the fibers and the CPU time for the simulations.

It is also found that the radial distribution of the showers measured in these conditions is rather narrow, that is 90% of the superluminar portion of hadronic showers is contained in a radius of 5 cm.

*4.2 Prototype studies*

The first prototype was built in the fall of 1991. Its small dimensions (20 cm long, $5 \times 5\,\text{cm}^2$ in cross section) made it an electromagnetic calorimeter. The Pb-to-fiber ratio was 4 and two photomultipliers were used, one each 10 cm in depth. Bombarded by 10 and 30 GeV electrons, it yielded a resolution $\sigma/E = 24\%/\sqrt{E} \oplus 3\%$. The pulses in the counting room were back to zero in 10 ns. The performances of the same prototype were also tested with 200 GeV/n sulfur ions with 0, 1 or 2 interaction lengths of tungsten in front. The comparison with the simulation (see fig. 13) gave confidence in this rather complicated Monte-Carlo program.

This is the reason why in spring 1992, the second prototype was built: a "small" ($90 \times 22 \times 22\,\text{cm}^3$) hadronic calorimeter with a ratio Pb-to-fiber of 10. The fibers were distributed to different photomultipliers in such a way that by choosing different photomultipliers, it was possible to work with four different fiber densities (1/40, 1/20, 1/13 and 1/10). Bombarded by 200 GeV/n sulfur ions it showed a resolution $\sigma/E$ in the range 7-9% dominated by the leakage of the showers (and in full agreement with the simulations). When moving the beam sideways, a significant change in output was observed only when closer to the side of the detector than 2 cm, indicating that the radial dimension of the detectable part of the shower is as small as predicted.

Pulse shapes were exactly the same as for the first prototype, back to zero in 10 ns after cables.

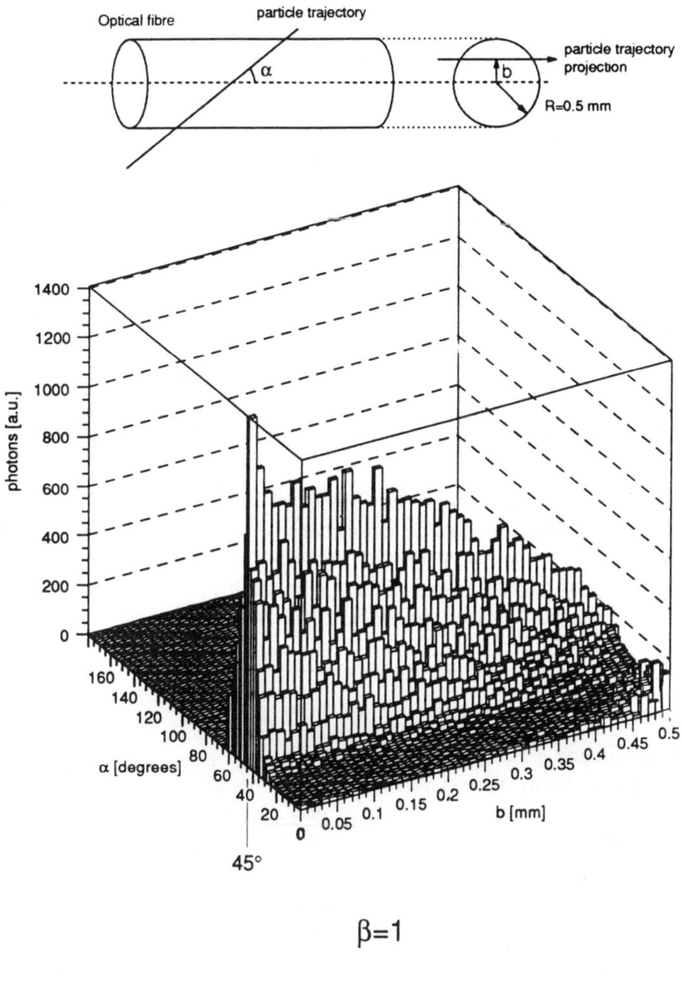


*Fig. 12. Top) Geometrical parameters of an optical fibre-particle trajectory system. Bottom) Number of photons produced by Cherenkov effect and collected at the extremity of a fibre (1 mm diameter and numerical aperture of 0.37) intercepted by a charged particle (Z=1, $\beta$=1).*

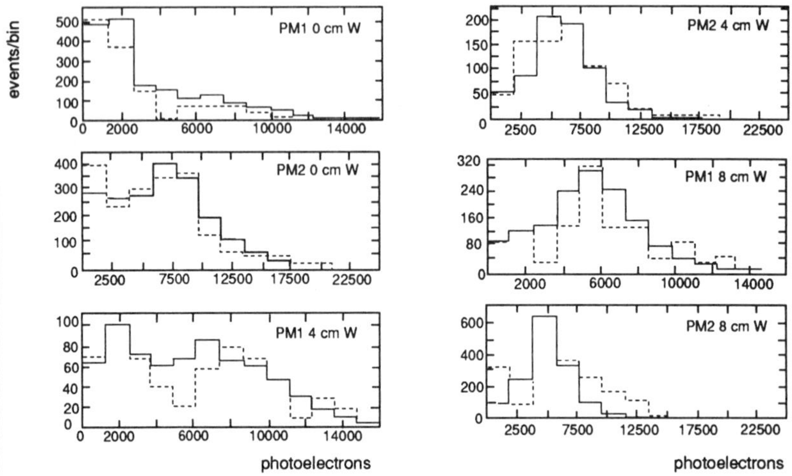


*Fig. 13. Comparison between the measured (full lines) and simulated (dotted lines) responses of the first prototype calorimeter. Zero, four and eight cm of tungsten (0, 1, 2 interaction lengths) were put in front of the calorimeter module. Incident particles were 200 GeV/n sulfur ions. PM1 and PM2 are the photomultipliers looking at the first and second half of the module in depth, respectively.*

The conclusions that can be derived from these studies are:

a) The response of such a calorimeter is very fast.
b) The quartz optical fibers are extremely radiation resistant.
c) The lateral shower dimensions of TeV particles are of the order of a few cm.
d) The Monte-Carlo program reproduces very well the experimental results.

*4.3 Further R&D to adapt the NA38 ZDC to the needs of Multi-TeV colliders*

The constraints on calorimeters for a multi-TeV collider due to radiation damage are a little bit less demanding than for NA38/50. In addition, the time response of the calorimeter and the effective shower size are ideal. However, a number of questions are still to be studied and are the subject of an R&D effort started in 1992 by the LAA Project at CERN, INFN Bologna, Italy and CRN Strasbourg, France.

In particular, compensation, resolution, linearity at low energies and different projectiles, spatial resolution for jets, optimal geometry and cost will be studied in detail. A special effort will be dedicated to improve the collection efficiency for photons as this relates directly with the energy resolution.

Simulation studies have started on a $40 \times 40 \times 200\,\text{cm}^3$ prototype with a fibre-to-lead filling ratio of 1:10 and uniform distribution of fibres inside the lead. Results on spatial and energy resolution, and on the signal variation as a function of the incident angle are shown in Figs. 14, 15 and 16, respectively.

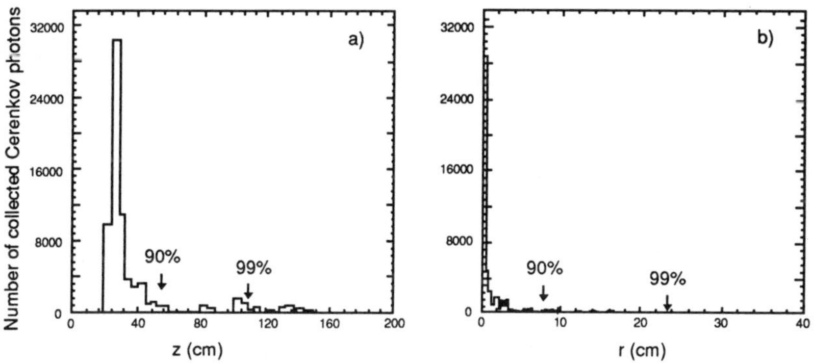


*Fig. 14. Longitudinal (a) and lateral (b) development of a 160 GeV proton shower in the lead/quartz fibre calorimeter described in the text.*

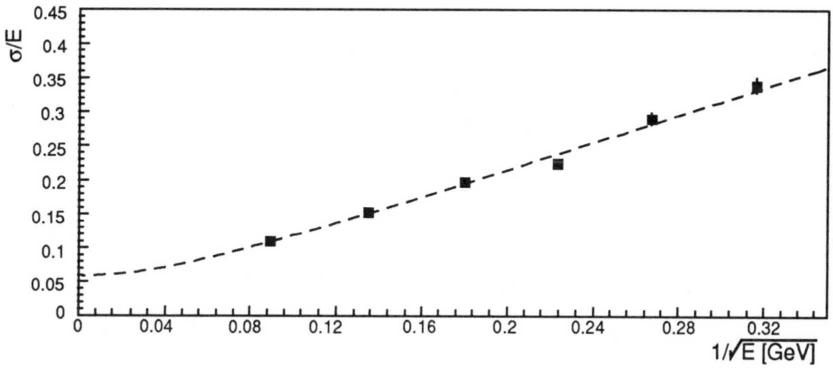


*Fig. 15. Hadronic energy resolution for the lead/quartz fibre calorimeter described in the text. The fit (dashed line) is: $\sigma/E = 100\%/\sqrt{E} \oplus 6\%$.*

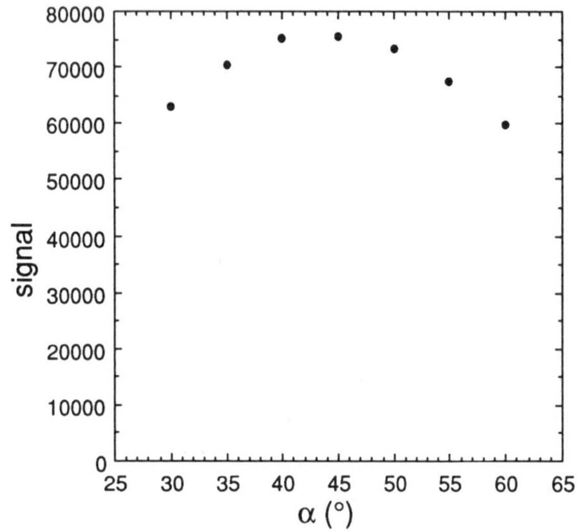


*Fig. 16. Signal versus particle incident angle $\alpha$ for the lead/quartz fibre calorimeter described in the text.*

## 5. Conclusions

Present calorimeter technologies allow the construction of hermetic, compact and precise calorimeters which meet the challenging environment of very high energy hadron colliders.

The main technical limitation appears to be the signal speed which imposes a maximum bunch crossing frequency of about 100 MHz. Other limitations at very high luminosities come from the pile-up of minimum bias events, which translates into an effective threshold on the detectable energy.

The conclusion is that the physics goals of ELOISATRON may be completely fulfilled in a staging scenario, where increasing luminosities allow to reach unexplored energies, while leaving a detailed analysis of each energy range to dedicated lower luminosity runs.

**References.**


[1] PYTHIA:
H.-U.Bengtsson and T.Sjöstrand, Computer Physics Commun. 46 (1987) 43.
JETSET:
T.Sjöstrand, Computer Physics Commun. 39 (1986) 347,
T.Sjöstrand and M.Bengtsson, Computer Physics Commun. 43 (1987) 367.

[2] J. Seguinot et al., "Liquid xenon ionization and scintillation studies for a homogeneous vector-electromagnetic calorimeter", Proceedings Int. Conf. on Liquid Radiator Detectors, Tokyo, 7-10 Apr. 1992 (Phys. Soc. Japan - Tokyo, 1992), p. 410.

[3] P. Lecoq et al., "Results on new scintillating crystals from the "Crystal Clear" Collaboration", preprint CERN-PPE/92-205.
V. Peskov and A. Zichichi, "A new scintillator: $LiBaF_3$, preliminary tests", preprint CERN-PPE/91-139.
A.V. Antipov et al., "Investigation of new inorganic materials for homogeneous electromagnetic calorimeters", preprint IHEP 92-81.
S. Anderson et al., "Further results on cerium fluoride crystals", preprint CERN-PPE/93-23.

[4] R. DeSalvo et al., Nucl. Instr. and Meth., A279 (1989) 467.
D. Acosta et al., Nucl. Instr. and Meth., A294 (1990) 193.
D. Acosta et al., Nucl. Instr. and Meth., A302 (1991) 36.
D. Acosta et al., Nucl. Instr. and Meth., A302 (1991) 36.
D. Acosta et al., Nucl. Instr. and Meth., A305 (1991) 55.
D. Acosta et al., Nucl. Instr. and Meth., A308 (1991) 481.
D. Acosta et al., Nucl. Instr. and Meth., A309 (1991) 143.
D. Acosta et al., Nucl. instrum. and Meth., B62 (1991) 116.
D. Acosta et al., Nucl. instrum. and Meth., A314 (1992) 431.
D. Acosta et al., Nucl. instrum. and Meth., A316 (1992) 184.
D. Acosta et al., Nucl. instrum. and Meth., A320 (1992) 128.
R. DeSalvo, "Dream on a Supercollider Spaghetti Calorimeter", Proceedings of the Workshop on Calorimetry for the Supercollider, March 13-17, 1989, Tuscaloosa, Alabama, World Scientific (ed. R. Donaldson, M.G.D. Gilchriese), page 383.
R. DeSalvo, "Trigger and data acquisition prospects for the Spaghetti Calorimeter", CERN/LAA-HC-90-002, presented at the Workshop on Perspectives for New Detectors in Future Supercolliders, Erice 17-24 October 1989.
J.M. Seixas, "Application of signal processing techniques on the analysis of a lead /scintillating fiber calorimeter in short acquisition time environments", preprint CERN-PPE/91-165.
A. Contin et al., Nucl. instrum. and Meth., A315 (1992) 344.
B. Bencheikh et al., Nucl. instrum. and Meth., A315 (1992) 354.
G. Anzivino et al., Radiat. Phys. Chem. 41 (1993) 283.

[5] Preliminary results.

[6] "Solenoidal detector collaboration SDC, technical design of a detector", SDC 92-201, and SSCL SRT 1215.
P. de Barbaro et al., Nucl. instrum. and Meth., A315 (1992) 317.
S. Han et al., "Radiation hardness tests of tile/fiber calorimeter structures for SDC", preprint FERMILAB FN 596 (1992).
D.G. Underwood et al., "Some fiber-tile optical studies for SDC electromagnetic calorimeter", preprint ANL HEP CP 92-94.

[7] B. Aubert et al., Nucl. instrum. and Meth., A321 (1992) 467.

[8] B. Aubert et al., Nucl. instrum. and Meth., A309 (1991) 438.
B. Aubert et al., "Performance of a liquid argon electromagnetic calorimeter with a cylindrical accordion geometry", preprint CERN-PPE/92-129.

[9] E. Radermacher, "Calorimetry with warm liquids: summary and prospects for LHC", CERN 90-10, Vol. III, p. 341.
M. Pripstein, "Developments in warm liquid calorimetry", preprint LBL-30282 (1991).
M. Hoff et al., "A warm liquid forward calorimeter for the SSC - a new technology with new challenges", preprint LBL-30402 (1991).

[10] Y. Galaktionov et al., Nucl. instrum. and Meth., A317 (1992) 116.

[11] M.C. Abreu et al., CERN/SPSLC 91-55, November 6, 1991.

[12] B. Grosdidier, Thése Université Luois Pasteur, Strasbourg, France, July 1992.

# DEVELOPMENT OF A 13.5 T $Nb_3Sn$ DIPOLE MAGNET FOR FUTURE PARTICLE ACCELERATORS

Q.T. Lin, G.F. Luan, W.J. Shen* ,L.Y. Wang & A. Zichichi
*World Laboratory, Laussane, Switzerland*
* *SSC Laboratory, Taxes, U.S.A.*

The development of a 13.5 Tesla superconducting Nb3Sn dipole magnet with 50 mm bore is under consideration at World Lab as a basic research program for future particle accelerators. In order to reach a field of 13.5 Tesla, the magnet will be designed with a Nb3Sn superconductor cooled with 4.3 K helium. In this paper the main parameters of the dipole and the uniformities of its field harmonics are presented.

## 1. Introduction

The desire to investigate elementary particles and to search for new phenomena has led physicists and engineers to build particle accelerators for higher and higher energies starting from small rectifier generators and cyclotrons about 60 years ago, over betatrons, linacs and synchrotrons, to the big accelerators and storage rings in the TeV and multi-TeV range existing today or presently under construction. The biggest one in the world, at present under design considerations, is the ( 100 + 100 ) TeV ELOISATRON Collider with a luminosity of $10^{34}$ $cm^{-2}s^{-1}$ in Italy (see Table.1). The development of 13.5 Tesla Nb3Sn superconducting dipole magnets will be a key issue to realize such a proton collider.

Table 1. List of Superconducting high energy particle accelerators

| Institute | FNAL | DESY | BNL | CERN | SSCL | INFN |
|---|---|---|---|---|---|---|
| Name | E/Double | HERA | RHIC | LHC | SSC | ENL |
| Energy [Tev] | 1.0 | 0.8 | 0.1/N | 7.7 | 20 | 100 |
| Diameter [km] | 2 | 2 | 2 | 9 | 25 | 273 |
| Particle | p/p-bar | p/e | N/N | p/p | p/p | p/p |
| Luminosity [$cm^{-2}s^{-1}$] | $10^{30}$ | $10^{31}$ | $10^{31}$ | $10^{34}$ | $10^{33}$ | $10^{34}$ |
| B in Dipole [Tesla] | 4.5 | 4.68 | 3.45 | 10 | 6.6 | 12 |
| #Dipoles | 774 | 416 | 288 | 1760(x2) | 8652 | |
| Completion [year] | 1986 | 1990 | (1996) | (2002) | (2003) | |

In this paper, the conceptual design of a 13.5 tesla $Nb_3Sn$ superconducting dipole magnet, the analysis of its field uniformity and its realization with Nb3Sn superconducting cables are presented.

## 2. Magnet design

The design goal for the Nb3Sn superconducting dipole magnet is to realize a nominal field of 13.5 Tesla in a 50 mm aperture surrounded by an iron yoke configuration. The superconducting dipole (shown in Figure 1) has 4 layers of Nb3Sn coils which are wound with four different Nb3Sn cables. The coils will be fabricated using the "wind and react" technology and are vacuum impregnated with epoxy. The reaction process for Nb3Sn requires to keep the coils in an oven at 700 $^{0}$C for one week. The cable will be insulated with a Mica-Alumina Silica 1/2 lapped tape which is 0.12 mm thick. The four layers of coils will be collared with an elliptical Aluminium-alloy collar in the middle plane, surrounded by two vertically split iron halves and finally an outer retaining Al-alloy cylinder of high prestress. The main parameters of the Nb3Sn dipole is listed in the Table 2.

Table 2. Main Parameters of 13.5 T, 50 mm Aperture Dipole

| ITEM | | | 1st layer | | 2nd layer | | 3rd layer | | 4th layer | |
|---|---|---|---|---|---|---|---|---|---|---|
| Coil Dimension | I.D. (mm) | | 50 | | 82 | | 114 | | 146 | |
| | O.D.(mm) | | 80 | | 112 | | 144 | | 176 | |
| | q | | $74^{o}$ | | $66.75^{o}$ | | $35^{o}$ | | $20^{o}$ | |
| Bmax at 4.3 K | | (T) | 14.28 | | 12.65 | | 10.34 | | 8.81 | |
| Location | | | X | Y | X | Y | X | Y | X | Y |
| (mm, mm) | | | 6.9 | 24.2 | 16.3 | 38.0 | 50.6 | 35.5 | 73.6 | 27.5 |
| Nb3Sn Cable Size | High* | (mm) | 15 | | 15 | | 15 | | 15 | |
| | Thick-ness | L(mm) | 1.696 | | 1.1996 | | 0.8118 | | 0.6727 | |
| | | S(mm) | 1.231 | | 0.8783 | | 0.6427 | | 0.5581 | |
| Current | | (A) | 5500 | | 5500 | | 5500 | | 5500 | |
| Current Density (A/mm$^2$) | | | 250 | | 385 | | 550 | | 650 | |

* Including insulation.

## 3. Analysis on field uniformity

The design has been performed with the PE2D package[1] that computes and optimizes analytically the magnetic field and the multipoles. The objectives of the optimization were to obtain the highest possible central field, low sextupole and decapole coefficients, and four layers with the same current margin. The maximum field at the conductor for a bore field of 13.5 T are shown in the Table 2. The magnetic flux lines in the coils and in the iron yoke are shown in Fig 2. Many efforts have been spent to reduce the change in sextupole from low field to the operating field. The harmonics field uniformities are summarized in the Table 3. The sextupole variation from low current to the field of 14.7 T at 6000 A/turn is 2.4 units ( Figure 3 ).

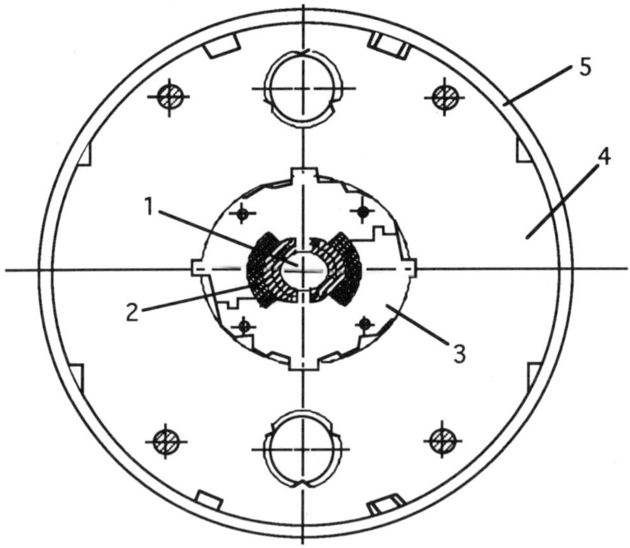


Fig. 1. Cross section of the 13.5 T Nb3Sn dipole magnet:
1) beam pipe, 2) Nb3Sn layers, 3) Al-alloy collar, 4) iron yoke, 5) Al-alloy loop

Table 3. Field Harmonics Uniformities ( at R = 1.00 cm )

| Current (A) | Bo (T) | b2/bo | b4/bo | b6/bo |
|---|---|---|---|---|
| 300 | 0.97069 | -1.1879e-03 | -4.6544e-05 | -1.2593e-05 |
| 600 | 1.9359 | -1.1577e-03 | -4.5869e-05 | -1.4156e-05 |
| 900 | 2.8301 | -8.8494e-04 | -4.4383e-05 | -1.3105e-05 |
| 1200 | 3.6335 | -5.2920e-04 | -4.2890e-05 | -1.2578e-05 |
| 1500 | 4.3965 | -2.6184e-04 | -4.2692e-05 | -1.4222e-05 |
| 1800 | 5.1349 | -9.9540e-05 | -4.3434e-05 | -1.4536e-05 |
| 2100 | 5.8486 | -2.7064e-05 | -4.4816e-05 | -1.3935e-05 |
| 2400 | 6.5473 | 5.3514e-07 | -4.6305e-05 | -1.4721e-05 |
| 2700 | 7.2383 | 1.3103e-05 | -4.7384e-05 | -1.5428e-05 |
| 3000 | 7.9255 | 2.0814e-05 | -4.8079e-05 | -1.5948e-05 |
| 3300 | 8.6105 | 2.6900e-05 | -4.8684e-05 | -1.5132e-05 |
| 3600 | 9.2939 | 3.2282e-05 | -4.9958e-05 | -1.6221e-05 |
| 3900 | 9.9762 | 3.7136e-05 | -4.9923e-05 | -1.5619e-05 |
| 4200 | 10.658 | 4.4241e-05 | -4.4985e-05 | -5.9942e-06 |
| 4500 | 11.339 | 4.6671e-05 | -5.9728e-05 | -3.6142e-05 |
| 4800 | 12.019 | 5.2939e-05 | -4.5805e-05 | -5.9794e-06 |
| 4950 | 12.359 | 6.2655e-05 | -5.0456e-05 | -2.2206e-05 |
| 5100 | 12.699 | 5.3167e-05 | -5.9939e-05 | -3.1044e-05 |
| 5250 | 13.039 | 5.8984e-05 | -4.0487e-05 | 4.6713e-07 |
| 5400 | 13.379 | 6.5153e-05 | -4.5590e-05 | -1.5506e-05 |
| 5700 | 14.059 | 6.9559e-05 | -4.3612e-05 | -7.5068e-06 |
| 6000 | 14.733 | 6.9323e-04 | -5.5636e-05 | -3.1704e-05 |

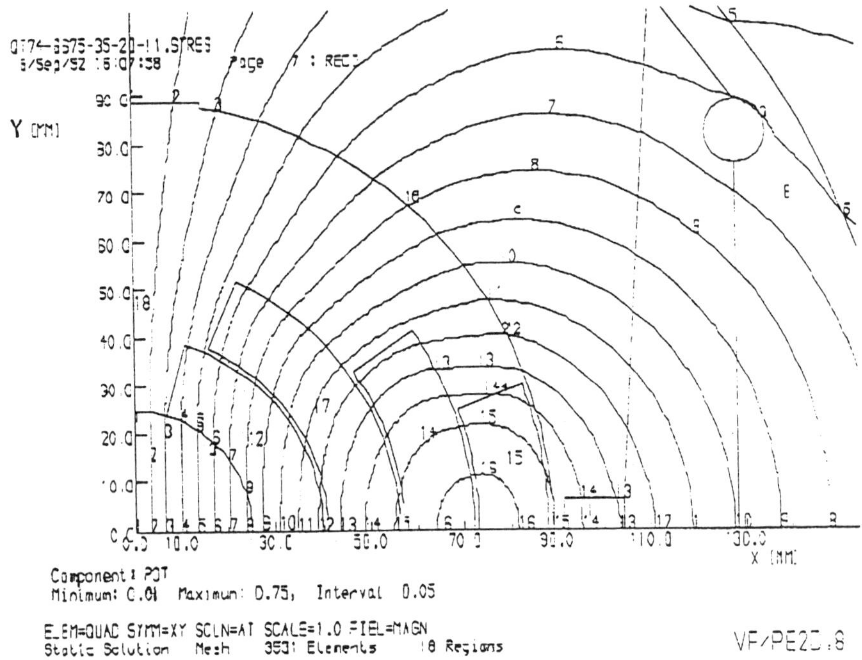


Fig. 2 Magnetic flux lines in the coils and in the iron

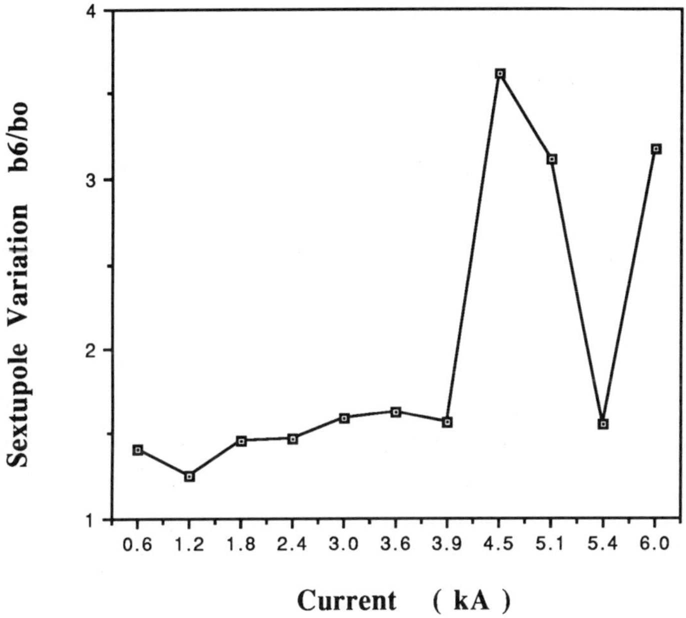


Fig. 3. Sextupole variation

**4. Development of Nb3Sn superconducting cables**

The proposed 13.5 T dipole magnet is based on the Nb3Sn current density values of 763 A/mm$^2$ at 14.28 T, 1176 A/mm$^2$ at 12.65 T, 1527 A/mm$^2$ at 10.34 T and 1985 A/mm$^2$ at 8.81 T in the non copper part of the superconducting wire. These values, which have been already obtained in 25 μ filaments fabricated by the powder technique on a laboratory scale ( see Table 4 ), represents an increase of 30 % in current density on the commercially available Nb3Sn and constitutes a new technical challenge for superconductor supplies. LBL has obtained some Nb3Sn short samples manufactured by IGC and Teledyne Wah Chang, which are specified to reach the current density values of 745 A/mm$^2$ at 13.5 T and 1550 A/mm$^2$ at 10.5 T at a temperature of 4.35 K[2].

Table 4 Resent laboratory data for $Nb_3Sn$

| Process | Jc, 10T, 4.2K ($A/mm^2$) | Filament size (μm) |
|---|---|---|
| Powder route | 1800 | 25 |
| Internal diffusion | 1200 | 45 |
| Sn in Nb tubes | 1500 | 15 |
| Improved bronze route | 1600 | 1 |
| Jelly roll | 1400 | 9-10 |
| European Collaboration Desiderata | 1600 | ≤5 |

## 5. Conclusion

A conceptual design study of a $Nb_3Sn$ superconducting dipole magnet for future high luminosity colliders has been carried out at World Lab. With a four-layer coil, it will generate a central field of 13.5 T at the nominal current of 5500 A. The sixtupole variation from low current to 14.7 T field at 6000 A/turn is 2.4 units. The R & D work on the $Nb_3Sn$ superconducting cables for this dipole magnet has to concentrate on increasing critical current density to 1000 $A/mm^2$ at 13 Tesla at a temperature of 4.35 K.

## Acknowledgements

The authors thank Dr.H. Leutz for his helpful discussions and the careful reading of the manuscript.

## References

1. Vector Fields Ltd, *The PE2D Reference Manual*, VF-06-90-24.

2. D.Dell'Orco, R. Scanlan, C.E. Taylor, *LBL-32072*,**UC-414** (to be presented in the 1992 ASC Conference).

# RADIATION AT HADRON COLLIDERS: SCALING TO THE HIGHEST ENERGIES

Donald E. Groom

*Lawrence Berkeley Laboratory, Berkeley, CA 94720, USA*

The ELOISATRON is proposed to operate at $\sqrt{s} = 200$ TeV with a luminosity of $\mathcal{L} = 10^{34}\,\text{cm}^{-2}\text{s}^{-1}$. Even higher luminosities are possible from a machine point of view, but the radiation and data acquisition problems are unprecedented even by LHC standards. We attempt quantitative assessment of some of the radiation levels at such a machine—in the tunnel, in the final focus system, and in a detector. The first step is to extrapolate the *p-p* inelastic cross section, the height of the pseudorapidity plateau, and the mean transverse momentum; these are thought to adequately characterize particle production at the interaction point. Extrapolation of neutron yield in cascades to ELN beam-loss situations is also straightforward. Neutron flux in the tunnel is nearly 6 times worse than at the LHC, but radiation in the final focus system is about an order of magnitude more severe. Detector levels are found by scaling results obtained for the SSC, where it was found that the ionizing dose ($D$) or neutron fluence ($\phi$) can be described by

$$D \text{ (or } \phi) = \frac{A}{r^2 \sin^{2+\alpha}\theta} .$$

Here $A$ depends on the process and exposure time, $\alpha$ is less than unity, and $r$ is the distance from the interaction point. It is found that detector radiation levels are about a factor of two higher than at the LHC if the ELN luminosity is limited to $10^{34}\,\text{cm}^{-2}\text{s}^{-1}$.

## 1. Overview of the problems

As part of the planning for the SSC, studies were made of the likely machine vacuum [1], neutron flux in the tunnel as a result of beam-gas collisions [1,2,3], heating due to debris from interaction point (IP) collisions being absorbed in the front collimator and final-focus quadrupoles [1], and radiation levels in detectors [4]. The results underscored the obvious: that the new era of high-energy, high-current, high-luminosity colliders would bring with it radiation situations fundamentally different that those to which we are accustomed. In the first place, proton synchrotron radiation desorbs gas from the vacuum pipe (or inner liner), so that cryopumping does not automatically result in an ultrahigh vacuum. Secondly, current losses are dominated by IP collision losses, which means that IP collisions are the dominant radiation source in the interaction hall. Under nominal SSC operating conditions, IP collisions result in the same energy deposition rate in the region (mostly in the final-focus quads) as would total loss of one beam into the region every 6 days.

In this report, I try to extrapolate even further, to the energy and luminosity of the proposed ELOISATRON (ELN). To say the machine parameters are not yet final is something of an understatement, so for definiteness we define the ELN in Table 1. For our present purposes it has a beam energy of 100 TeV and a luminosity of $10^{34}\,\text{cm}^{-2}\text{s}^{-1}$.

Three main subjects will be addressed:

1. *The accelerator tunnel.* The extrapolation to 100 TeV is not difficult, because the center of mass energy of the *p*-gas system is only 433 GeV (for hydrogen), where the cross sections are well measured. The problem has to do with effective gas pressure. On the basis of the SSC studies, I think the estimated $10^{-9}$ torr is highly imaginative, unless something very clever is done with liners.

**Table 1**

Quantities related to radiation levels at the Tevatron, LHC, SSC, and ELN. In the first group are operational or design parameters; in the case of the ELN they define the the machine for purposes of this report. The second set of entries are derived quantities and extrapolated particle production parameters. The third group have to do with the neutron flux in the main ring tunnel, and the final group have to do with a $4\pi$ detector in an interaction hall. Complete descriptions are given in the text. Many of the Tevatron entries are for reference only, since actual radiation levels are not dominated by IP collisions.

| Quantity | Tevatron | LHC | SSC | ELN |
|---|---|---|---|---|
| $E_b$ (TeV) | 0.90 | 7.7 | 20 | 100 |
| MR circumference (km) | 6.28 | 26.7 | 87.1 | 356 |
| $\mathcal{L}_{ave}$ ($cm^{-2}s^{-1}$) | $2 \times 10^{30}$ | $1.7 \times 10^{34}$ | $1 \times 10^{33}$ | $1 \times 10^{34}$ |
| Protons/ring | $6 \times 10^{11}$ ($\bar{p}$) | $1.0 \times 10^{14}$ | $1.5 \times 10^{14}$ | $7 \times 10^{14}$ |
| $\sqrt{2m_p E_b}$ (GeV) | 41 | 120 | 194 | 433 |
| $\sigma_{pp}(\sqrt{2m\, E_b})$ (mb) | 42 | 48 | 51 | 58 |
| $\sqrt{s}$ (TeV) | 1.8 | 15.4 | 40 | 100 |
| $\sigma_{inel}(\sqrt{s})$ (mb) | 56 | 84 | 100 | 130 |
| $H$ (charged particles per unit $\eta$) | 3.9 | 6.2 | 7.5 | 10.6 |
| $\langle p_\perp \rangle$ (GeV/$c$) | 0.46 | 0.55 | 0.60 | 0.70 |
| Plateau edge $\eta_0$ (5% $\sqrt{s}$) | 3.9 | 5.4 | 6.1 | 7.2 |
| Interactions/s | $1.1 \times 10^{5}$ | $1.4 \times 10^{9}$ | $1.0 \times 10^{8}$ | $1.3 \times 10^{9}$ |
| Interactions/yr | $1.1 \times 10^{12}$ | $1.4 \times 10^{16}$ | $1.0 \times 10^{15}$ | $1.3 \times 10^{16}$ |
| Relative interaction rates | $1 \times 10^{-3}$ | 14 | 1 | 13 |
| $\sqrt{s} \times$ rate (kW) | $3 \times 10^{-5}$ | 3.5 | 0.64 | 41 |
| Direct dose rate @ 10 cm (Gy/yr) | 2.2 | 45200 | 3800 | 69300 |
| Protons/cm (one ring) | $9.6 \times 10^{5}$ | $3.8 \times 10^{7}$ | $1.7 \times 10^{7}$ | $2 \times 10^{7}$ |
| $N_{H_2}\, \sigma_{pp}\, c$ ($s^{-1}$ $torr^{-1}$) | 88 | 101 | 109 | 123 |
| Beam-gas lifetime@$10^{-9}$ torr $H_2$ (hr) | 3160 | 2740 | 2550 | 2250 |
| Loss rate @$10^{-9}$ torr $H_2$ ($cm^{-1}s^{-1}$) | 0.08 | 3.79 | 1.87 | 2.42 |
| Loss rate$\times E_b$ @$10^{-9}$ torr ($\mu$W/cm) | 0.01 | 4.7 | 6.0 | 39 |
| Loss rate around ring (W) | 0.008 | 12 | 52 | 1380 |
| Neutrons $cm^{-1}$ (per collision) | 6.4 | 39 | 88 | 340 |
| Flux @$10^{-9}$ torr ($cm^{-2}\, s^{-1}$) (2 m from 1 ring) | 0.54 | 150 | 165 | 824 |
| Flux @$10^{-8}$ torr ($cm^{-2}\, yr^{-1}$) (2 m from 2 rings) | $1.1 \times 10^{8}$ | $3.0 \times 10^{10}$ | $3.3 \times 10^{10}$ | $1.7 \times 10^{11}$ |
| "Final focus neutron flux" ($cm^{-2}yr^{-1}$) | $2.4 \times 10^{9}$ | $1.9 \times 10^{14}$ | $2.9 \times 10^{14}$ | $1.5 \times 10^{15}$ |
| $A$ for neutron flux @ 1 m ($cm^{-2}yr^{-1}$) ($\alpha = 0.67$) | $7.4 \times 10^{8}$ | $1.7 \times 10^{13}$ | $1.5 \times 10^{12}$ | $3.0 \times 10^{13}$ |
| Relative neutron flux | $5 \times 10^{-4}$ | 11 | 1 | 20 |
| $A$ for dose rate from photons @ 1 m (Gy $yr^{-1}$) ($\alpha = 0.93$) | 0.057 | 1360 | 124 | 2610 |
| $A$ for dose rate from hadrons @ 1 m (Gy $yr^{-1}$) ($\alpha = 0.89$) | 0.013 | 320 | 29 | 610 |
| Cavity neutron flux scale factor ($\alpha = 0.50$) | $5 \times 10^{-4}$ | 11 | 1 | 20 |

2. *Final-focus optics* (and possible front collimator). This is an exceedingly difficult problem. Until recently, none of the codes described particle production in the diffraction region. While one knows that 95% of the collision energy goes into these magnets, its distribution depends on the momentum distribution of the debris from the IP. The number of neutrons from the two ends is comparable to the number produced around the entire tunnel.

3. *The detector.* This is regarded as a solved problem, except for modest extrapolations of cross section, average number of particles per pseudorapidity interval, and mean value of the transverse momentum.

We first revisit our extrapolations of these quantities, with 100 TeV as the target energy and with some new data. After a discussion of the energy scaling of hadronic cascades, we apply the results to the three situations described above.

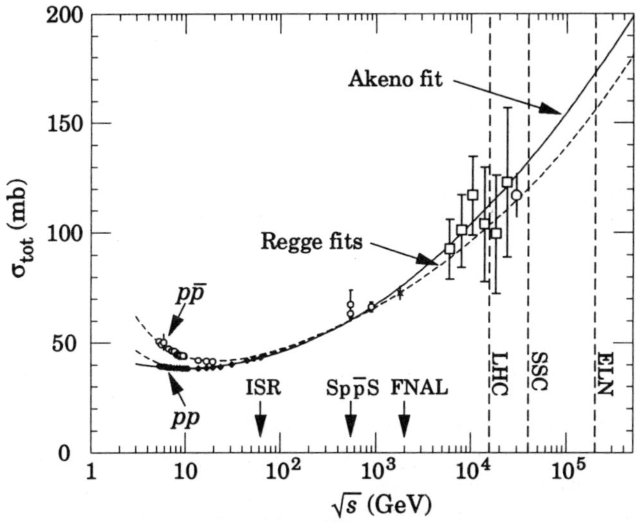


FIG. 1. Data and extrapolations of the *p-p* and $\overline{p}$-*p* cross sections. References to accelerator data prior to 1985 are given in Table II of Ref. 5. The Fly's Eye result (circle at 30 TeV), the final CØ cross section at 1.8 TeV, and the cosmic ray results from the Akeno detector (squares at high energy) are from Refs.6, 7, and 8, respectively. The dotted curves are recent Regge fits to the accelerator data [9], and the solid curve, $\sigma_{\rm tot}^{pp}$ (mb) $= 38.5 + 1.37 \ln^2(\sqrt{s}/10\ {\rm GeV})$, is a best fit to all of the *pp* data shown exclusive of Fly's Eye. The Akeno result at 40 TeV is $(133 \pm 10)$ mb, with additional uncertainty due to conversion of *p*-air cross sections to *p-p* cross sections.

## 2. The *p-p* cross section

The available *p-p* and $\overline{p}$-*p* cross section measurements are shown in Fig. 1. The data include a final value from the CØ experiment at 1800 GeV and new cosmic ray data from the Akeno detector.

Conventional wisdom says that about 1/4 of the total cross section is elastic, 1/4 diffractive, and 1/2 inelastic. In the present analysis we are dependent on such parameters as the number of charged particles per pseudorapidity interval per inelastic event, and in this context inelastic includes diffractive. We take 3/4 of the total cross section as the appropriate inelastic cross section. The Regge fits (of the form $c_1\, s^{e_1} + c_2\, s^{-e_2}$) would differ very little from the Akeno plus accelerator $\ln s$ fits if the cosmic ray data had been used in the former, so we take the Akeno parameterization for our present purposes. The inelastic cross sections listed in Tables 1 and 2 are 3/4 of the values obtained using the form given in the figure caption.

Most of the elastic cross section is exponential in the 4-momentum transfer $t$, which means it is Gaussian in both $\theta_x$ and $\theta_y$. For the SSC parameters the width of the distribution is only slightly greater than the angular spread of the beam due to the emittance at a high-luminosity IP, so elastic scattering has no noticeable effect or importance. The same is likely true at the ELN, depending upon the chosen emittance.

Although 3/4 of the Akeno fit to the total cross section is almost exactly 100 mb at 40 TeV, this value really amounts to a definition in the SSC studies: the average collision rate at the nominal SSC luminosity ($10^{33}\,\mathrm{cm}^{-2}\mathrm{s}^{-1}$) is taken as $10^8\ \mathrm{s}^{-1}$, and the "year" is defined as $10^7$ s to allow for the realities of machine operation. There is no particular rationale for these numbers, but comparisons with the SSC should be made to the nominal rate of $10^{15}\ \mathrm{yr}^{-1}$, not to the intermediate numbers.

Similarly, scaling to the LHC or ELN needs to be done in a way which is consistent with this if the numbers are to be compared. That is, reference inelastic cross sections should lie on a reasonable smooth curve which includes the lower-energy data. The numbers from the Akeno formula numbers are no more defensible than others being used, but do satisfy the "smooth curve" requirement.

### 3. Height of the pseudorapidity plateau

The the number of charged particles produced in an inelastic $p$-$p$ collision per pseudorapidity interval ($dN_{\mathrm{ch}}/d\eta$) is a fairly constant function of $\eta$, which we call $H$ in this paper. There is a dip in the middle which may become more pronounced with increasing energy, so one should take $H$ to be about 10% higher than $dN_{\mathrm{ch}}/d\eta|_{\eta=0}$.

The data are summarized in Fig. 2. Sp$\bar{\mathrm{p}}$S data and earlier references are given in Ref. 10, and the Tevatron point is taken from Ref. 11. In interpreting these data, care must be taken to understand the trigger, that is, the definition of an inelastic collision. If single-diffractive events can trigger an event, then a higher number of triggers is used to normalize $dN_{\mathrm{ch}}/d\eta$, and the result is smaller. The UA5 data are reported both ways, and differ by about 10%. We show the "inelastic" case. My understanding of the Tevatron results is that a double-ended trigger was used, so a point 10% lower than their reported result is shown.

Just as was the case with cross sections, there is no particularly dependable way to extrapolate to higher energies. We use power-law growth shown in the figure because it describes the CDF result (which was not included in the fit) and because it is a little more conservative. The adopted values of $H$, 10% higher than the values from the graph, are given in Table 1.

Some care must be taken. In the first place, the various production codes give seriously different values for $H$, and it even changes between code versions. It is important to normalize code results to experimental measurements whenever possible, and to report the assumed value of $H$ along with any simulation results. It is also possible that the curve

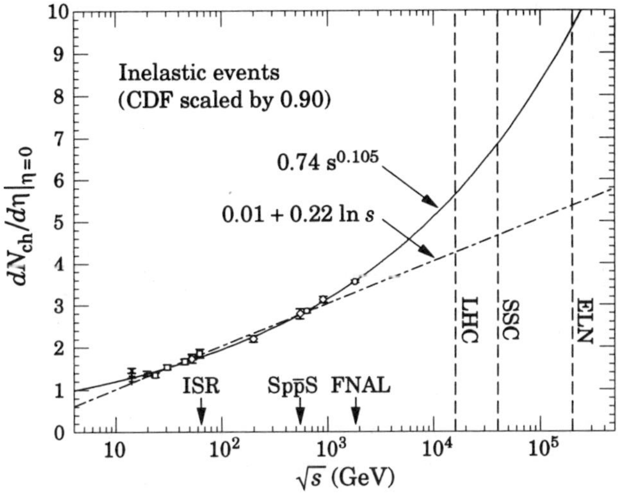


FIG. 2. Data and extrapolations of $dN_{ch}/d\eta|_{\eta=0}$.

actually does have some kind of "kink" after which it turns sharply upward [12]. The value $H(200\text{ TeV}) = 13$ used elsewhere in these proceedings might not be unreasonable.

### 4. The mean transverse momentum

The available data and all simulations indicate that the momentum distribution at a given $\eta$ is an $\eta$-independent function of the transverse momentum, which we write as $f(p_\perp)$. Measured mean values of $\langle p_\perp \rangle$ are shown in Fig. 3. Radiation level predictions at the new accelerators depend upon the assumed value, but usually scale as $\langle p_\perp \rangle^\alpha$, where $\alpha$ is a little less than one. In the absence of a good model, we do not show a curve, but rather indicate then assumed values at the new accelerators with crosses; they are also given in Table 1. As with the cross sections and pseudorapidity plateau heights, these values are not defensible, but their ratios are.

Several functional forms have been used to describe $f(p_\perp)$. During the work on Ref. 4, these were numerically folded with various energy-dependent functions (all power-law fits) to find radiation levels. In all cases the approximation $f(p_\perp) = \delta(p_\perp - \langle p_\perp \rangle)$ gave results which in the worst cases were in error by 6% or so. Since this was much better than some of the input assumptions (like the constancy of $H$), the approximation was used throughout. We continue to do so.

Since $p = p_\perp / \sin\theta = p_\perp \cosh\eta$, the mean charged particle energy per pseudorapidity interval is $H \langle p_\perp \rangle \cosh\eta$, and the charged particle energy in the interval $-\eta_0 < \eta < \eta_0$ is $2H \langle p_\perp \rangle \sinh\eta_0$. Using either this formula or the Tevatron and UA5 data points, one finds that about 5% of the charged particle energy lies in the region in which $H$ is reasonably constant—the pseudorapidity plateau region. This seems like a good definition of the pseudorapidity plateau width at the new machines. It is given in Table 1 for our extrapolated parameters.

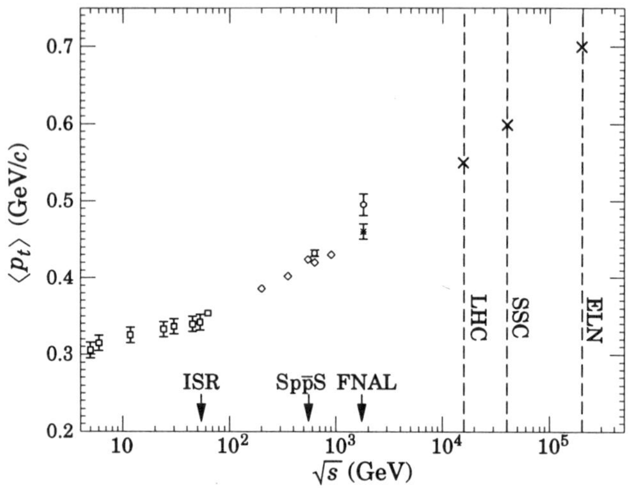


FIG. 3. Experimental measurements of $\langle p_\perp \rangle$. Crosses show extrapolated values.

## 5. Scaling hadronic activity

Two features of high-energy hadronic cascades have been long known to shielding specialists: (a) in a high-energy hadronic cascade in a given material (incident $E \gtrsim 10$ GeV), the relative abundance and spectrum of each hadronic species responsible for most of the energy deposition is independent of the energy or species of the incident hadron, and (b) because $\pi^0$ production bleeds off more and more energy into the electromagnetic sector as the energy of the incident hadron increases, the level of this low-energy activity rises less rapidly than the incident energy, and in fact rises very nearly as $E^m$, with $m < 1$. It is possible to demonstrate this approximate power law behavior by means of an induction argument and to obtain an estimate of the exponent $m$ in terms of other cascade parameters [13,14]. According to the HETC and FLUKA simulations over the range 3 GeV $< E <$ 20 TeV, $0.80 \lesssim m \lesssim 0.85$. At the present level of analysis there is no clear correlation of the slope with material, incident hadron species, or incident energy.

Both of these (related) observations have direct applicability to the problem of understanding radiation problems at the new energies. The first says that the radiation is indistinguishable from the radiation produced at any other proton accelerator, except for magnitude. This is shown in Fig. 4, where simulation results at 875 GeV and 20 TeV are superimposed—without the caption, one cannot tell which is which. The same will be true at 100 TeV.

The second observation permits extrapolation to situations such as neutron yield in 100 TeV beam-gas collisions. Two examples of such simulations are shown in Fig. 5. Although the extrapolation in this figure seems quite extreme, similar results are found in a large number of simulations extending to 20 TeV; it is just difficult to summarize all of them on a single plot. The normalizing factor for the spectra shown in Fig. 4 corresponds to $m = 0.84$. If our estimate that $m$ lies between 0.80 and 0.85 is correct, then the error in scaling the level of hadronic activity such as radioactivation or neutron production from 100 GeV to 200 TeV is about 20%, and the relative error in comparing the LHC, SSC, and ELN is far less.

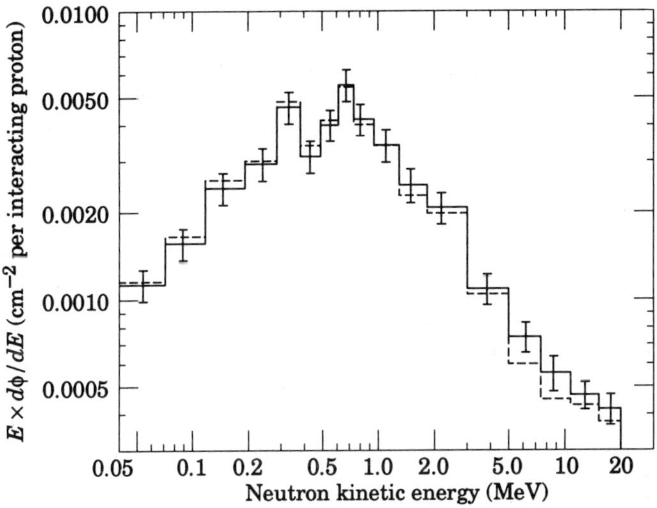


FIG. 4. Neutron spectra at 875 GeV and 20 TeV in the Tevatron tunnel. The solid histogram is an HETC/MORSE simulation at 875 GeV, and the dashed histogram is a simulation at 20 TeV, normalized to emphasize the identity of the spectral shapes.

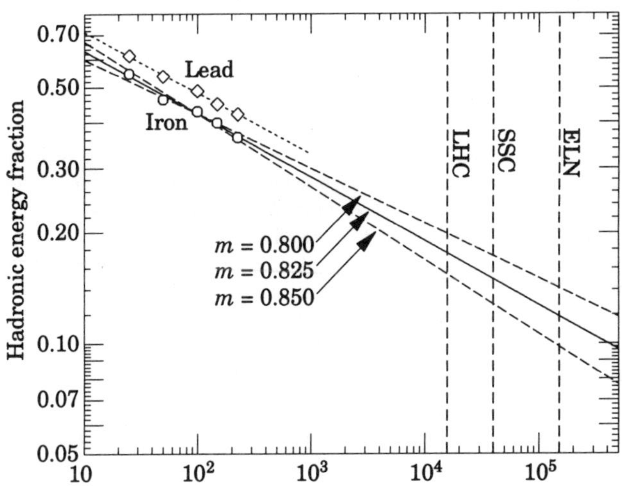


FIG. 5. The mean hadronic fraction $F_h$ as simulated with HETC for two calorimeter plate materials. The fitted slopes are indistinguishable, but the hadronic fraction (non-$\pi^0$ energy fraction) is lower in iron because of less neutron production. Slopes obtained with other simulations, some extending to 20 TeV, are in the same range.

**6. Neutron flux in the accelerator tunnel**

The neutron flux in the main ring tunnels of the new accelerators is expect to be mainly the result of beam-gas collisions. Synchrotron radiation desorbs gas from the vacuum pipe, and the molecules appear to make several crossings before sticking to the walls again. A series of experiments were conducted at the BNL synchrotron light source in the course of preparing the SSC CDR [1]. These included mass spectroscopic measurement of the desorbed gas. Molecular hydrogen dominated. Carbon monoxide could be identified ab about 1/20 of the hydrogen partial pressure, and carbon dioxide was thought to be present at an even smaller pressure. For purposes of this discussion, we will assume only molecular pressure, since the other gases contribute to the losses at only the 10% level.

The gas kinematics are such that pressure is not defined, so by "pressure" we will mean the pressure at 300 K which corresponds to the correct number density. Care must also be taken to remember the factor of two relating proton density to molecule density.

The number (molecule) density and partial pressure at 300 K for the $i$th species are related by

$$N_i = 3.54 \times 10^{16} P_i \ \mathrm{cm^{-3} torr^{-1}} ,$$

so that the interaction rate (reciprocal lifetime for a single proton) with an interaction cross section $\sigma_{pp}$ is

$$\begin{aligned} N_{\mathrm{H_2}} \sigma_{pp}\, c &= 212\, (\sigma_{pp}/100 \text{ mb})\, (P_{\mathrm{H_2}}/1 \text{ torr})\ \mathrm{s^{-1}} \\ &= \frac{(\sigma_{pp}/100 \text{ mb})\, (P_{\mathrm{H_2}}/10^{-9} \text{ torr})}{1310 \text{ hr}} . \end{aligned}$$

This quantity may be multiplied by the number of protons per unit length in the ring to find the loss rate per unit length. These and related quantities are given in the 3rd group of entries of Table 1.

The debris from a collision hits the beam pipe at near-grazing incidence, with the neutrals and negative particles hitting the outside with respect to the center of the ring. About 1/3 of the energy is carried by $\gamma$'s from $\pi^0$ decay, and additional 1/3 of the remaining energy is converted to $\gamma$'s when the high-energy hadrons have their first collision near the interior surface of the beam pipe. More than half of the collision energy is thus deposited in electromagnetic shower, and therefore within about a Moière radius of the beam pipe. However, according to Table 1 the power is little more than 6 times that at the SSC, and it is far below the energy deposit which could produce a quench. It is also more than three orders of magnitude below the power of the synchrotron radiation.

In fact, the only worrisome radiation in the tunnel is from neutrons. These are produced at the low-energy end of the cascades and diffuse out of the magnets. During the conceptual design of the SSC, these neutrons were studied experimentally (in the Tevatron tunnel) and with simulations [2,3,4].

In both the experiment and the simulations, the beam-gas collision was with a gas-leak target which was treated as a point source. The neutron flux at the tunnel wall was obtained as a function of distance along the tunnel, and the reported quantity was the *integral* of the flux along the tunnel. The units (neutrons/cm) look funny, but when this quantity is multiplied by the loss rate one recovers the normal flux in $\mathrm{cm^{-2}s^{-1}}$. The integrals given in the Table for the Tevatron and SSC are from the HETC simulations. From the ratio

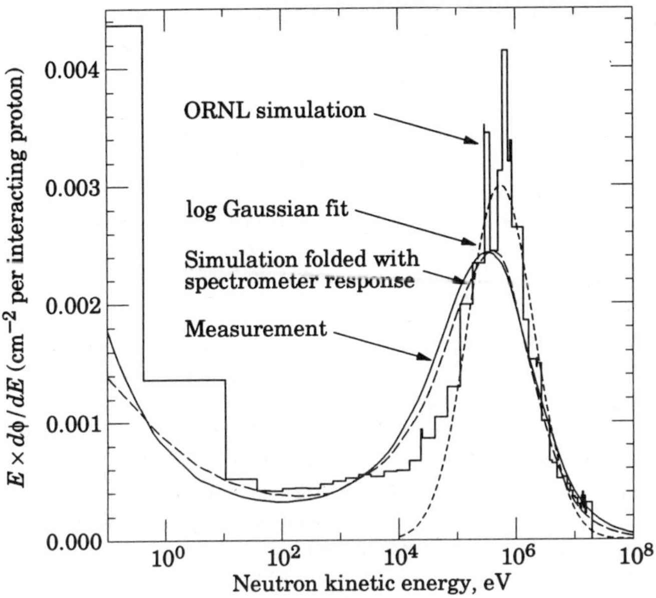


FIG. 6. Neutron spectra at 875 GeV in the Tevatron tunnel. The solid histogram is an HETC/MORSE simulation. The solid curve is the deconvolution of Bonner sphere spectral data taken in the Tevatron tunnel. The dashed curve is the 875 GeV simulation result as convoluted with the Bonner sphere response and then deconvoluted with the same program as was used for the data.

one obtains an effective power of 0.84 for the energy scaling (see Fig. 5); this power law is used to obtain the LHC and ELN entries. A Tevatron-like tunnel was assumed; further discussion of the geometry and scaling is given in Refs. 2 and 4, but any corrections are minor.

The flux near the tunnel wall from just one ring, assuming an effective pressure of $10^{-9}$ torr, is given in the Table. Although this reference pressure is often given, the SSC studies of gas desorption led to the conclusion that the SSC beam tube pressure would be about an order of magnitude greater. Accordingly, we multiply these numbers by $2\times10\times10^7$ to obtain the annual flux from both rings at $10^{-8}$ torr, given in the last row of this group of entries.

The spectra obtained in these studies are shown in Figs. 4 and 6. Fig. 4 has already been used to show that the spectrum has no memory of the energy of the original hadron. In Fig. 6 we show the simulated spectrum, the simulated spectrum as processed through the response of the Bonner sphere spectrometer used to make the measurements, and the deconvoluted experimental spectrum. As can be seen, experiment and simulation are in reasonable agreement.

In the simulation shown in Fig. 6, neutrons were transported down to thermal energies. The flux due to thermals is about the same as the flux in the "1 MeV" peak. Since it is only neutrons above about 160 keV which damage semiconductors, the fluxes reported in

Table 1 are for neutrons above the valley separating the two regions (40 keV).

The simulations also showed that about 80% of the flux is from neutrons reflected from the tunnel walls. The scaling with tunnel diameter and scaling to the inside of a $4\pi$ detector are discussed in Ref. 4.

The effects of beam losses due to lost particles are *not* included in the above discussion. If they are more serious than the beam-gas losses, then the machine will have a prohibitively short beam lifetime. However, they will occur at scrapers, insertion regions, *etc.*, and in these locations the situation will be much worse than described here. In addition, the pressure around the ring is unlikely to be constant. Both observations are in accord with observations at the Sp$\overline{\text{p}}$S which indicate that radiation levels vary by at least an order of magnitude with distance along the tunnel.

## 7. The final-focus optics

As mentioned in the introduction, an adequate description of energy deposit in the final focus quadrupoles (and a possible front scraper or collimator) is hampered by the absence of a good particle production in the forward region. Recent work by J. Ranft and N. V. Mokhov is very promising [15,16], and improvements may be expected in the near future.

In the meantime, we can learn a little from the energetics of the problem. At the SSC, 640 W of particles are produced at an IP. More than 90% of this energy goes into the final focus optics and a possible front collimator, half into each end. Of this energy, 2/3 is in the form of hadrons which can cascade to produce neutrons. The geometry is unclear, but suppose most of the cascades occur in the first 20 m of magnets, so that over this region the average energy loss rate is $\frac{1}{2} \times \frac{2}{3} \times 640/2000$ W/cm. From the Table, we see that the average loss rate in the ring at $10^{-9}$ torr is 6.0 $\mu$W/cm, so the loss rate (and, as a result, neutron production rate) are $1.8 \times 10^4$ times greater than in the ring. The single-ring contribution the fluence at the tunnel wall is $1.65 \times 10^9\,\text{cm}^{-2}\text{yr}^{-1}$, so if the shielding around the final focus quads has some resemblance to the tunnel, the average flux in this region should be $2.9 \times 10^{13}\,\text{cm}^{-2}\text{yr}^{-1}$. Entries for the other colliders can be found in the Table.

In reality the flux inside the shielding will be greater (because it is packed tightly around the magnets, not in the form of a tunnel) and the flux outside substantially less (because of the shielding), but our calculation underscores the problem. In Ref. 2 it is noted that about 4/5 of the flux in the tunnel has been reflected once or more from the concrete walls, so that the relevant quadrupole magnet "source term" for neutrons in the interaction hall may be obtained by dividing the entries in the Table by five.

## 8. Radiation levels in the detector

Detector radiation levels were throughly studied by an SSC Task Force [4], have been reported in numerous conference proceedings, and are even summarized in the *Review of Particle Properties* [17]. I therefore consider this to be a solved problem, and give only a brief treatment here.

Since the original report, the only major change was the discovery that the electromagnetic dose estimate for calorimeters was high by a factor of three due to a geometric error [18]. The report also had some major omissions. Having developed a methodology for describing the spatial dependence of cascade-associated quantities (dose at electromagnetic shower maximum, neutron flux at hadronic cascade maximum, albedo neutron numbers), it failed to include radioactivation. A first pass at this inclusion has now been made [19], and further work is in progress.

The report also failed to address the difficult problem of neutrons in the interaction hall. This was mainly because the only particle source considered (besides reflection) was the interaction point. The neutrons in the hall, which may be a major source of occupancy in the outer muon system, come mostly from the final-focus optics region and through cracks. (I have tried to address the neutron flux in the final-focus region in the last section.) The report considered only neutrons which could cause displacement damage in silicon ($E_n > 160$ keV), while drift tube occupancy induced from the gamma rays from $(n, \gamma)$ reactions by thermal neutrons may be important. Significant progress is being made in understanding these problems via codes which include detailed geometry [16,20], but they remain outside the scope of this paper.

Finally, I have gradually learned to present the basic argument about how various cascade-associated quantities scale with angle and distance from the IP with a little more clarity (they all go almost as $r^{-2}\sin^{-3}\theta$). In retrospect, the summary given in SSC-SR-1033 (Chapter 5) seems a little obtuse.

For the present purposes, I will address (a) dose from particles coming directly from the IP (because of the relevance to inner tracker discussions elsewhere in these proceedings, (b) the scaling of cascade-related effects to the ELN, and (c) neutrons inside the detector.

*(a) Direct particle dose*

We have already observed that $dN_{\text{ch}}/d\eta \approx H$. It then follows that the charged particle flux across a unit area normal to the IR radius vector at an angle $\theta$ to the beam line is

$$\frac{dN_{\text{ch}}}{da} = \frac{dN_{\text{ch}}}{d\eta}\frac{d\eta}{d\Omega}\frac{d\Omega}{da} = H\left(\frac{1}{2\pi\sin^2\theta}\right)\left(\frac{1}{r^2}\right) = \frac{H}{2\pi r_\perp^2}\ ,$$

where $r_\perp^2$ is the perpendicular distance from the beam line. The event rate is $\mathcal{L}\sigma_{\text{inel}}$, and a particle deposits $dE/dx$ as it goes through a thin absorber. (Thin enough that cascade processes are unimportant.) With a suitable conversion of MeV/g to joules/kg = Gy (2 MeV/g = $3.20 \times 10^{-10}$ Gy), we have

$$\begin{aligned}\dot{D} &= \frac{H\mathcal{L}\sigma_{\text{inel}}}{2\pi(r_\perp/1\ \text{cm})^2}\,\frac{dE}{dx}\\ &= 0.38\left(\frac{H}{7.5}\right)\left(\frac{\text{Rate}/(10^{15}\ \text{yr}^{-1})}{(r_\perp/1\ \text{cm})^2}\right)\ \text{MGy/yr}\ .\end{aligned}$$

Direct particle doses calculated from this formula are shown in Fig. 7. The assumed event rates are taken from Table 1. Note that elsewhere in these *Proceedings* the LHC luminosity has been taken as $1.0 \times 10^{34}\,\text{cm}^{-2}\text{s}^{-1}$; our value ($1.7 \times 10^{34}\,\text{cm}^{-2}\text{s}^{-1}$) is taken from Ref. 17.

*(b) Cascade-related quantities*

A hadron entering the solenoid coil or calorimeter initiates a cascade with its associated ionizing dose and neutron production. The scale length for cascade development is a interaction length, so the energy deposit and neutron production are distributed over about a meter. Neutrons produced earlier in the cascade have a higher probability of leaking out of the front face as "albedo neutrons."

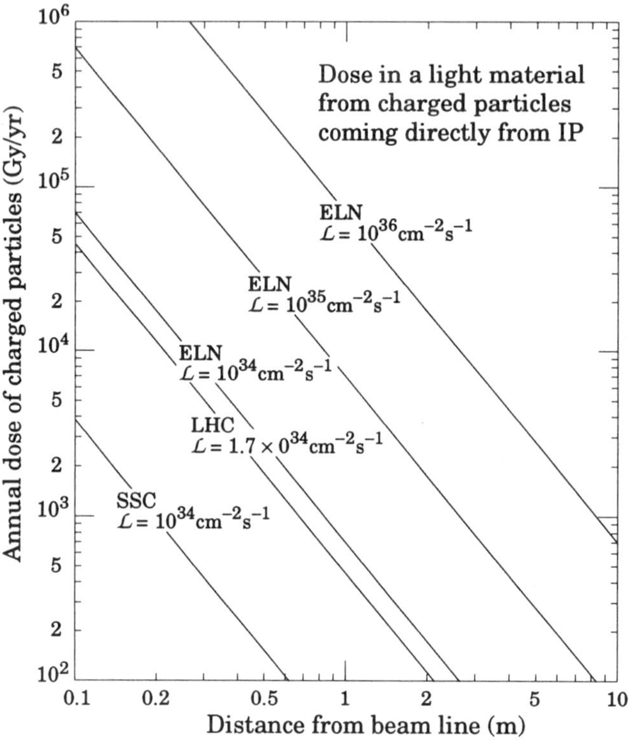


FIG. 7. Dose as a function of perpendicular distance from the beam line due to charged particles coming directly from the beam line. $dE/dx = 2$ Mev $\text{g}^{-1}\text{cm}^{-2}$ is assumed. The other parameters (inelastic cross section, luminosity, and pseudorapidity plateau height) can be found in Table 1.

To some level of approximation, quantities such as the ionizing dose and neutron fluence near cascade maximum are proportional to the incident energy. In this case the scaling with distance and angle is very simple, as outlined in Fig. 8. However, because a higher-energy cascade is longer, the dose (flux) at cascade maximum is reduced for purely geometrical reasons. In addition, the fraction of cascade energy involved with hadronic processes decreases with energy (see Fig. 5 and Section 5). As a result, radiation levels at cascade maximum increase a little less rapidly than linearly with energy. While there is considerable scatter in the results, the Monte Carlo simulations suggest that a quantity such as neutron flux at shower maximum is described reasonably well by a power law, with the flux proportional to $E^{0.67}$. (Unlike the situation in Section 5, this is a purely empirical observation.) As a result, the neutron flux is proportional to $1/\sin^{2.67}\theta$ rather than $1/\sin^3\theta$. In general, flux and dose-related quantities in a calorimeter go as $1/\sin^{2+\alpha}\theta$, where $0.5 \lesssim \alpha < 1$.

We have assumed that $\pi^0$'s going into a given solid angle carry half as much energy as do the "stable" hadrons, or, almost equivalently, that there are just as many photons (from $\pi^0$decay) as hadrons, but that they have half the mean momentum. Most of the energy in

$$\left.\begin{array}{l}\dfrac{dN_{\text{ch}}}{d\eta} = \text{Const} \\[2ex] \dfrac{d\eta}{d\Omega} = \dfrac{1}{2\pi \sin^2\theta}\end{array}\right\} \frac{dN_{\text{ch}}}{d\Omega} = \frac{\text{Const}}{\sin^2\theta}$$

$$\left.\begin{array}{l}\dfrac{dN_{\text{ch}}}{d\Omega} = \dfrac{\text{Const}}{\sin^2\theta} \\[2ex] E \approx p = \dfrac{p_\perp}{\sin\theta}\end{array}\right\} \frac{dE}{d\Omega} = \frac{\text{Const}}{\sin^3\theta}$$

$$\left.\begin{array}{l}\dfrac{dE}{d\Omega} = \dfrac{\text{Const}}{\sin^3\theta} \\[2ex] \dfrac{d\Omega}{da} \propto \dfrac{1}{r^2}\end{array}\right\} \begin{array}{l}\text{Neutron fluence} \\ \text{or ionizing dose}\end{array} = \frac{K}{r^2 \sin^3\theta}$$

FIG. 8. Logic flow for the scaling in a cascade process resulting in neutron fluence or ionizing dose. A flat pseudorapidity plateau means that the particle flux per unit solid angle is proportional to $1/\sin^2\theta = \cosh^2\eta$. Since the mean energy (momentum) of the particles at $\theta$ is equal to $\langle p_\perp\rangle / \sin\theta$, the energy flow into $d\Omega$ is proportional to $1/\sin^3\theta$, or the energy flow into unit area is proportional to $1/r^2 \sin^3\theta$. A number of dose and fluence related quantities are related to the local energy flow, but the proportionality constant must be found via a simulation.

an electromagnetic cascade is deposited in about 15 radiation lengths, which in a lead-plate calorimeter can be 10 cm or so. Even though the $\pi^0$'s carry only have half the energy, they produce a very much larger ionizing dose because of the short shower development length.

We can summarize the radiation levels produced in most cascade processes with the statement

$$\text{Dose or fluence} = \frac{A}{r^2}\cosh^{2+\alpha}\eta = \frac{A}{r^2\sin^{2+\alpha}\theta}\,,$$

where $A$ is proportional to $\sigma_{\text{inel}}\,\mathcal{L}_{\text{ave}}\,H\,\langle p_\perp\rangle^\alpha$, and $\mathcal{L}_{\text{ave}}$ is the average luminosity over a $10^7$ s year. (The appearance of $\langle p_\perp\rangle$ instead of an integral over a momentum distribution was discussed in Section 4.) It remains only to tabulate values of $A$ for relevant processes at the new accelerators. This is done in Table 1 for a uranium/scintillator calorimeter. The neutron levels should be halved for a lead/scintillator calorimeter. As an example of the table's usage, suppose the problem is to find the maximum ionizing dose near the edge of a calorimeter 4 m from the IP at $\eta = 3$, operating at the $10^{34}\,\text{cm}^{-2}\text{s}^{-1}$ ELN:

$$\dot{D} = 2630\cosh^{2.67}\eta/4^2\ \text{Gy/yr} = 78\,000\ \text{Gy/yr}$$

The model also assumes that particle flux on the calorimeter does not change much over the transverse dimensions of a cascade—there are no transverse leakage corrections. This approximation breaks down at small angles It is also assumed that the particle flux comes from the IP. At wide angles, there can be a substantial contribution from albedo from the large-$|\eta|$ regions, but the dose here is very low already.

*(c) Neutrons inside the detector*

Backscattered (albedo) neutrons mostly originate near the surface of the calorimeter, and hence the number produced grows only as a very small power of the incident momentum ($\alpha \approx 0.5$). They tend to scatter once before being absorbed, so the flux inside the cavity is about twice as great as the flux from the neutrons diffusing out from the cascades. The

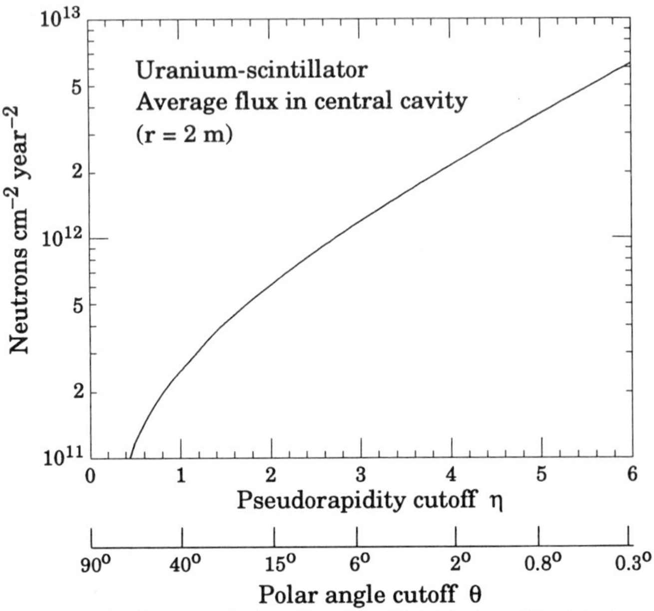


FIG. 9. Neutron flux in the central cavity of an SSC detector. The detector consists of a uranium/lead calorimeter with a 2 m radius inside cavity. It covers the range $-\eta_0 \leq \eta \leq -\eta_0$. Scaling factors for other accelerators are given Table 1, and the flux should be reduced by a factor of about two for a lead/scintillator electromagnetic calorimeter.

flux appears to vary by a factor of two inside the cavity, and the dominant sources are the large-$|\eta|$ endcap regions.

In Ref. 4 we give the number of albedo neutrons as a function of $\eta$, but the relevant quantity is the integral of this distribution over the $\eta$ coverage of the central detector. We show this integral in Fig. 9 for the SSC, for a detector consisting of a uranium/scintillator calorimeter with a 2 m radius inside cavity. The calorimeter extends to $\pm\ \eta_0$. Scale factors relative to the SSC for the other new colliders are given in Table 1; within rounding errors these are the same as the scale factors for the neutron flux at cascade maximum. Once again, all fluxes should be reduced by a factor of two for a lead calorimeter on the inside of the sphere. The flux is for the "1 MeV" peak ($E_n > 40$ keV).

What about other geometries? We have also shown that in the limit of isotropic flux, the neutron flux scales inversely as the square of a characteristic dimension of the cavity, which can be defined a little more analytically as the mean chord divided by the volume.

The SDC detector is expected to have an average annual neutron fluence of about $2.3 \times 10^{11}\,\mathrm{cm}^{-2}$ in the tracking region. A similar detector at the $10^{34}\,\mathrm{cm}^{-2}\mathrm{s}^{-1}$ ELN would have a fluence of $4.6 \times 10^{12}\,\mathrm{cm}^{-2}$.

## 9. Discussion

The SSC event rate will be hundreds of times higher than the highest collider rate with which we actually have experience. In the years since the SSC CDR, we have gotten used to the daunting radiation levels, trigger rates, data-handling requirements, *etc.*, without actually experiencing them in laboratory. But since the physics requirements are equally daunting (a 300 GeV Higgs will be produced in only $10^{-10}$ to $10^{-9}$ of the collisions), we have coveted even more luminous machines. In this study, we were asked to consider an ELOISATRON operating at $10^{35}\,\mathrm{cm^{-2}s^{-1}}$ or even $10^{36}\,\mathrm{cm^{-2}s^{-1}}$, but I have restricted the present report to $10^{34}\,\mathrm{cm^{-2}s^{-1}}$—the reader can multiply by 10 or 100 if he wishes.

Scaling to the ELN is not difficult. The particle production parameters do not need to be extrapolated much further from their LHC and SSC extrapolations (on log paper), and we know with fair certainty how cascades scale. But the main result is that the $10^{34}\,\mathrm{cm^{-2}s^{-1}}$ ELN is not very different than the LHC, except in a few places. The neutron flux in the tunnel is about 6 times greater, and the final focus system has radiation problems about 9 times worse than at the LHC (which are bad enough already). Problems inside the detector are only worse by a factor of two.

The main benefit of this study, however, has little to do with whether ELOISATRON is actually built or not. The exercise of extrapolation, revisiting old calculations, and re-thinking old conclusions has led to new and productive ways of thinking about the machines actually under construction. I am grateful for the opportunity to have participated in this exploration.

## References


1. Conceptual Design of the Superconducting Super Collider, SSC Central Design Group Report SSC-SR-2020 (1986).
2. T. A. Gabriel, F. S. Alsmiller, R. G. Alsmiller, Jr., B. L Bishop, O. W. Hermann, and D. E. Groom, "Preliminary Simulation of the Neutron Flux Levels in the Fermilab Tunnel and Proposed SSC Tunnel," SSC Central Design Group Report SSC-110 (1987).
3. R. G. Alsmiller *et al.*, Nucl. Instrum. Meth. **A313** (1992) 357-366.
4. "Radiation Levels in the SSC Interaction Regions," ed. by D. E. Groom, SSC Central Design Group Report SSC-SR-1033 (June 1988).
5. M. Block and R. N. Cahn, Rev. Mod. Phys. **57** (1985) 563-598.
6. R. M. Baltrusaitis *et al.*, Phys. Rev. Lett. **52** (1984) 1380;
   R. M. Baltrusaitis *et al.*, Proc. 19th Inter. Cosmic Ray Conf. (La Jolla) **6** (1985) 5.
7. N. A. Amos *et al.*, Phys. Rev. Lett. **68** (1992) 2433-2436.
8. M. Honda *et al.*, Phys. Rev. Lett. **70** (1993) 525.
9. A. Donnachie and P. V. Landshoff, Phys. Lett. **B 296** (1992) 227-232.
10. K. Alpgård etal., Phys. Lett. **112B** (1982) 193;
    G. J. Alner *et al.*, Z. Phys. **C33** (1986) 1.
11. F. Abe *et al.*, Phys. Rev. **D41** (1990) 2330.
12. J. Ranft, private communication (1993).
13. D. E. Groom, "Energy Scaling of Low-Energy Neutron Yield, the $e/\pi$ Ratio, and Hadronic Response in a Calorimeter," *Proc. of the Workshop on Calorimetry for the Superconducting Super Collider,* Tuscaloosa, Alabama, 13–17 March 1989, ed. R. Donaldson and M. G. D. Gilchriese, World Scientific (June 1990), 59–75.
14. T. A. Gabriel, D. E. Groom, P. K. Job, N. V. Mokhov, and G. R. Stevenson, "Incident Energy Dependence of Hadronic Activity," *Proc. of the Workshop on Simulating Radiation Environments,* Santa Fe, NM, 11–15 January, 1993, ed. by L. Waters and A. P. T. Palounek (to be published). This is a draft of a paper to be submitted to Nucl. Instrum. Meth.

15. J. Ranft, "Hadron Production at Supercolliders: DTUJET," *Proc. of the Workshop on Simulating Radiation Environments,* Santa Fe, NM, 11–15 January, 1993, ed. by L. Waters and A. P. T. Palounek (to be published).

16. N. V. Mokhov, "MARS12 Code System," *Proc. of the Workshop on Simulating Radiation Environments,* Santa Fe, NM, 11–15 January, 1993, ed. by L. Waters and A. P. T. Palounek (to be published).

17. K. Hikasa *et al.*, "Review of Particle Properties," Phys. Rev. **D46**, S1 (1 June 1992).

18. D. E. Groom, H. Hirayama, and W. R. Nelson, "Erratum to SSC-SR-1033," SSC Laboratory Report SSCL-285 (19 July 1990).

19. D. E. Groom, p. 390 in "Extension of SSC-SR-1033 Approach to Radioactivation in LHC and SSC Detectors," *Proc. II Inter. Conf. on Calorimetry in High Energy Physics,* Capri, Italy, 14–18 October 1991, ed. by A. Ereditato, World Scientific (1992).

20. L. Waters, "LAHET Calculations for the SSC," *Proc. of the Workshop on Simulating Radiation Environments,* Santa Fe, NM, 11–15 January, 1993, ed. by L. Waters and A. P. T. Palounek (to be published).

# MATERIALS IN THE RADIATION ENVIRONMENT OF HIGH-LUMINOSITY COLLIDERS

H. SCHÖNBACHER

*CERN*
*CH - 1211 Geneva 23, Switzerland*

ABSTRACT

This paper gives an overview of radiation effects to organic polymers, metals and inorganic materials including semiconductors. It is shown that in the high-radiation areas of high-luminosity colliders such as LHC, SSC and ELN-34 extremely careful selection of semiconductor and plastic materials is required in order to avoid radiation damage. For the highest energies and luminosities such as ELN-35 and ELN-36 according to our present knowledge and available technologies only inorganic materials and metals will survive in the highest radiation areas. Further, in these areas, access will be restricted respectively impossible due to very high induced radioactivity.

## 1. Introduction

Material selection for future high-luminosity colliders and detectors is a challenge to design engineers and physicists. Many operational and safety aspects need to be taken under consideration. One of the key parameters is radiation. Since several years the high-energy physics community has become aware of the problem and a large number of R&D projects are carried out in this direction on many places. This report will give an overview of radiation effects on the materials concerned in general without going into detail of ongoing research. In particular it will be concentrated on the limitations when talking about colliders at the highest energies and highest luminosities such as the ELOISATRON[1] *.

## 2. Semiconductors

From experience with existing high-energy particle accelerators, it is known that components based on semiconductors are very sensitive to radiation. For example all electronics has been removed from the tunnel of the CERN 450 GeV Super Proton Synchrotron after 6 years of operation at a dose of about $10^2$ Gy (1 Gy = 100 rad) [2]. Semiconductors in future colliders and superdetectors will have to operate at doses from charged particles exceeding this by three to four orders of magnitudes. Therefore only specially selected, radiation hard components can be used as available from military and to some extend from space applications.

The degradation in semiconductor devices is caused by ionization and displacement effects. Ionization processes affect the electronic structure of the atoms causing electron-hole pair generation and resulting in charge accumulation within the insulator and at the interface with the silicon. Displacement damage reduces the charge efficiency and increases the leakage current and noise in the bulk silicon. The degradation can be described by the reduction of the minority carrier lifetime, carrier removal and reduction of the carrier mobility [3)].

Examples of different radiation effects in semiconductors dependent on radiation type and energy are given in Figs. 1 and 2. Fig. 1 shows the maximum clock frequency of a PAL 16 R8 programmed binary counter which was irradiated to 1.2 MeV gammas ($^{60}Co$), to 2 MeV electrons and to 1 MeV neutrons in a high-energy particle accelerator field [4)]. Fig. 2 shows the calculation of non ionizing energy deposition from different displacement cross sections in silicon by electrons, protons and neutrons at different energies[5)].

Due to the above explained difference in radiation effects of the radiation types and energies, the damage must be related either to total dose (ionization) or to fluence (displacements, point defects, clusters). Fluence to dose conversion of the particle or photon fluence for various energies to the absorbed dose is shown in Fig. 3. The characteristics reflect the stopping power relation for electrons and protons, the cross sections for neutrons and the mass attenuation coefficient for photons for various energies. It must be noted that apart from electrons above 1 MeV the conversion factors are very dependent on energy and may be an important source of error if the energy spectrum is not well known[6)].

At present enormous research efforts are made to find semiconductor detectors and electronics to cope with the high radiation fields to be expected for the LHC and SSC design luminosities. From a summary presented at this Workshop[7)], the perspectives are promising and it appears possible to find the components which can cope with the radiation field of ELN-34.

As regards to the highest luminosities such as ELN-35 and ELN-36, the use of semiconductors will have to be **excluded** according to our present knowledge and technologies.

## 3. Organic Materials

As organic polymers age under the influence of high-energy radiation, they can suffer from loss of their mechanical, electrical, chemical and optical properties. The major cause of changes in the macroscopic mechanical properties of polymers results from scission of polymer chains and/or from crosslinking between neighbouring polymer chains.

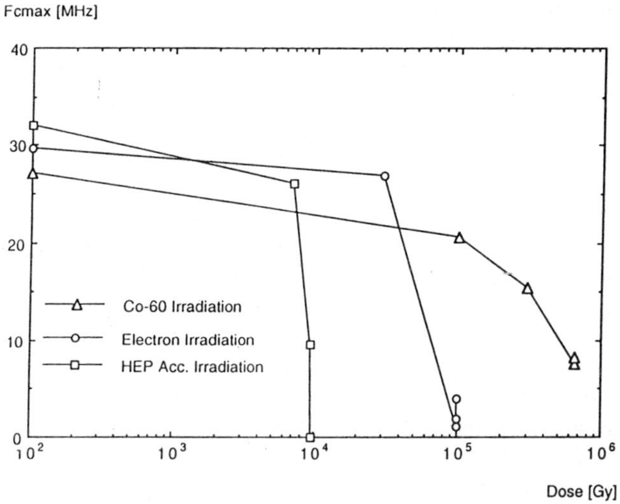


Fig. 1 Example of difference in radiation effects from 2 MeV Electrons, 1 MeV neutrons from a HEP accelerator and a Co-60 gamma source (Maximum clock frequency of a PAL 16R8 programmed as a 6 bit binary counter, Ref. 4).

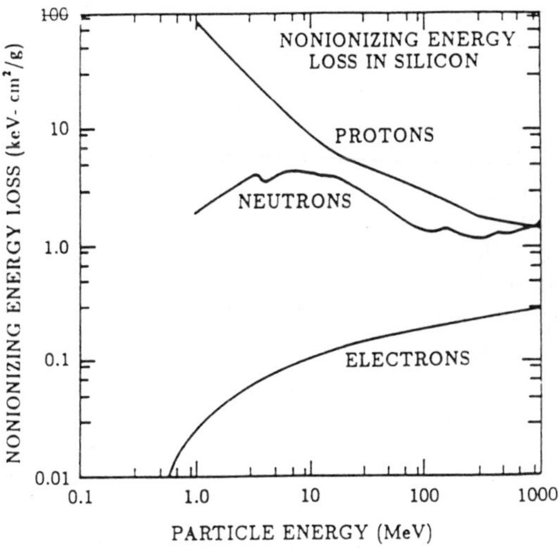


Fig. 2 Calculation of nonionizing energy deposition in silicon by electrons, protons and neutrons (Ref. 5).

When these materials are irradiated in the presence of oxygen, the rate of degradation can be strongly influenced by the dose rate. This increases the difficulty in predicting long term (low dose rate) radiation degradation behaviour. In many cases, lower dose rates result in comparatively higher mechanical degradation at the same dose. Therefore, if the material is in service over a long time at low dose rates it may degrade faster than expected from short time tests.

This effect is particularly pronounced in some currently thermoplastics, such as polyethylene (see Fig. 4) [8]. Table 1 gives a review of a number of organic materials which are frequently used in high-energy particle accelerator and detector construction in increasing order of radiation resistance[9]. Organic materials which should be avoided because their radiation resistance is below $10^5$ Gy are:

- Butylrubber, Perfluoro ethylene-propylene (FEP), unfilled Phenolic resin, Polyacryl (Plexiglas), Polyamide (Nylon), unfilled Polyester resin, Silicon rubber, Polytetrafluoroethylene (Teflon, PTFE), Fluorinated copolymer (Viton)

Further, halogen containing materials such as PVC, Hypalon and Neoprene should be avoided due to the production of corrosive gases in the case of fire.

A number of organic materials are available for application in the LHC, SSC or ELN-34 radiation environment. For application in colliders at the highest energies and luminosities such as ELN-35 and ELN-36, a very careful material selection is required and no organic materials shall be used in areas with radiation levels exceeding $10^8$ Gy.

## 4. Ceramics[10]

Exposure of ceramic materials to ionizing radiation affects the crystal lattice of the ceramic by displacement of atoms within the lattice, conversion of atoms to different species by nuclear reactions, and ionization and electronic excitation.

A variety of effects are produced in ceramic materials as a result of lattice displacement. The introduction of interstitial atoms into the lattice produces swelling which may proceed to the point of causing the material to disintegrate into powder. Defects in the lattice reduce the thermal conductivity of the ceramic. Thermal spikes produced particularly near the end of the path of the particle through the lattice lead to localized regions of high temperatures.

The conversion of atoms in the lattice to other species by reactions with neutrons produces impurity atoms. This is generally not of great importance except in the cases where gases are formed, such as the neutron reaction with beryllium which produces helium. Such gases in the lattice

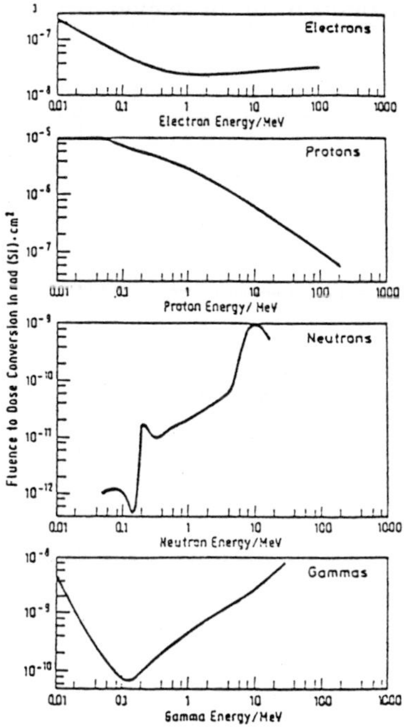


Fig. 3 Fluence to ionisation dose conversion for electrons, protons, neutrons and gammas as a function of their energies (Ref. 6).

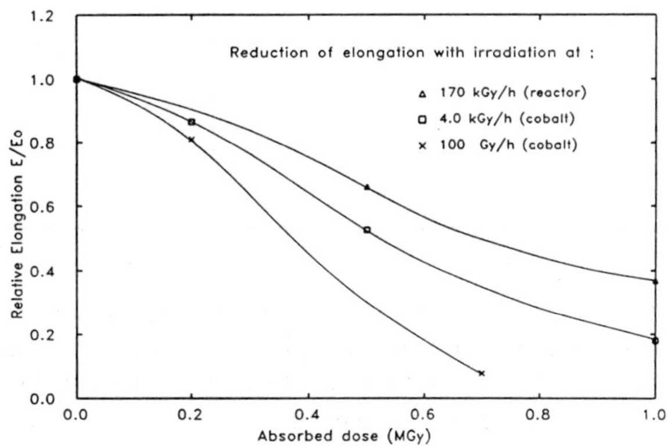


Fig. 4 Reduction of elongation at break of cross-linked polyethylene cable insulation material irradiated at three different dose rates (Ref. 8).

**Table 1: Radiation dose limits in air for commonly used organic materials[9)]**

| Organic Materials | Radiation Dose Limits |
|---|---|
| Epoxy resin, cured at ambient temperature (Araldite D)<br>Chlorosulfonated PE (Hypalon, CSP)<br>Cross-linked PE (XLPE)<br>Ethylene-acrylate rubber (EAR)<br>Ethylene-propylene rubber (EPR)<br>Ethylene vinyl acetate (EVA)<br>Halar (CTFE)<br>Polyethylene terephthalate (Hytrel, PETP)<br>Polyethylene (Lupolen, PE)<br>Polychloroprene (Neoprene)<br>Polyolefin<br>Polyvinyl chloride (PVC) | $10^5$ - $10^6$ Gy |
| Epoxy resins (Araldite B, Araldite F, Epikote)<br>Epoxy Novolac<br>Epoxy resin, aromatic hardener<br>Paints based on epoxy or polyurethane resins<br>Polyimide (Kapton)<br>Polyether-etherketone (Peek) | $10^6$ - $10^7$ Gy |
| Polyphenylene sulfide (Ryton, PPS)<br>Polyphenylene oxide (PPO)<br>Inorganic filled resins:<br>- Epoxy, aromatic hardener<br>- Phenolic<br>- Polyester<br>- Polyimide<br>- Polyurethane<br>- Silicone | $10^7$ - $10^8$ Gy |

may nucleate to form gas pockets or bubbles and cause severe swelling. All these effects occur in frequently used structural or insulating materials such as aluminium oxide ($Al_2 O_3$) or beryllium oxide ($Be0_2$) at very high doses exceeding $10^{10}$ Gy and neutron fluences above $10^{21}$ n/cm$^2$ (E > 0.1 MeV).

Ionization and electronic excitation produced by the passage of fast particles or gamma radiation through the lattice do not affect the general engineering properties of ceramics to any great extent. However, electrons stripped from atoms in the lattice may be trapped at defects to form colour centers, or regions of altered optical absorption characteristics. This effect can be of great importance in applications where optical glasses and optical glass fibres are utilized. For this reason radiation damage in optical fibres used for data transmission may occur at a few Gy; special radiation hard fibres are however available which could be used up to 1 MGy[11)].

An interesting application of quartz fibres for calorimetry up to 10 MGy was presented at this Workshop[12)].

## 5. Metals[10, 13)]

The mechanism of radiation effects in metals is essentially related to neutrons which in a primary collision with an atom create vacancies. The neutrons continue to bounce about knocking atoms out of place until its energy is dissipated. A secondary effect results from the energy imparted to the knocked-on-atom. This atom collides with other atoms in the lattice knocking them out of position and imparting energy to them. Finally, these "knocked-on" atoms lose all energy and come to rest in an interstitial position in the lattice. Hence, two types of point defects are formed by the collision between neutrons and elastic atoms: vacancies and interstitials.

A further secondary effect of neutron bombardment is the change in structure caused by "thermal spikes". It is postulated that in some instances the energy of the knocked-on atoms can be dissipated locally as a thermal pulse, which can produce extremely high temperatures for very short periods of time. The area affected by such a "spike" could contain interstitial vacancies, misoriented regions, and dislocation loops.

Fission processes produce a variety of foreign atoms, the effect of these atoms on interstitial positions will also contribute to property changes. In certain materials, the same effect can be achieved by the transmutation of the basic material to atoms of another element. Of all transmutation products the inert gases cause some of the most notable effects in metals by swelling which is produced by retained gas.

As in the case of ceramics, these effects occur only at very high doses exceeding $10^{10}$ Gy and $10^{21}$ n/cm$^2$ (E > 0.1 MeV).

## 6. Summary and conclusions

Table 2 summarizes for the groups of materials discussed in this paper the damage mechanisms and the radiation types which cause the damage. Fig. 5 shows for the same materials the radiation dose ranges and neutron fluences where they undergo no damage, mild to moderate damage and

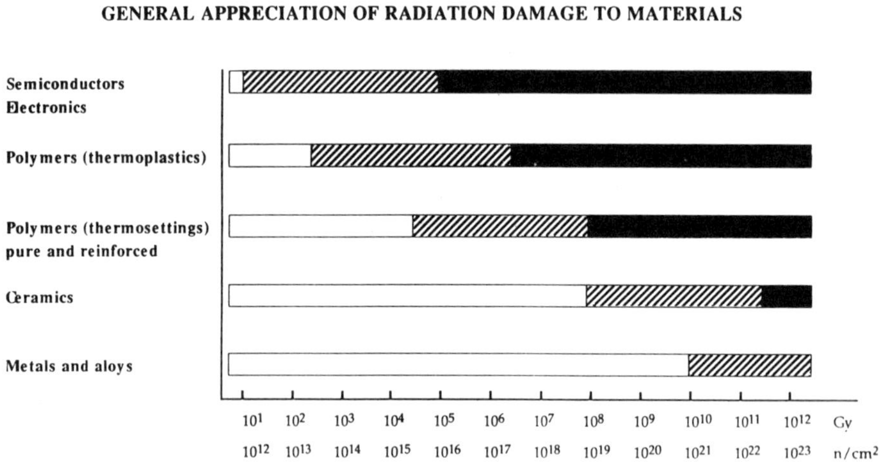


Fig. 5 General appreciation of radiation damage to materials. Shown are ranges where materials undergo no damage (blank) mild to severe damage (shaded) or destruction (black).

**Table 2: Radiation effects to materials**

| Material | Damage mechanism | Damage due to |
|---|---|---|
| Semiconductor s | Photoelectric interaction<br>Electron-Hole Pairs<br>Displacement of atoms | Total dose<br><br>Neutrons, HEP |
| Polymers | Main and side chain rupture Cross-linking and degradation, gas evolution, radical production, reaction with environment | Total dose |
| Ceramics | Displacements,<br>trapped charge carriers,<br>thermal peaks<br>colour centres | Neutrons, HEP<br><br><br>Total dose |
| Metals | Displacements<br>Dislocations<br>Nuclear reactions producing clusters (Helium), voids and bubbles | Neutrons, HEP |

and where they are usually unusable. From this figure it needs particularly noted that:

- all semiconductor devises with mild to moderate damage shown in dose ranges above $10^3$ Gy and $10^{13}$ n/cm$^2$ are non standard products and need to be specially selected for radiation hard technologies;
- most of the thermoplastic polymers are very sensitive to oxygen induced damage during irradiations at low dose rates.

From the doses to be expected in the central area of a supercollider detector[14, 15] it can be concluded that:

1. For ELN 34 one may cope with its radiation environment by extremely careful selection of semiconductor and plastic materials.

2. For ELN-35 radiation damage will exclude semiconductors and will restrict the choice to very few plastic materials.

3. For ELN-36 only ceramics and metals will survive in the high-radiation areas of the detector.

In all cases the very high induced radioactivity in the highest radiation areas will restrict or exclude access to these zones.

## 7. References


1. A. Zichichi, *The INFN ELOISATRON Project*, August 1989.
2. H. Schönbacher, F. Coninckx, *Dose to the CERN 450 GeV Super Proton Synchrotron and estimate of radiation damage,* Nucl. Instr. and Meth, in Physics Research **A288** (1990) 612.
3. J.E. Gover and T.A. Fischer, *Radiation hardned microelectronics for accelerators,* IEEE Trans. on Nuclear Science, **35** (1988) 160.
4. L. Larsen, T. Massam, H, Schönbacher, F. Wulf, *Radiation tests on service electronics for future multi-TeV detectors,* LAA Report in preparation.
5. R.L. Pease, A.H. Johnston, J.L. Azarewicz, *Radiation Testing of Semiconductors for Space Electronics, IEEE 1990 International Nuclear and Space Radiation Effects Conference,* Reno Nevada 16 July 1990, Short Course.
6. D. Bräunig, F. Wulf, W. Gaebler, A. Boden, *Irradiation test guidelines for radiation hardness of electronic components,* HMI-B 390, TN 53-10, Berlin 1982.
7. F. Wulf, D. Braünig, A. Boden, D. Fritsch, *Electronic components in a high-luminosity environment,* Paper published in the Proceedings of this Workshop.
8. H. Danielsson, H. Schönbacher and M. Tavlet, *Estimate of lifetime of cables in the radiation environment of LEP 200,* CERN TIS Division Report CFM 91-13 (1991).
9. P. Beynel, P. Maier, H. Schönbacher and A. Stolarz-Izycka, M. Tavlet, *Compilation of radiation damage test data, Part I: Cable insulating materials,* CERN 79-04 (1979), 2nd edition, CERN/89-12 (1989), *Part II: Thermosetting and thermoplastic resins,* CERN 79-08 (1979), *Part III: Materials used around high-energy accelerators,* CERN 82-10 (1982).
10. J.F. Kircher and R.E. Bowman, *Effects of Radiation on Materials and Components,* Reinhold Publishing Corporation, New York, 1964.
11. B. Leskovar, *Radiation effects on optical data transmission systems, Radiation effects at the SSC,* M.G.D. Gilchriese, Editor, SSC-SR-1035, LBL, BERKELEY, CA, June 1988.
12. A. Contin, *Calorimetry,* Paper published in the Proceedings of this Workshop.
13. W.F. Sheely, *Radiation Effects,* Metallurgical Society Conferences, **Vol. 37**, Published by Gordon and Breach, Science Publishers, Inc., 1967.
14. D. Groom, *Radiation at hadron colliders: scaling to the highest energies*. Paper published in the proceedings of this Workshop.
15. H. Leutz, *Particle detection at Eloisatron,* Paper published in the Proceedings of this Workshop.

# Electronic Components in a High Luminosity Environment

F. Wulf, D. Bräunig, A. Boden, D. Fritsch
Hahn-Meitner-Institut Berlin GmbH
Glienicker Straße 100
D-1000 Berlin 39

ABSTRACT

New high-luminosity colliders and detectors require a large number of different electronic components. Depending on the distance and the angel of the interaction point of the beam, the total ionizing dose and neutron fluence for a one year operation with a luminosity of $10^{34}$ cm$^{-2}$ s$^{-1}$ is in the range of 100 Gy(Si) to 1 MGy(Si) and $10^{12}$ to $10^{17}$ n/cm$^2$, respectively. The degradation of the electronic devices due to the basic radiation effects -total ionizing dose effects and displacement damage- are reviewed. The complexity of the degradation mechanisms of bipolar integrated circuits will be discussed by using the test results of 3 device types fabricated in Low Power Schottky, Advanced Low Power Schottky and Fast technology. The devices from four manufactures are compared. This paper considers only the degradation of electronic components, but the general degradation mechanisms of semiconductor detectors are related to the same basic radiation effects. The evaluation of electronic components for future detectors requires a well defined radiation test procedure. Specifications for military and space applications can be used, but must be adapted to the radiation environment of high-energy and high-luminosity colliders and detectors. A summary of the radiation hardness of standard electronic devices will be given. The radiation hardness of various materials, passive and active electronic devices and mechanical components are summarized at the end of this paper.

## 1. Introduction

New high-luminosity colliders and detectors require a large number of different electronic components. Depending on the distance and the angel of the interaction point of the beam, the total ionizing dose and neutron fluence for a one year operation with a luminosity of $10^{34}$ cm$^{-2}$ s$^{-1}$ is in the range of 100 Gy(Si) to 1 MGy(Si) and $10^{12}$ to $10^{17}$ n/cm$^2$, respectively. The beam crossing time of 15 ns hence follows a maximum clock rate of about 66 MHz for the first trigger level. Under these requirements radiation hard advanced bipolar and CMOS technology must be used as front-end-electronic.

The degradation of electronic devices are caused by displacement damage and total ionizing dose effects TID , also often called total dose effects. The displacement damage is related to the interactions of neutrons, protons and electrons with the matter. These interactions generate energy levels in the bandgap of the semiconductor and scattering centers. Electromagnetic quanta as well as electrons, high energy particles and protons generate electron-hole-pairs in the target material. These two effects lead to the degradation of electronic devices as well as of particle detectors.

In general bipolar devices are more sensitive to displacement damage and MOS-structures are sensitive to TID effects. The typical behaviour of MOS-devices due to TID effects are reviewed in the proceedings of the ECFA Study Week[1], Large Hardron Collider Workshop[2] and other publications[3,6]. In this paper the prompt and delayed degradation of advanced bipolar technologies will be discussed.

The characterization of the radiation hardness of electronic devices requires a well defined test procedure. Only under this assumption a comparison of the degradation of tested devices is possible. This restriction must also be taken into account when the data of various test reports are summarized.

## 2. Basic Mechanisms of Radiation Effects on Electronic Devices and Circuits

### 2.1 *Displacement Damage*

From a physical point of view the interaction of radiation with matter is characterized by the probability of creating a displaced atom that can change the material's parameter of the semiconductor, for instance the minority carrier lifetime as the most sensitive quantity. This is caused by the introduction of additional recombination centers induced by radiation. Unfortunately there is no direct correlation between the probability in terms of a cross-section and the amount of created active centers. This becomes intuitively evident if one realizes that the most probably created defects - the vacancy and the interstitial - are unstable at room temperature and these defects are mobile and tend to create more complex defects, for instance divacancies, A-centres or clusters. In addition the formation of more complex defects depends on the mobility of these simple defects which in turn is a function of the charge state of ,say, the vacancy. The charge state on the other hand depends on the energetic position of the Fermi level in the semiconductor and this can be varied by a number of means: doping, temperature, charge injection etc. In this way there is a variety of interdependencies that makes a prediction quite complicated.

#### 2.1.1 *The Concept of the Damage Factor*

By generating additional electrically active centers within the semiconductor material some properties of the material can be changed, which are directly accessible to the experiment. For instance, the minority carrier lifetime $\tau$ is defined as

$$\tau = \frac{1}{\sigma \cdot v_{th} \cdot N_t} \tag{1}$$

where $\sigma$ is the capture cross-section, $v_{th}$ the thermal velocity and $N_t$ the concentration of electrically active centers in $cm^{-3}$. In a very simple way the addition of radiation induced defects $N_{trad}$ can be described by

$$N_t + N_{t\,rad} = \frac{1}{\sigma \cdot v_{th}} \cdot \frac{1}{\tau_1} \tag{2}$$

where $\tau_1$ is the post-radiation lifetime ($\tau_1 < \tau_0$, $\tau_0$ being the initial lifetime). This yields the relation

$$\Delta\left(\frac{1}{\tau}\right) = \frac{1}{\tau_1} - \frac{1}{\tau_0} = \sigma \cdot v_{th} \cdot N_{trad} \tag{3}$$

On the other hand the following holds

$$N_{t\,rad} = \sigma_d \cdot N_A \cdot \Phi \tag{4}$$

with $\sigma_d$ the total displacement cross-section, $N_A$ being the number of lattice atoms per unit volume and $\Phi$ the particle fluence per unit area. Combining Eq. (3) and Eq.(4) yields

$$\Delta\left(\frac{1}{\tau}\right) = \underbrace{\sigma \cdot v_{th} \cdot \sigma_d \cdot N_A}_{k_\tau} \cdot \Phi = k_\tau \cdot \Phi \tag{5}$$

$k_\tau$ is called the damage factor or damage coefficient and is commonly used as a very powerful tool to describe the behaviour of a semiconductor device after irradiation. It is at least required that $k_\tau$ is independent on the fluence and this is fulfilled for fluencies not so high that the wave function of the damage centers do overlap. Extensive experimental work is carried out in order to determine the damage factor of Si. Here especially for neutrons a bulk of knowledge exists, but the more abundant species in space, electrons and protons, are also present [4,5,6]. Tab.1 gives the values of $k_\tau$ for a number of resistivities and injection levels (this quantity is the ratio of excess carriers to majority carriers) and electrons, protons and neutrons at a specified standard energy.

| | Substrate Resistivity [Ω cm] | | | | |
|---|---|---|---|---|---|
| injection level | | LOW(<1E-2) | | HIGH(>1) | |
| | n-type | | | | |
| | 1 | (0.6-3)E-7 | | ≈5E-8 | |
| **3-MeV-electrons** | 10 | (2-10)E-8 | | ≈1E-8 | |
| | p-type | | | | |
| | 1 | (1-4)E-8 | | (2-8)E-9 | |
| | 10 | (0.5-2)E-8 | | (1-4)E-9 | |
| | 100 | 3E-9 | | ≈6E-10 | |
| injection level | | 1E-3 | | 1E-1 | |
| | n-type | | | | |
| | 1 | (2-10)E-5 | | (1-5)E-5 | |
| **20-MeV-protons** | 10 | - | | ≈5E-6 | |
| | p-type | | | | |
| | 1 | (1-3)E-5 | | ≈1E-6 | |
| | 10 | - | | ≈5E-6 | |
| injection level | | 1E-5 | 1E-3 | 1E-1 | 1 |
| | n-type | | | | |
| | 1 | 1E-5 | 5E-6 | 2E-6 | 1.5E-6 |
| | 10 | 6E-6 | 3E-6 | 1.5E-6 | 1E-6 |
| **1-MeV-neutrons** | 100 | 1E-5 | 2.5E-6 | 5E-7 | 3E-7 |
| | p-type | | | | |
| | 1 | 8E-6 | 2E-6 | 5E-7 | 3E-7 |
| | 10 | 8E-6 | 2E-6 | 5E-7 | 3E-7 |
| | 100 | 2.5E-6 | 1.5E-6 | 5E-7 | - |

Tab.1 The damage factor $k_\tau$ in $cm^2/s$ for 3 MeV-electrons, 20 MeV-protons and 1 MeV-neutrons for Si and various resistivities.

To cover the whole energy range, Fig.1 gives the energy dependence of the damage factor for the individual radiation species, where the damage factor is normalized to that value at the standard energies.

The minority carrier lifetime represents the most sensitive semiconductor quantity, but in some cases the apparent change of doping (carrier removal) or the change of mobility are requested. In a way similar to the approach for the lifetime damage factor the following definitions are established:

**carrier removal**

$$n = n_0 - \left(\frac{\Delta n}{\Delta \Phi}\right) \cdot \Phi \tag{6}$$

and the change in **carrier mobility**

$$\frac{1}{\mu} = \frac{1}{\mu_0} + k_\mu \cdot \Phi \tag{7}$$

For 1 MeV-neutrons in Si the following values hold[6] : $(\Delta n/\Delta\ \Phi)=6\ (ns)^{-1}$ and $k_\mu=3\cdot10^{-19}$ Vs/n. The use of the damage factor concept enables the application engineer to check quantitatively the irradiation induced degradation of a certain device parameter, which is a function of e.g. the minority carrier lifetime $\tau$. This includes, for instance, the DC-current gain of a bipolar transistor or the spectral yield of a solar cell or other important properties of bipolar devices. From Tab.1 it is obvious that p-type Si is more resistant against displacement damage than n-type material and that has some impact on hardening of solar cells.

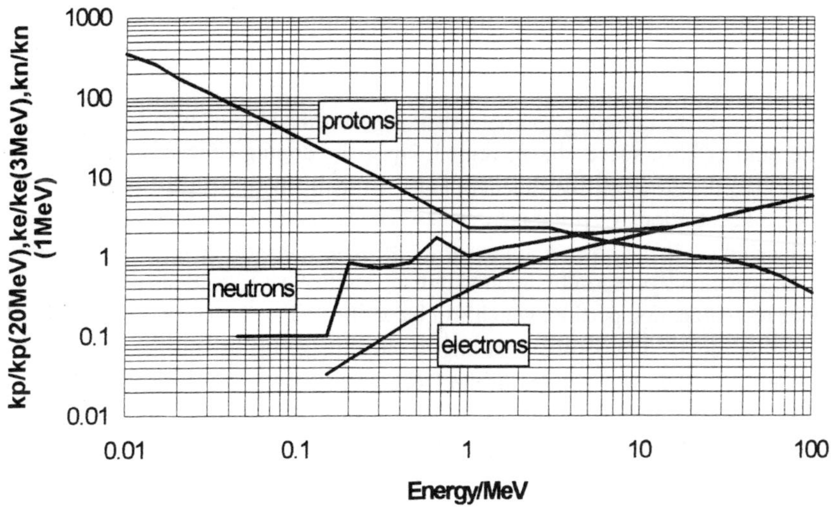


Fig.1 The ratio of damage factors for electrons (3 MeV), protons (20 MeV), neutrons (1 MeV) according to $k_\tau(E)/k_\tau$ (at standard energy).

Let us assume as a simplified example a **bipolar pnp-transistor** and the common-emitter DC-current gain β according to the well-known WEBSTER-equation. Leaving only the irradiation sensitive parts and arranging for the change in reciprocal β we arrive at

$$\Delta\left(\frac{1}{\beta}\right) = \frac{1}{\beta_{irr}} - \frac{1}{\beta} = \underbrace{\frac{1}{2}\cdot\frac{w_B^2}{D_{pB}}\cdot\Delta\left(\frac{1}{\tau}\right)}_{(1)} + \underbrace{\frac{N_{DB}\cdot w_B\cdot w_{EB}}{2\cdot D_{pB}\cdot n_i}\cdot\exp\left(\frac{-q\cdot V_{EB}}{2\cdot k\cdot T}\right)\cdot\Delta\left(\frac{1}{\tau}\right)}_{(2)}$$

$$\left(+\underbrace{\frac{N_{DB}\cdot w_B}{2\cdot D_{pB}\cdot n_i}\cdot\frac{A_s}{A_j}\cdot\exp\left(\frac{-q\cdot V_{EB}}{2\cdot k\cdot T}\right)\cdot\Delta(s)}_{(3)}\right) \quad (8)$$

3

Here $\beta_{irr}$ is the current gain after irradiation, $w_B$ the base width and $w_{EB}$ the depth of the emitter-base space charge region, $N_{DB}$ the doping concentration of the base region, $D_{pB}$ the diffusion coefficient of holes within the base (= $L_{pB}^2/\tau$), $n_i$ the intrinsic carrier concentration, $V_{EB}$ the emitter-base voltage and kT/q the thermal voltage. The term labeled (1) is the portion governed by the base transport factor, (2) represents the volume space-charge recombination within the emitter-base junction and term (3) is the surface recombination portion, which will be omitted here. $\Delta(1/\tau)$ was introduced before and can be replaced by $k_\tau \cdot \Phi$ according to the damage factor concept. Therefore it is possible to describe the change in the DC-current gain as a function of fluence and emitter current ( ~ $\exp(-qV_{EB}/kT)$ ). Regarding the relative importance of terms (1) and (2), term (2) is commonly negligible at VEB-voltages >500 mV compared to term (1) and Eq.(8) reduces to

$$\Delta\left(\frac{1}{\beta}\right) = \underbrace{\frac{1}{2}\cdot\frac{w_B^2}{D_{pB}}}_{t_{tr}}\cdot k_\tau\cdot\Phi \quad (9)$$

where $t_{tr}$ is just the transit time of holes across the base region. This quantity is related to the cut-off frequency by $f_{tr}=1/(2\pi t_{tr})$. As a first result, Eq.(9) suggests, that high-frequency transistors are much more radiation resistant than their low-frequency counterparts and this is true as a rule of thumb: reduce base width to get radiation hardness. There is a clear order from power- and small-signal transistors at the weak end and high-speed digital devices at the rad-hard regime. In addition most experimental results show a linear relationship between change of reciprocal gain and fluence as would be expected from Eq.(9). A second rule of thumb is, that a transistor should operate in the high collector current range, where the $\beta(I_C)$ characteristic is almost flat.

A great variety of electrical parameters of semiconductor devices depends on the minority carrier lifetime and all these quantities are affected in a analogous way. The efficiency of solar cells, for examples, depends on the diffusion length and any degradation results in loss of yield, most pronounced in the infra-red regime. In other cases, where the leakage current is of importance, an increase is a direct result of displacement damage. At higher fluencies the appearant reduction of doping concentration (carrier removal) results in increased bulk resistances, which is specifically important for power or majority carrier devices. The reduction of mobility, at high fluencies, is most pronounced at low temperatures in particular, when scattering at defects dominates. In order to get a feeling of the vulnerability of current devices most of modern devices withstand a neutron fluence of $10^{14}$ $cm^{-2}$, but $10^{15}$ $cm^{-2}$ is a critical fluence for most of the digital ICs. Bipolar circuits are generally more sensitive. Here the radiation species are neutrons, but the range of fluence can easily be translated to other radiation species by using Tab.1 and Fig.1.

### *2.1.2 Test Considerations*

The only parts in a space environment affected by displacement damage are solar cells and low-frequency bipolar transistors. The particles under concern for this type of damage are protons and electrons and both are heavily ionizing as well. Testing is accomplished by a step-stress approach and the radiation sources, mostly used, are van de Graaff accelerators and high energy protons from cyclotrons and linear accelerators. When neutrons are used they come from nuclear reactors and neutron generators. In the case of a spectral distribution of the radiation species it is advisable to calculate the equivalent fluence at the standard energies as mentioned in section 2.1.1 according to Fig.1 and Tab.1 by using the following Eq.(10)

$$k_{stand} \cdot \Phi(E_{stand}) = \sum_i k(E_i) \cdot \Phi(E_i) \qquad (10)$$

Here, all the individual spectral contributions are summed up in their effectiveness in producing displacement defects.

### 2.2 *Total Ionizing Dose Effects*

Total ionizing dose effects on electronic devices and circuits, have been studied for more than 25 years, but up to now the complex interactions are not fully understood. Different models are used to predict the behaviour of the prompt and delayed radiation response of electronic devices. A brief review of the basic effects will be discussed.

The TID has the strongest impact on MOS devices and circuits which use passivation layers in the active area of the transistors like advanced bipolar technology. Therefore, the following description is limited to these technologies.

Some preliminary explanations about the interaction of ionizing particles and electromagnetic quanta with the insulating layers of the MOS technology are necessary to understand the degradation mechanisms of semiconductor devices. The first effect is the transfer of energy to the matter quantified by the absorbed dose and the second step is the generation of electron-hole pairs and their impact on the insulating material.

#### 2.2.1 *Dosimetry*

High energy particles, such as electrons, protons and electromagnetic quanta interact with matter in different ways but they always lose energy in the irradiated volume. The **absorbed dose**, often called **dose** or **total dose**, gives the mean energy absorbed per unit mass of irradiated material at the point of interest [7,8].

$$D = \frac{dE}{dm} = \frac{1}{\rho} \bullet \frac{dE}{dV} \tag{12}$$

E : energy [ J ]  m: mass [kg]  $\rho$ : density [kg/ dm$^3$]  V : volume [dm$^3$]

The SI unit of dose is the gray (Gy) :

$$1\ \text{Gy} = 1\ \text{J/kg} \tag{13}$$

Due to the different atomic properties of the target material, the absorbed dose depends on the material of interest and therefore the specific material should always be referenced in parentheses of the unit rad or Gy (e.g. rad(Si), Gy(Si)). The use of the gray has been strongly advocated by the various standard organizations but the rad is still used by many experimenters and in many applications.

$$1\ \text{Gy} = 100\ \text{rad} \quad \text{with the definition of: } 1\ \text{rad} = 100\ \text{erg/g} \tag{14}$$

The material of interest to the MOS technology is silicon (Si) and silicon dioxide ($SiO_2$). The stopping power in Si and $SiO_2$ is nearly the same for particle energies above 1 keV[9]. Therefore the specified material for MOS devices is silicon (rad (Si) or Gy (Si)).

The **absorbed dose rate** (also often called **dose rate**) D' is the time rate of the absorbed dose [8].

$$D' = \frac{dD}{dt} \tag{15}$$

### 2.2.2 *Generation of Oxide Charges and Interface States in $SiO_2$*

#### 2.2.2.1 *Charge Generation and Recombination*

For the practical solution of charge generation it can be supposed that the absorbed energy in the material is totally used to create electron-hole (e-h) pairs. The energy needed to create one e-h pair in $SiO_2$ is $E_{e\text{-}h}$ = 18 eV±1eV [10,11]. More accurate sets of measurements and analysis have shown a value of $E_{e\text{-}h}$ = 17eV ±1eV [12], but the value of $E_{e\text{-}h}$ = 18 eV is still used. The typical ionizing energies and e-h density generation factors $g_0$ of some semiconductor materials are summarized in Table 2.

| **Material** | **$E_{e\text{-}h}$ [eV]** | **$g_0$ [e-h/(Gy·cm$^3$)]** |
|---|---|---|
| Silicon Si | 3.6 | $4.0 \cdot 10^{15}$ |
| Silicon dioxide $SiO_2$ | 18 | $8.2 \cdot 10^{14}$ |
| Gallium arsenide GaAs | ~ 4.8 | $\sim 7 \cdot 10^{15}$ |
| Germanium Ge | 2.8 | $1.2 \cdot 10^{16}$ |

Table 2 Electron - hole (e-h) pair generation energies and pair density generation factors of different materials [13]

Immediately after the first phase of e-h pairs generation there is a stage of recombination. Some of the electrons and holes generated initially recombine in less than picoseconds. Only a fraction of the primary generated e-h pairs remains in the oxide. The recombination rate is dependent on the particle energy and type and the oxide field. Two basic models are used to describe the ionization and recombination process of e-h pairs in insulators:

(i) the **columnar model** , where the e-h pairs are very close together [14-19]

and

(ii) the **geminate model**, where the charge density N along the ionizing track is smaller [20-22].

These models are determined by two parameters:

(i) the **thermalization distance $r_t$** which characterizes the initial separation between the hole and its corresponding electron after reaching their thermal energy

and

(ii) the **mean separation distance $\lambda$** between the e-h pairs.

In $SiO_2$ the thermalization distance is in the order of 5-10 nm. The separation distance is inversely proportional to the charge generation per cm. The generation of e-h-pairs is proportional to the electronic stopping power dE/dx.

$$N = \frac{\rho}{E_{e\text{-}h}} \bullet \frac{dE}{dx} \tag{16}$$

$$\lambda = \frac{1}{N} \tag{17}$$

N: charge density [$cm^{-1}$] $\rho$ : density [$g/cm^3$]
dE/dx: electronic stopping power [$eV/g/cm^2$]

For electrons, X-rays and Co-60 irradiation, the geminate model describes the experimental results sufficiently. Only for low energy radiation with a stopping power greater than 10 $MeV/g/cm^2$ the effects of the columnar and geminate model overlap and a modified model must be used [18]. An intermediary model has been developed by Brown and Dozier[23,24]. This model is valid only for sufficiently large electric fields ($E_{ox}$> 0.5MV/cm), otherwise the diffusions effects of the carrier cannot be neglected.

In summary, the charge generation and recombination of holes in silicon dioxide is a very complex phenomenon depending on the radiation type (particle or quanta), its energy and the oxide field. Especially for low energy X-rays the effects of multilayer structures with different densities must be taken into account. In practice X-ray and Co-60 sources are used for the irradiation testing of electronic devices. Dozier et. al. [25] have summarized different results of radiation experiments to find an empirical solution for the fraction of holes which escape recombination after Co-60 and 10 keV X-ray irradiations as a function of the applied oxide field. The following empirical formulas for the fractional hole yield $f_h$ have been found:

$$f_h(E_{ox}) = \left(\frac{0.5}{E_{ox}} + 1\right)^{-0.7} \qquad \text{Co-60} \tag{18}$$

$$f_h(E_{ox}) = \left(\frac{1.35}{E_{ox}} + 1\right)^{-0.9} \qquad \text{X-ray} \tag{19}$$

$E_{ox}$ : oxide field [MV/cm]

*2.2.2.2 Transport of Charges in the Oxide*

The phenomena of charge generation and recombination are essentially completed after some picoseconds. After this time, the excess electron and holes are free to undergo transport in the $SiO_2$ layer depending on the electric field. Electrons have a high mobility of $\mu_e$=20 $cm^2/Vs$ at 300 K and an increasing mobility at lower temperatures

($\mu_e$=40 cm$^2$/Vs at 114 K )[26-29]. The maximum velocity is about $1 \cdot 10^7$ cm/s [30,31]. After some picoseconds the electrons are swept out of the oxide ($d_{ox}$< 100 nm) and are collected at the positive electrode. The mobility of the holes is $1.6 \cdot 10^{-5}$ cm$^2$/Vs at 300 K and is strongly temperature dependent. The thermal activation energy is determined [32-34] to 0.16 eV .

*2.2.2.3 Hole and electron trapping*

Under the influence of the electrical field in the oxide the holes move to the $SiO_2$/Si or $SiO_2$/gate interface. At these interfaces a fraction of them are trapped in deep hole traps. The remaining holes recombine with the electrons of the gate or from the substrate contact. Trapped oxide charges $N_{ot}$ at the $SiO_2$/Si interface have the strongest effect on the threshold voltage shift of MOS transistors. The distance of the charge centroid of the $N_{ot}$ from the $SiO_2$ /Si interface is in the range of 3-13 nm [35-46], with a typical value of 4-5 nm [47]. The range of the hole trap density is $10^{12}$ to $10^{13}$ cm$^{-2}$ [47-49]. The conditions under which MOS devices are produced have a strong impact on the initial hole trap density $N_{t\infty}$.

*2.2.2.4 Hole trapping factor*

After the basic mechanisms of generation of electron-hole pairs and trapping of positive charges have been examined, the buildup of the oxide charge density $N_{ot}$ as a function of the accumulated dose, oxide field and trap density can be summarized by a simplified model. Assuming that the oxide contains only one type of hole trap $N_{t\infty}$ with a capture cross section of $\sigma_h$, the trapping rate may be described by a first order kinetic equation.

$$\frac{d\,\Delta N_{ot}}{d\,t} = \left(N_{t\infty} - \Delta N_{ot}\right) \cdot I_{inj} \cdot \frac{\sigma_h}{q} \tag{20}$$

$$I_{inj} = q \cdot g_0 \cdot f_h\left(E_{ox}\right) \cdot d_{ox} \cdot D' \tag{21}$$

| | |
|---|---|
| $I_{inj}$ : | generated current by the ionizing irradiation in the $SiO_2$ [A/cm$^2$] |
| $N_{t\infty}$ : | initial defect density [cm$^{-2}$] |
| $g_0$ : | e-h density generation factor (table 2) [e-h /(Gy • cm$^3$)] |
| $f_h(E_{ox})$: | fraction hole yield factor Eq. 18,19 |
| $d_{ox}$ : | oxide thickness [cm]     $D'$: dose rate [Gy/s] |
| $\sigma_h$ : | capture cross section [cm$^2$]     q : elementary charge [As] |

Providing that ionizing irradiation does not generate new hole traps in the oxide (which was found to be true up to a maximum dose of $10^5$ Gy(Si) [50]) and the annealing of $N_{ot}$ during the exposure time can be neglected, the following result is found:

$$\Delta N_{ot}(t) = N_{t\infty}\left(1 - \exp\left(-Q_{inj} \cdot \frac{\sigma_h}{q}\right)\right) \quad (22) \text{ with } \quad Q_{inj} = \int I_{inj} \cdot dt \quad (23)$$

Radiation hardness of different technologies or devices can be expressed by the hole trapping factor A introduced by Freeman et. al. [51]. Using the slope of the function at the time t = 0, the hole trapping factor can be defined as the ratio of the trapped charge density $\Delta Q_{ot}$ to the density of injected charges $\Delta Q_{inj}$. The trapping factor is then the product of the capture cross section and the initial trap density.

$$\left|\frac{d\,\Delta N_{ot}}{d\,t}\right|_{t=0} = q \cdot g_0 \cdot f_h(E_{ox}) \cdot d_{ox} \cdot D' \cdot \frac{\sigma_h}{q} \cdot N_{t\infty} \quad (24)$$

$$A = \frac{\Delta Q_{ot}}{\Delta Q_{inj}} = \sigma_h \cdot N_{t\infty} \quad (25) \qquad \text{with } \Delta Q_{ot} = q \cdot \Delta N_{ot} \quad (26)$$

Multiple capture cross sections and trap types have been found [52], but in practice this formula is sufficient to characterize the hole trapping behaviour of a MOS technology.

### *2.2.2.5 Annealing of Trapped Charges*

Most of the trapped holes anneal with time. Two models are used for the description of the logarithmic time dependence of the annealing of $N_{ot}$ after irradiation. The thermal emission model assumes the annealing of $N_{ot}$ by thermal emission of holes from the traps in the oxide to the valence band. The annealing behaviour of $N_{ot}$ as a function of the temperature and oxide field has been examined in great detail [53-56]. McLean´s tunnel model [57] assumes that the positive charges recombine with electrons tunneling from the substrate into the $SiO_2$. McWhorter et. al. [58] have derived a new model for predicting hole annealing by combining the tunnel and thermal model.

A good description of the annealing behaviour of $N_{ot}$ as a function of temperature magnitude and direction of the oxide field is important for characterizing of dose rate effects and the long term stability of irradiated MOS devices.

### *2.2.2.6 Generation of Interface States*

In addition to hole trapping in silicon dioxide, there is a buildup of interface states $N_{it}$. The prompt and delayed generation process is more complicated than the buildup of oxide charges. Many models have been proposed to explain the nature and generation process of these defects[3].

### 2.3 *Radiation Effects on MOS Devices and Circuits*

Basic effects of ionizing radiation on MOS structures have been discussed in the previous sections. These primary effects of trapped oxide charges and generated interface states cause the following changes in MOS transistor characteristics and CMOS device performance:

**MOS transistor**

| | | |
|---|---|---|
| - | negative threshold voltage shift | $\Delta V_{Th}$ |
| - | decrease of the surface mobility | $\mu_0$ |
| - | increase of surface recombination velocity | $S_0$ |
| - | decrease of the transconductance | $g_m$ |
| - | increase of source drain leakage current | $I_L$ |
| - | decrease of drain to source breakdown voltage | $V_{BR}$ |
| - | increase of noise | N |

**CMOS circuits :**

| | | |
|---|---|---|
| - | decrease of the output voltage | $V_{out}$ |
| - | decrease of the max. output current | $I_{os}$ |
| - | increase of the propagation delay time | $t_p$ |
| - | increase of the quiescent current | $I_{DD}$ |

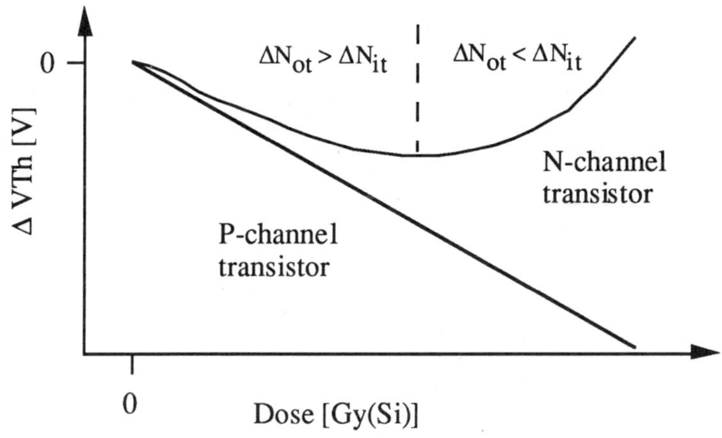


Figure 2 Schematic threshold voltage shift $\Delta V_{Th}$ of N-and P-channel transistors versus absorbed dose

Total dose response of the threshold voltage shift of N- and P-channel transistors due to the number of trapped charges $N_{ot}$ and generated interface states $N_{it}$ is different, as shown in Figure 2. Provided that the total amount of charges is trapped at the $SiO_2/Si$ interface, the threshold voltage shift $\Delta V_{Th}$ can be defined as follows :

$$\Delta V_{Th} = -\frac{d_{ox} \cdot q}{\varepsilon_{SiO_2}} \cdot (\Delta N_{ot} \pm \Delta N_{it}) \tag{27}$$

with the permittivity of silicon dioxide being $\varepsilon_{SiO_2}$, the elementary charge q, and the oxide thickness $d_{ox}$.

The threshold voltage shift is proportional to the number of trapped oxide charges and the number of charged interface states. The fermi level controls the charge state of the $N_{it}$. In the case of the N-channel transistor, the interface states are negatively charged (acceptor states in upper half of the band gap) and therefore the minus sign in Eq.27 is valid. For P-channel transistors, the interface states are positive charged (donator states in the lower half of the band gap) and must be added to the oxide charge (plus sign). Therefore, under the same test conditions, the threshold voltage shift of P-channel transistors is stronger compared to the shift of N-channel transistors.

If the magnitude of $\Delta V_{Th}$ of the P-channel transistor reaches the supply voltage, the transistor cannot be switched into the **on**-state, which leads to functional failure.

Negative threshold voltage shift of the N-channel transistor turns the transistor on. This increases the leakage current and the transistor cannot be switched into the **off**-state, thus generating function failures. For the N-channel transistor, the balance of $\Delta N_{ot}$ and $\Delta N_{it}$ leads to a turn around effect of $\Delta V_{Th}$ (Fig. 2). This so called **rebound** [53] or **superrecovery effect** [59] can increase the threshold voltage above its pre-irradiation value.

The mobility degradation $\mu_i$ is related to the Coulomb scattering by trapped charges [60-65]. The reduction of $\mu_0$ was first fitted only by the interface state density [63], but recent work include the trapped oxide charges as well [64,65].

$$\mu_i = \frac{\mu_0}{1 + \alpha_{it} \cdot \Delta N_{it} + \alpha_{ot} \cdot \Delta N_{ot}} \tag{28}$$

with the fitting values being $\alpha_{it} = 3.9 \cdot 10^{-12}$ $cm^2$ and $\alpha_{ot} = 0.7 \cdot 10^{-12}$ $cm^2$ at 300 K and $\alpha_{it} = 3.4 \cdot 10^{-12}$ $cm^2$ and $\alpha_{ot} = 1.3 \cdot 10^{-12}$ $cm^2$ at 77K [65].

Transconductance $g_m$, cut off frequency $F_T$ and the output short current $I_{os}$ is proportional to the surface mobility $\mu_i$. Also the increase of the propagation delay $t_p$ time of a CMOS inverter is related to the mobility degradation due to the reduction of $I_{os}$ [63].

$$\frac{\Delta t_p}{C_{load}} = \frac{0.5 \cdot V_{out}}{I_{os}} \qquad \text{with } V_{out} = V_{DD} - 0.5 \tag{29}$$

where $V_{out}$ is the max. output voltage, $V_{DD}$ the drain voltage and $C_{load}$ the load capacitor of the circuit

The generation of $N_{it}$ increases surface recombination velocity $S_0$ [62,63,66]. This characteristic is important for the storage time of CCD and DRAM. $S_0$ is proportional to $N_{it}$ and the square root of the product of the hole and electron capture cross sections $\sigma_n$, $\sigma_h$:

$$S_0 = 0.5 \bullet ( \sigma_n \bullet \sigma_h )^{1/2} \bullet \upsilon \bullet k\,T \bullet N_{it} \tag{30}$$

where $\upsilon$ is the thermal velocity.

The dominant radiation effect on CMOS technologies is the subthreshold leakage current of the N-channel transistor [67,68], and for nonradiation hard technologies the parasitic leakage path between source and drain at the bird´s beak [69]. Parasitic field oxide leakage is also significant for the radiation hardness of advanced bipolar technologies using recessed field oxide or walled emitter structures [70-73].

## 3. Total Dose Study on LS-, ALS- and F-Technology Integrated Circuits

Bipolar components demonstrate a relatively high tolerance to the total dose environment. With the exception of linear microcircuits bipolar technology digital circuits are radiation hard to at least $10^3$- $10^4$ Gy (Si) [74,75]. But on the other hand they are sensitive to neutron radiation. Low Power TTL devices[75] start to degrade at levels of $4 \bullet 10^{12}$ n/cm$^2$. CMOS devices are less sensitive to neutron damage (CMOS[74,75] bulk $2 \bullet 10^{14}$ n/cm$^2$; CMOS/SOS $10^{15}$ n/cm$^2$ ), but much more sensitive to ionizing radiation. Advanced bipolar technology is therefore of great interest for the application in new detectors. One of the major advances in the bipolar technology have been made by the introduction of the so called recessed oxide. The recessed oxide is a field oxide which extends from the surface into the silicon as deep as the active area of the device. This oxide provides lateral dielectric isolation, acts as a diffusion stop and reduces the junction capacitance. This allows smaller feature size and together with the reduced capacitance an increase of the speed. The impact of the recessed oxide on the radiation hardness of this type of devices is studied in great detail in the last years [70,71,72,76].

### 3.1 Test Procedure

Four devices were irradiated at room temperature with 2.5 MeV electrons. One device is used as a control device without any irradiation but it undergoes the same handling procedure. The dose rate was varied depending on the dose step from 0.05 Gy(Si)/s to

320 Gy(Si)/s. The electron beam is monitored by Faraday cups and the number of electrons per $cm^2$ is converted into an absorbed dose by the following equation:

$$D\ [Gy(Si)] = 2.5 \cdot 10^{-10} \cdot \Phi_e\ [e/cm^2] \qquad (31)$$

The ionizing effect of the 2.5 MeV electron irradiation is compared to the radiation effects of a Co-60 source [77].

The devices were biased under the expected worst case condition during irradiation and the supply current ICC was monitored. The electrical measurement was performed about 5 minutes after exposure. The annealing behaviour was studied 48 h after the end of last exposure for the first time and again after some weeks. During this time the devices were stored at 22 °C ± 1 °C without bias, that means all pins were grounded.

The electrical measurements were performed by using the Gen Rad 1732 digital test system, evaluating the DC-characteristics according to the data sheets. The functional tests were done with the best and worst case test conditions. An additional set up was used to measure the propagation delay time and the fall and rise time at one output stage. To reduce the number of data the worst value of each input or output characteristic was selected for the statistical analysis.

It should be noted, that in many cases tests are performed by using the worst case test conditions for the input and output characteristics. This can give significantly different test results compared to other bias conditions. The use of the standard worst case test conditions allows the designer to perform the worst case analysis of a new application in the usual way and the irradiation data is not limited to a dedicated application.

## 3.2 Test Results

### 3.2.1 LS - Technology

74LS241 and 74LS244 octal buffers and line drivers from Motorola were irradiated up to 0.7 MGy(Si) and 0.8 MGy(Si) before the first parameter exceeds the specification limit. All inputs of the devices were connected via 470 Ω to the supply voltage $V_{DD}$= 5V. The irradiation dose steps are in the order of 1, 3, 5, 10 for each decade, but only steps with significant changes in one of the 18 measured characteristics (IIL, IIH, VOL, VOH, IZL, IZH, IOS, ICC, tpLH, tpHL), together with the function test, using different parameter settings, are stored in the standard data files. The complete test results are published in the data compilation report [78].

The input low current IIL showed an increase by a factor of 11 at the max. dose of 0.7 / 0.8 MGy(Si) (VDD=5.25 V with an input voltage of Vin=0.5 V ). After 48 h as well as after 4900 h storing time nearly no annealing effect is observed. Fig. 3 depicts the input low current of the 74LS241/244 as an average value of 4 irradiated devices.

The high-level input current (IIH) , the tristate low- and high-level current (IZL, IZH), short-circuit output current (IOS), supply current (ICC) are nearly unchanged for both device types up to the maximum radiation level.

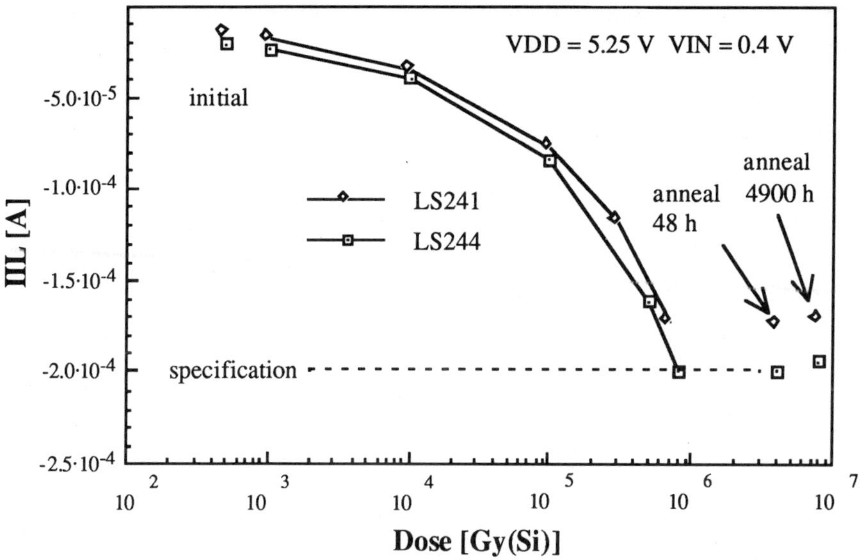


Figure 3 Input low current IIL as a function of total dose of the 74LS241/244 devices including the annealing after 48 h and 4900 h from Motorola.

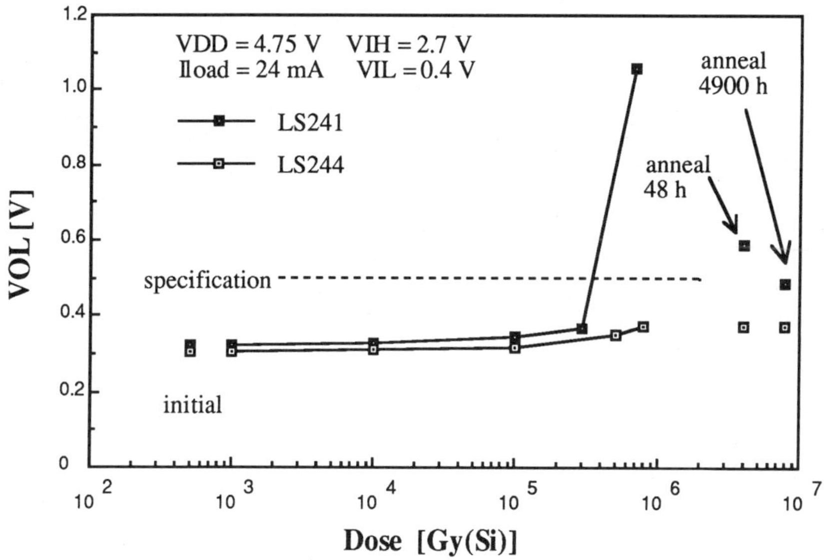


Figure 4 Low level output voltage VOL as a function of total dose of the 74LS241/244 devices including the annealing after 48 h and 4900 h from Motorola

In the case of the 74LS241 the low-level output voltage at VDD= 4.75 V with the minimum input voltage and maximum output current exceed the specification limit of 0.5 V at 0.7 MGy (Si) as shown in Fig.4. This value showed a strong annealing effect after 48 h but nearly no change after additional 4900 h. The VOL of the 74LS244 showed only a small increase after 0.8 MGy(Si) Fig. 4.

Increasing the supply voltage to 5.25 V and the low- level input voltage, the 74LS241 is again in the specification limit of 0.5 V. Also when the input voltage high level VIH is increased from 2.7 V to 3 V, keeping VDD= 4.75 V, the VOL stays below 0.5 V. Because the short-circuit output current showed no changes, this failure is due to the degradation of the input stage. In the case of the LS244 the propagation delay times, low-to-high level tpTH and high-to-low level output tpHL are nearly unchanged and showed above 0.1 MGy(Si) a slight increase of 25%. The propagation delay times of the LS 241 showed the same increase at 0.7 MGy(Si). Under some restriction in the performance, it should be possible to use this type of devices from Motorola up to 1 MGy(Si). The compared neutron dose to this level of electron irradiation is about $6 \cdot 10^{13}$ n/cm$^2$.

### 3.2.2 ALS -Technology

Two different devices, the 74 ALS 244 (octal buffer and line driver) and 54 ALS 151J (1 of 8 multiplexer), were irradiated up to 0.8 MGy(Si) and 2 kGy(Si), respectively, before the first functional failure was observed. The two different device types are coming from the same manufacturer with nearly the same date code. These devices showed complete different prompt radiation response. First the results of the 74ALS244 will be discussed. The high level input current IIH exceeds the specification limit at about 0.5 kGy(Si) and then decreases again to a value of $4 \cdot 10^{-7}$ A at 100 kGy(Si) (Fig. 5). This is a typical behaviour for ALS devices showing the dose window effect in the case of the IIH current [70,71]. The low-level input current IIL showed a confusing result. It starts with an initial value of $- 3 \cdot 10^{-5}$ A and changes to $+4 \cdot 10^{-5}$ A after 1 kGy(Si) and goes to nearly zero at 10 kGy(Si) and switches then to $-7 \cdot 10^{-7}$ A at 100 kG(Si). But the IIL current never exceeds the specification limit of 100 μA. The functional failure is caused by the increase of the low level output voltage VOL at about 600 kGy(Si) as illustrated in Fig. 6. The propagation delay time measured with a load capacitor of 12 pF exceed the specification limit of 10 ns at 500 kGy(Si) (Fig. 7). The quiescent current is continually decreasing from the initial value of 17 mA to 1.6 mA at 800 kGy(Si) and decreases further after the last exposure. All the other measured characteristics ( VOH, IZL, IZH, IOS) showed no significant changes after irradiation.

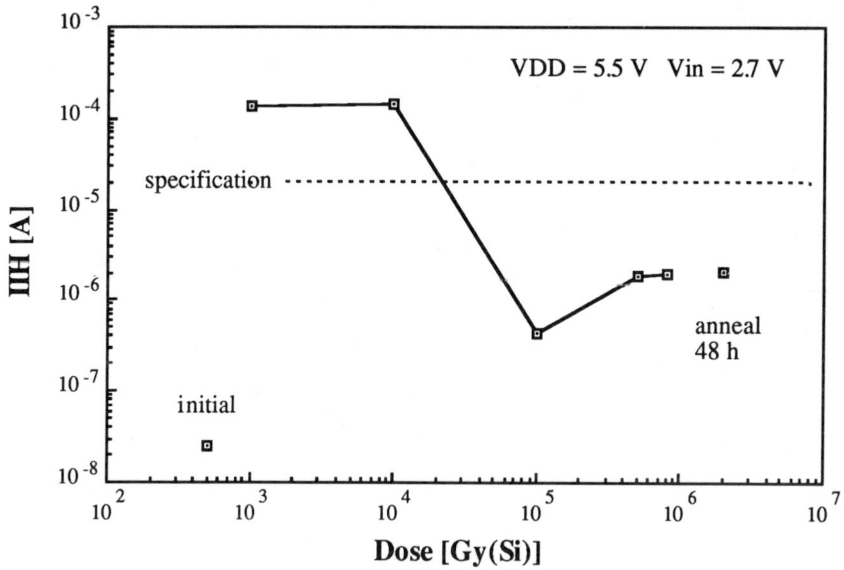


Figure 5 High level input current IIH as a function of dose of the 74ALS244 device. Positive voltage applied during irradiation

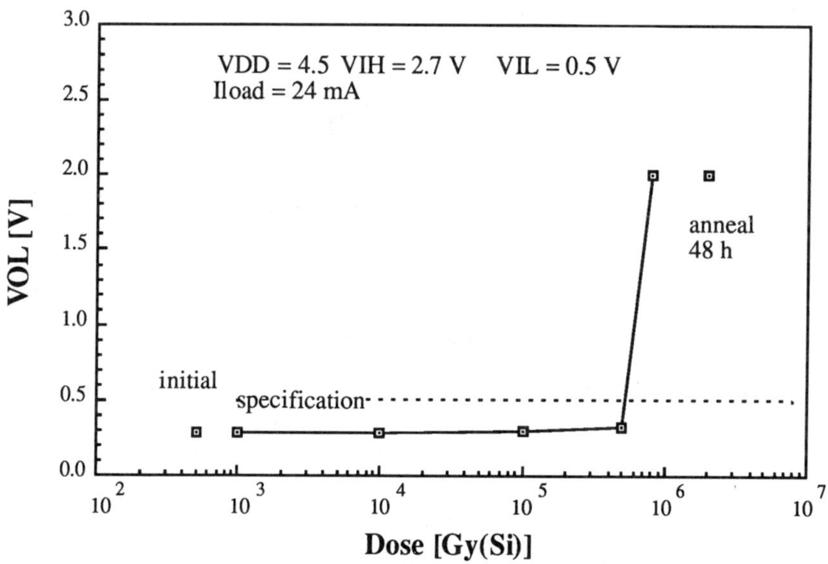


Figure 6 Low level output voltage VOL as a function of dose of the 74 ALS 244 devices under worst case test conditions.

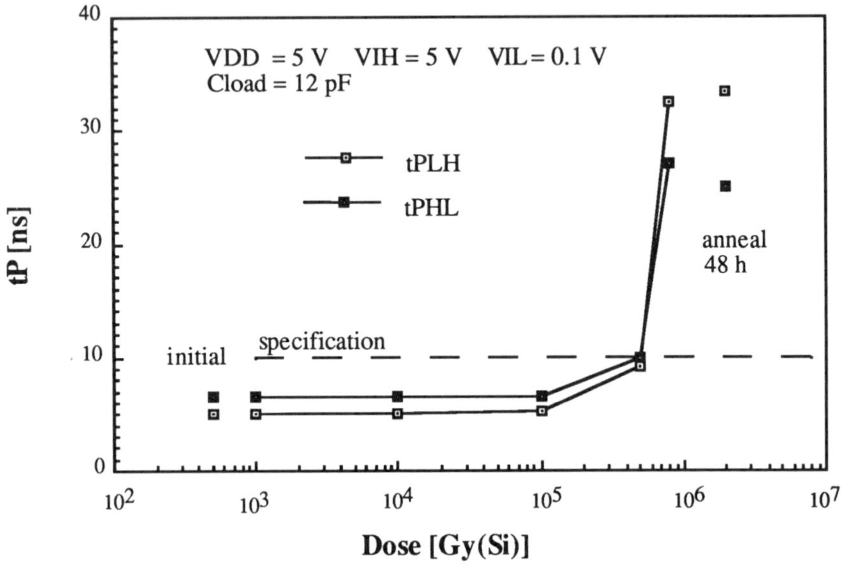


Figure 7 Increase of the propagation delay time tPLH and tPHL of the 74 ALS 244 devices as a function of the absorbed dose.

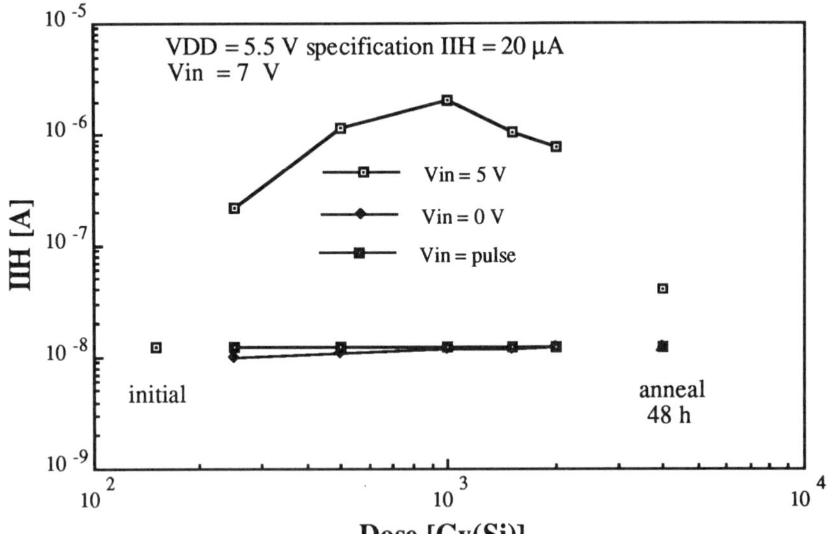


Figure 8 High level input current IIH of the 54ALS151 devices as a function of the absorbed dose using different bias conditions during irradiation.

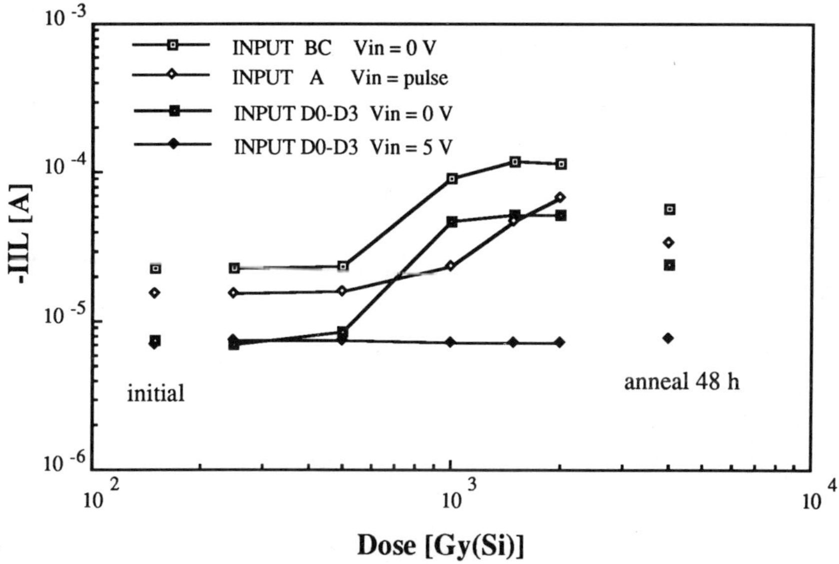


Figure 9 Low level input current IIL of the 54ALS151 devices as a function of the total dose. Different behavior of the inputs D0-D3 and ABC under positive, zero and pulse bias conditions during irradiation.

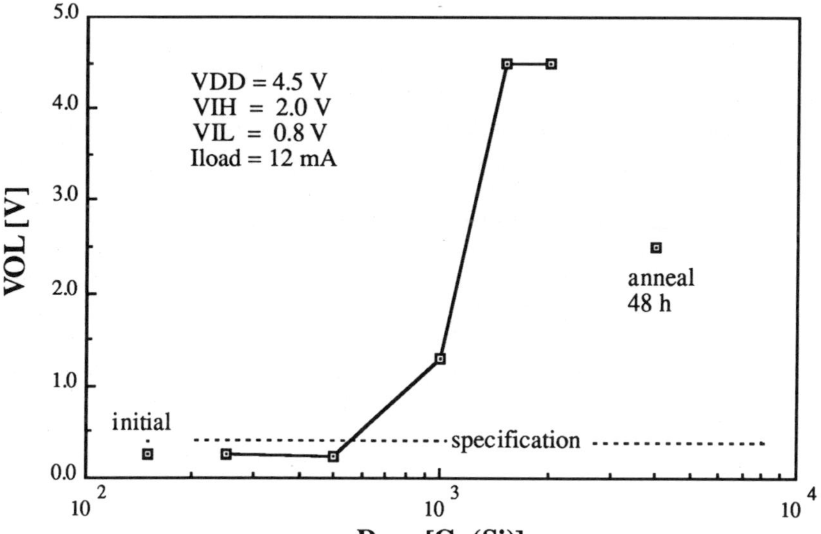


Figure 10 Increase of the low level output voltage VOL of the 54ALS151 devices as a function of the total dose measured using worst case test conditions.

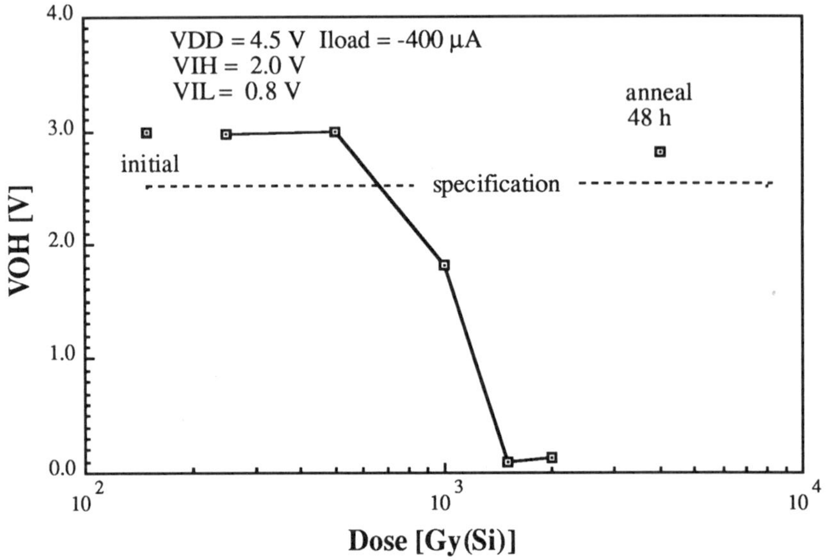


Figure 11 Decrease of the high level output voltage VOH of the 54ALS151 devices as a function of the total dose measured under worst case test conditions.

54ALS151 J devices are much more sensitive to radiation compared to the 74ALS244. The maximum IIH current is found at about 1 kGy(Si). At that level it just reached the specification limit (Fig. 8). Wether the inputs were grounded or pulsed (100 Hz, duty cycle of 50 %, VIH= 3 V, VIL = 0.5 V), no changes of the IIH current were observed. The low level input current IIL has an opposite bias dependence and the individual inputs have a different response to the radiation. There are two groups, the ABC inputs and the address inputs D0-D3. The increase of IIL is much stronger when the input is connected to ground, and the degradation of the ABC inputs is higher than of the D0-D3 inputs (Fig. 9). The dose dependence of the A-input is weaker when the input is forced by pulse voltage (100Hz, duty cycle of 50 % , VIH= 3 V, VIL =0.5 V). The IIH and IIL current are measured by VIN = 7 V and VIN = 0.5 V, respectively, and the IIH current is positive and the IIL current is negative. The bias dependence during irradiation is just in the opposite direction. This indicates that two different failure mechanisms must be involved in the degradation of the input stage. The functional failure is caused by the decrease of VOH and the increase of VOL (Fig. 10,11).

### 3.2.3 F-Technology

Two commercial nonradiation hard types 74F241 and 74F244 (octal buffer and line driver ) from Fairchild and Signetics were investigated. It has already been shown that the input stage of this technology is radiation sensitive due to the parasitic N- channel transistor [76]. The maximum positive voltage at the input during irradiation is the worst case bias condition. The difference in the IIH current of the Fairchild and Signetic devices F 241 is illustrated in Fig. 12. The specification limit of IIH is reached at a dose level of < 100 Gy(Si). The IIH current has a strong dose rate dependence which is indicated by the rapid annealing behaviour at RT (Fig. 12). The tristate current IZH, IZL at the high- and low- output voltage of the Fairchild devices showed nearly the same increase compared to the IIH current. In the case of the Signetic devices the IZH and IZL degrade much slower as illustrated in Fig. 13 and 14. This currents are correlated to the trapped positive charges in the soft field oxide. These charges anneal very fast, which has been observed. The functional test of the Fairchild devices failed at 50 Gy(Si) due to the high IZL and IZH current. The other characteristics (IIL, VOL, VOH, IOS, IL) are nearly unchanged up to the total dose of 100 Gy(Si). The devices from Signetics did not fail the functional tests until 300 Gy(Si).

The 74F244 from Signetics have been irradiated up to 1.8 kGy(Si) without any functional failure. The increase of the IIH current is in the same order compared to the 74F241 (Fig. 15). The IZL and IZH current (Fig. 16) of the 74F241 at 100 Gy(Si) are higher than from the 74F244 (Fig. 13, 15). Comparing the IZL and IZH values from the Fairchild 74F241 with the Signetics 74F244, the specification limit is reached at about 50 Gy(Si) and for the Signetics 74F244 at about 1 kGy(Si).

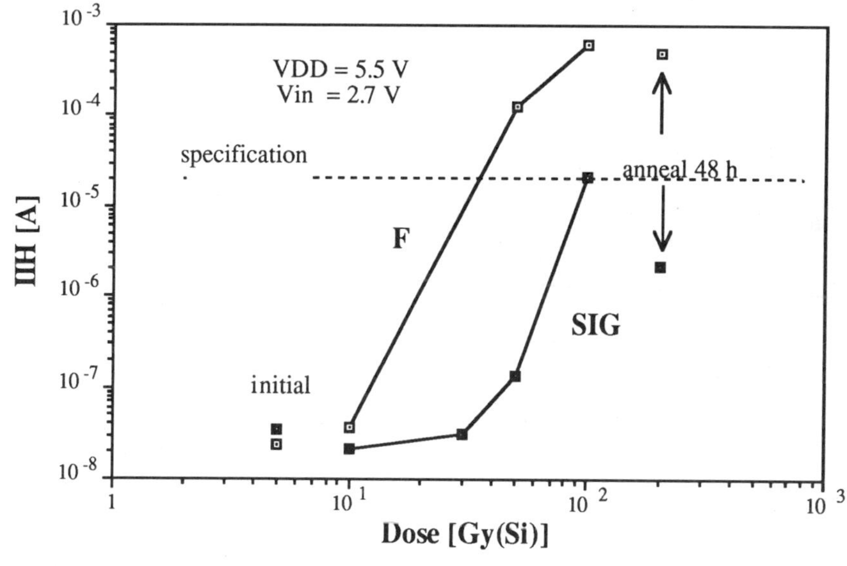


Figure 12 Increase of the high level input current IIH of the 74F241 devices from Fairchild and Signetics as a function of dose using positive voltage during irradiation.

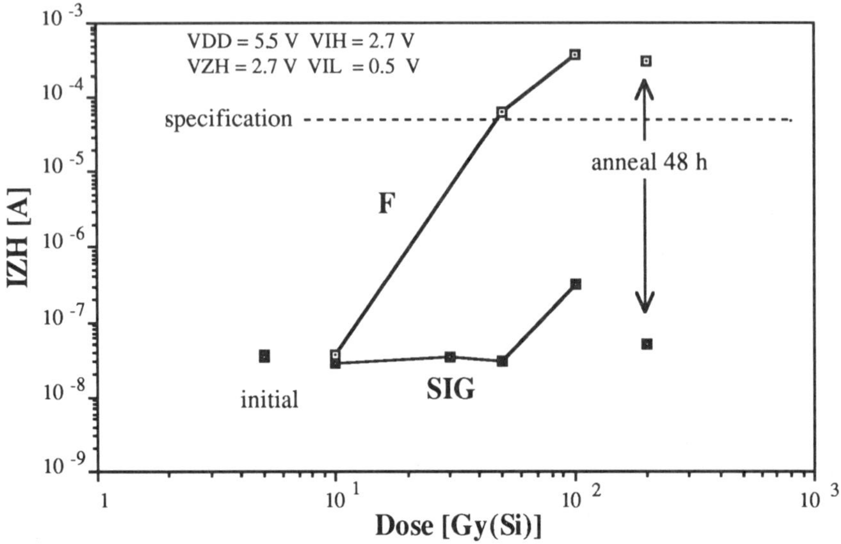


Figure 13 Increase of the tristate output high level current IZH of the 74F241 devices from Fairchild and Signetics as a function of dose.

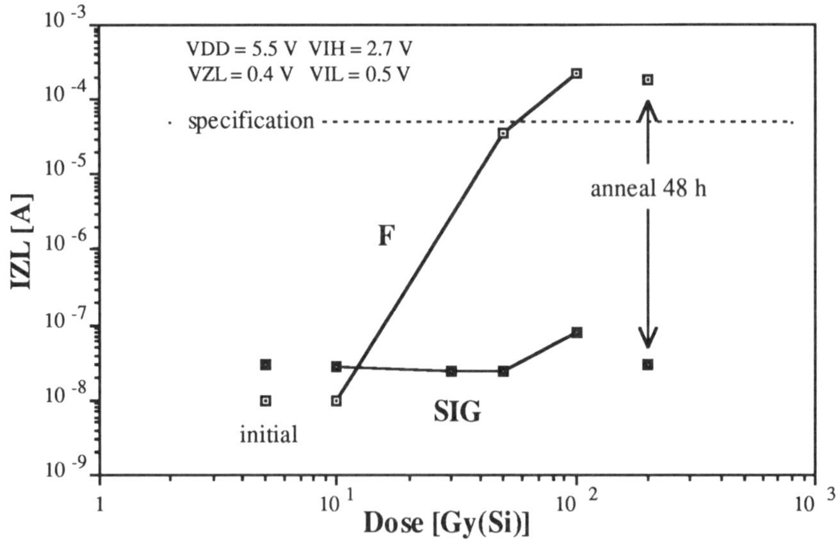


Figure 14 Increase of the tristate output low level current IZL of the 74F241 devices from Fairchild and Signetics as a function of dose.

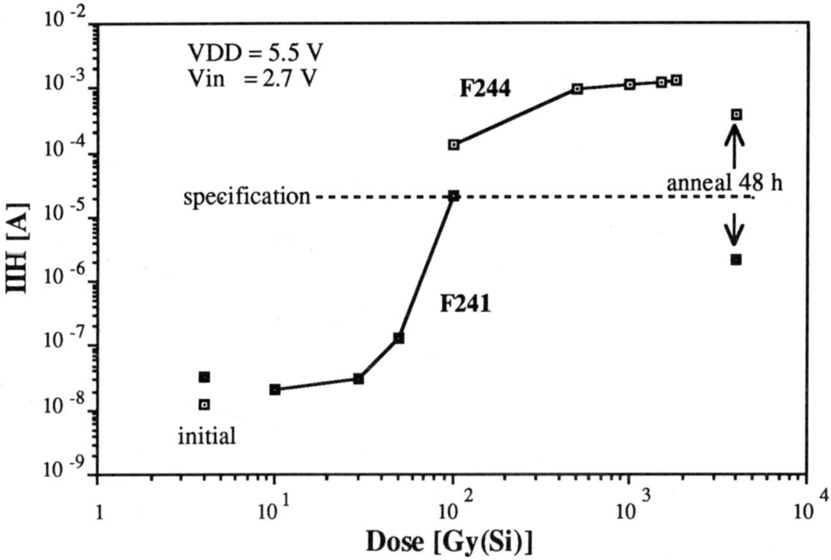


Figure 15 Increase of the high level input current IIH of the 74F241 and 74F244 devices from Signetics as a function of the total dose.

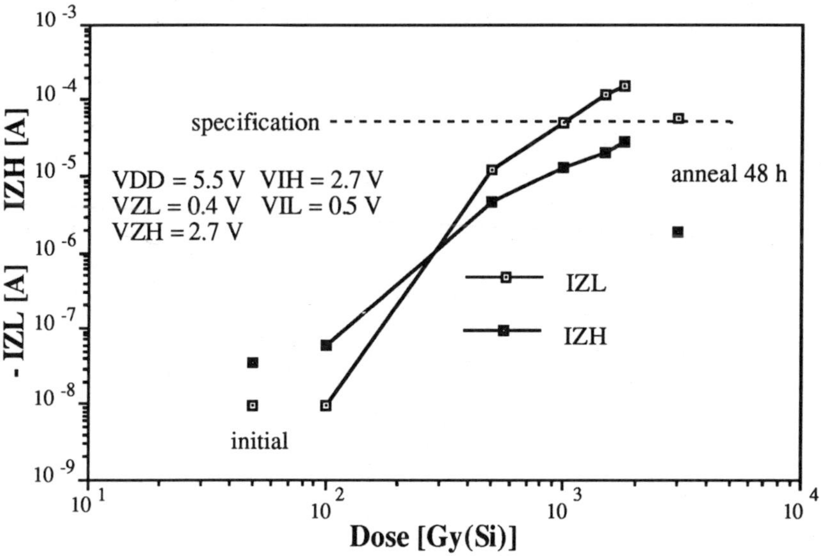


Figure 16 Increase of the tristate low and high level current IZL, IZH of the 74F244 devices from Signetics as a function of the total dose. The IZH is positive and the IZL is negative.

### 3.3. Discussion

The irradiation test on LS devices demonstrates again the total dose radiation hardness of this technology. The higher operational speed of the advanced ALS and F technologies is required for new applications in radiation environment (e.g. LHC or space). The F-technology seems to be more sensitive to ionizing effects than the ALS-technology.

The recessed oxide plays the dominant role in the degradation mechanisms of the F-technology [70]. The use of nested or walled emitter [71] for the ALS -technology have a strong impact on the radiation hardness. It has been shown, that there are improvements of the radiation hardness of this technology [71,72], but the improved process steps are not implemented in the standard and commercial devices.

It has been demonstrated by this investigation, that the radiation hardness of one technology from one manufacture is very strong dependent on the device type. Furthermore, all the other effects like bias conditions, dose rate etc. [79] have a strong impact on the prediction of the radiation hardness.

A wide range of total dose failure levels may be expected, even for a specific device type coming from one manufacturer. Normally little changes in design, layout and processing can result one or more orders of magnitude in the radiation hardness. Also a factor of three in the total dose failure level has been observed for logic circuits coming from the same wafer [70].

A good control of the process parameters can improve this large variation down to ±30% [76]. **Therefore, the traceability of the date code is not sufficient to guarantee the radiation hardness of commercial lots.**

### 3.4. Summary and Conclusion

Total dose failures due to parameter or functional test have been studied on bipolar digital microcircuits of different LS-, ALS-, F-technologies, device types and various manufacturers. The failure mechanisms of these devices were attributed to a) high leakage current of the characteristics IIH, IIL, IZL, IZH, for ALS and F technology, b) VOL and VOH in the case of ALS 151, c) VOL of the LS technology (Fig.17).

The observed failures are sensitive to the DC or pulse bias condition during irradiation, the dose rate and therefore the time of measurement after irradiation as well as to the total dose level for which the measurements are taken (see Fig. 8,9,15). A well defined test procedure for the worst case bias condition during irradiation and the evaluation measurement of the characteristics have to be established. This should be worked out together with the manufacturer of the devices, the application engineer and the independent test institute. The variation in the layout and process of commercial devices are done without any respect to the radiation hardness and therefore, the failure level of those parts could vary orders of magnitude from one date code to the other. Each production lot of devices should be carefully tested to determine the total dose failure level. The homogeneity across the wafers from on diffusion lot must be known. All this requires a close and confidential cooperation between the user and the manufacturer of the devices. This causes sometimes problems, but it seems the only practical way. Until radiation hardened designs and processes are fully implemented and controlled for example by statistical process control, any of bipolar digital microcircuits utilizing recessed oxides should be carefully evaluated before they are used in a radiation hard application. The trend in the advanced bipolar microcircuits is toward more use of recessed oxides, walled or nested emitter and closer spacing including trench isolation. This will result in even lower total dose failure levels unless some hardening techniques are implemented. It appears that in many cases a radiation hardness level of 10 kGy(Si) can be achieved if the known techniques are implemented.

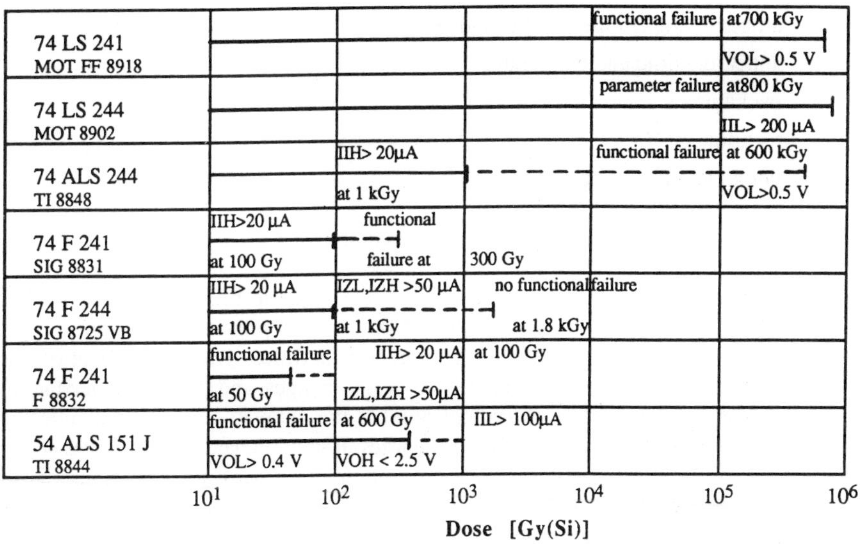


Figure 17 Summary of the degradation of the tested LS-, ALS- and F-devices.
——┤ no damage, – – –slight damage

## 4 Irradiation Qualification Tests

In the previous sections the basic effects of ionizing radiation on MOS devices have been briefly discussed. Many phenomena are in general understood, but the details are still part of research programs. The impact of processing steps on the radiation hardness of devices and circuits is not completely under control. Irradiation qualification tests must always be performed to verify the radiation hardness of a technology. For space and military applications many different device types from various technologies and manufacturers are used. Many evaluation and qualification tests have been performed over recent years. Depending on the properties of the device and the technology used radiation sensitivity is related to displacement damage and/or total dose effects. Irradiation tests must consider the dominant degradation effect of device types and technologies. The relative sensitivity due to the degradation effects is summarized in Table 3 [80,81]. The complex interdependencies of the radiation response of electronic devices depend on the following facts:

- type of radiation [25]
- dose rate [53,59,68,82-84,94,100-102]
- bias condition during and after irradiation [80,85,103,104]
- ambient temperature during irradiation [70]
- annealing time and temperature [53,69,84]
- packaging [87]
- technology [88,89]
- layout and circuit design [90]

Therefore a well defined irradiation test plan is required [77,91].

Standardizations of the test procedures for total dose testing are defined by MIL Standard Method 1019.4 [92] or ESA/SCC 22900 [93]. This provides the opportunity to compare irradiation test results from different test facilities. These specifications are continually updated to the implication of technology changes, and the new understanding of the radiation response [94,97]. A critical point is the delayed generation of interface states and the annealing behaviour of oxide charges. Radiation hard MOS devices show a different prompt and delayed response from nonradiation hard devices. Nonradiation hard devices degrade dominantly by hole trapping in the gate and the field oxide. Dose rate dependence of nonradiation hard devices is smaller compared with radiation hard devices [94]. Delayed generation of interface states is more pronounced in radiation hard devices, leading to rebound and superrecovery effects. These effects cannot be observed in short time annealing after irradiation at RT. High temperature annealing at 100°C over a period of 168 h with the same bias condition as used during irradiation has been recommended [94].

A new approach to radiation hardness assurance prediction is the qualified manufacturer list (QML) methodology [96,97].

| | Total Dose | Displacement | 0 = not, 3 = very sensitive<br>**Remarks** |
|---|---|---|---|
| **Diodes** | | | |
| Rectifier | 1 | 1 | Recombination and generation increase. Breakdown voltage increases slightly. At very high fluencies series resistances become pronounced. |
| Switching diode | 1 | 1 | The switching is even faster. No severe limitation. |
| Voltage reference | 0 | 1 | If temperature is compensated, the temperature coefficient changes to a positive sign. Zener-voltage increases or decreases. |
| **Transistors** | | | |
| Bipolar NF | 3 | 3 | Very sensitive, especially at low currents. Current gain decreases.Leakage currents increase by orders of magnitude. |
| HF | 1 | 1 | Less vulnerable compared to NF-transistor. Higher transit frequencies are advantageous. |
| Microwave | 1 | 1 | Thinning the base of the transistor improves hardness. |
| Power | 3 | 3 | Very sensitive especially at low currents. Both emitter yield and base transport factor degrade heavily. |
| JFET | 1 | 1 | Si and GaAs-FETs (J- and MES-) are commonly hard against all kind of damage, probably except for SEU. |
| MOSFET | 3 | 0 | Very sensitive to total dose, pMOS is partly harder than nMOS. Due to its anomalous annealing behaviour (rebound effect), the nMOS degradation is less predictable. For n- and pMOS, threshold voltage shifts to more negative values and leakage currents increase by orders of magnitude. |
| **Opto electronic** | | | |
| LED | 0 | 2 | GaAs is relatively hard, but the creation of non radiative centers limit the light output. Lenses darken. |
| Photo transistor | 3 | 3 | Very sensitive. Photo-sensitivity decreases very rapidly. |
| Opto coupler | 3 | 3 | Combination of LED and photo transistor deteriorates even worse compared to individual components. |
| Solar cell | 0 | 2 | Very sensitive against displacement, reducing the diffusion length and thus efficiency, especially in the long wave length region. |
| Optical waveguide | 3 | 1 | Extremely sensitive due to darkening. But hardened fibers are available. |
| CCD | 3 | 1 | Most sensitive devices, especially when they are operated at cryostatic temperatures. Slight improvement by using buried channel CCDs. |
| **Linear IC** | | | |
| OpAmp | 3 | 3 | Very sensitive, but hardened. OpAmps are available. The sensitivity is caused by low current operation, matching properties and DC-coupling. |
| Comparator | 3 | 2 | Sensitive, large failure range, depending on technology. |
| Reference source | 2 | 2 | Sensitive. |
| **Digital IC** | | | |
| TTL/ECL | 1 | 1 | Hard technology due to fast transistors, ECL even harder than TTL. |
| LSTTL | 1 | 1 | As TTL with minor deterioration. |
| CMOS/HCMOS | 3 | 0 | The commercial CMOS and HCMOS parts are the most vulnerable technology for total dose (only CCDs are more sensitive). The failure mechanism is similar n- and pMOS. |
| CMOS (rad hard) | 1 | 0 | Changes in technology and layout improve the performance to up to 1 kGy to 1 MGy. Very expensive. |
| SOS (rad hard) | 1 | 0 | Like CMOS (rad hard). Design rules are known to almost supress SEUs. |
| GaAs | 1 | 1 | Very hard against displacement and total dose.. |
| HMOS | 3 | 0 | Very sensitive, but hardened devices are available. |
| EPROMS | 3 | 1 | Very sensitive due to decharging of information. |

Table 3 Sensitivity and damage behaviour of device families

## 5. Radiation Hardness of Electronic Components

The prediction of the radiation hardness of electronic components due to total ionizing dose effects and displacement damage is a very complex task, as shown in the previous sections. The radiation sensitivity of different device types with respect to their susceptibility to the damage process is summarized in table 3.

Since many years, radiation tests have been performed on a large number of various device types from different manufactures. There are several dependences regarding the radiation hardness of an individual device such as layout, circuit design, bias condition during and after irradiation or dose rate effects. That means that the test conditions are not always the same. Under this consideration total dose failure levels of relevant technologies and device families are summarized in fig 18. This review of the sensitivity of the various technologies and device families shows the range at which the **first** failure has been observed (left margin of the bars) to the highest level at wich **all** devices failed (right margin of the bars). Fig. 18 contains irradiation test results from various test facilities published in recent years [74,98,99,105]. A careful selection by a lot acceptance test is always recommended when the radiation level exceeds the left margin of the bars.

As another example, the total dose radiation hardness data of different technologies from one manufacturer are shown in fig. 19. In addition to the data already shown at the workshop in Barcelona [1] and Aachen [2], the radiation hardness due to neutron damage of different device types is shown in fig. 20.

An extensive study for fusion reactors [106] has summarized a large number of radiation test results of electronic components and materials. Some results of the radiation hardness due to neutrons are shown in fig. 21. In table 4 the radiation damage effects of electronic components and materials are reprinted from this report, because the data are only available on microfilm and very hard to read. This information can assist in selecting components for use in high dose invironments. But, as explained before, the individual radiation hardness of a device or component depends strongly on the specific application requirements, design details and fabrication conditions and therefore this table can only be used as a guideline.

### Summary and Consequences

It has been shown, that the prediction of the radiation hardness of electronic components is a very complex task. There are radiation hard devices on the marked, but they are very expensive and the delivery time can be very long. Furthermore there are some limitations about the export licence for devices coming from US. Since some years also a few European manufacturers produce radiation hard devices. Semiconductor research institutes are developing radiation hard technologies. Therefore there exist extensive knowledge in this field and manufacturer resources are available to meet most of the requirements for the high luminosity collider and detectors. A radiation hardness assurance program should be established as soon as possible to define the test procedures and to organize the device procurement. This must lead into the definition of an approved part list (APL) for all experimentalists. The standardization of components and a close cooperation between the organization and the manufacturers or semiconductor research institutes can reduce the total cost significantly.

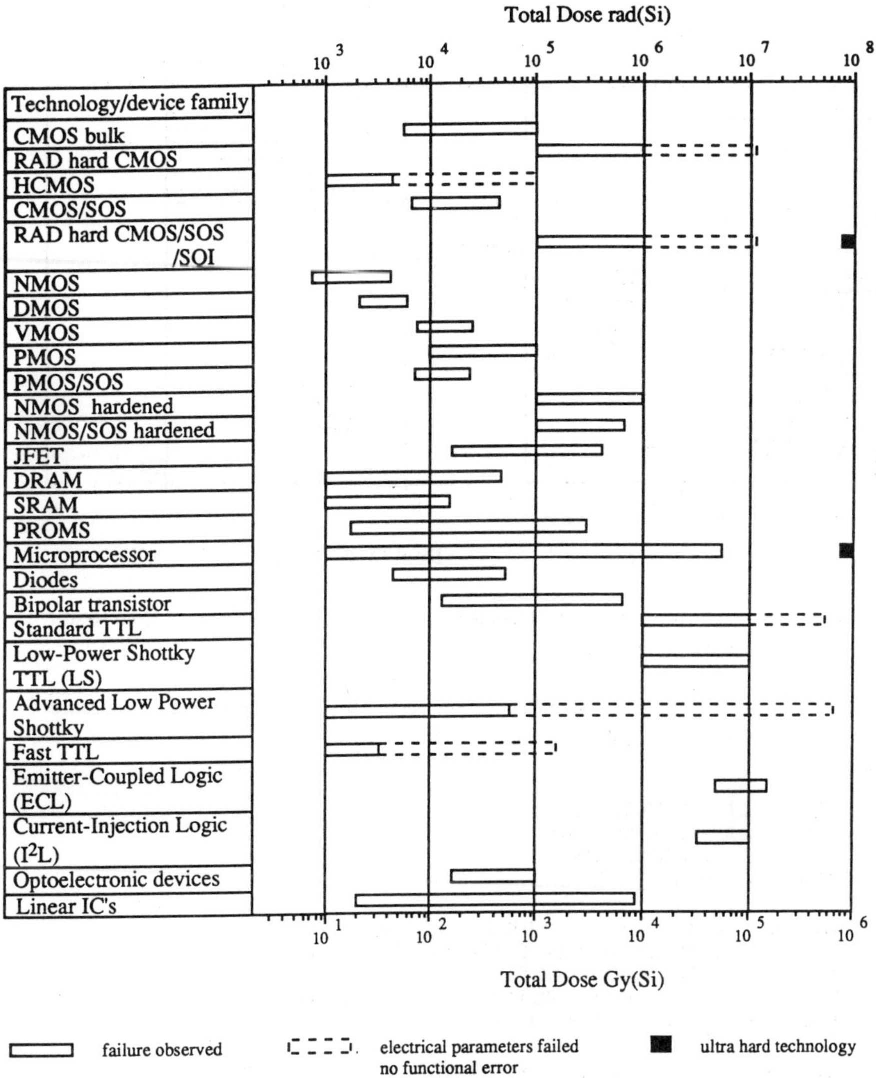


Fig. 18 Total dose failure of various technologies and device families. The left margin of the bars indicates the **first** observed failure and the right margin the highest dose level at which **all** devices failed.

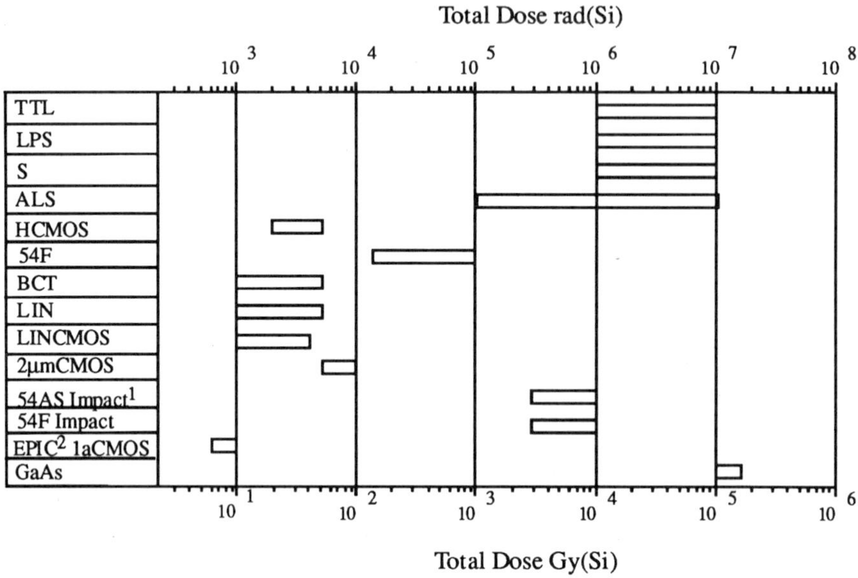


Note 1 Impact is a Trademark of Texas Instruments. (2µ Bipolar Oxide Isolated Nested Emitter Technology)

Note 2 Epic is a Trademark of Texas Instruments. (1µ Enhaced Performance implanted CMOS. Twin Well. Epitaxial Substrate)

TTL = ALL 54 FUNKTIONS
LPS = ALL 54LS FUNCTIONS
S = ALL 54S FUNCTIONS
ALS = ALL 54ALS FUNCTIONS

HCMOS = ALL 54HC AND 54HCT FUNCTIONS
54F = ALL 54F AND 54AS FUNCTIONS
BCT = ALL 54BCT FUNCTIONS
LIN = ALL UA, LM AND SN55 LINEAR FUNCTIONS
LINCMOS = ALL TLC FUNCTIONS

2µm CMOS = 54AC/ACT, GA AND STD CELL
EPIC 1a CMOS = 1M DRAM/VRAM

Fig. 19 Total dose failure range of device families from Texas Instruments

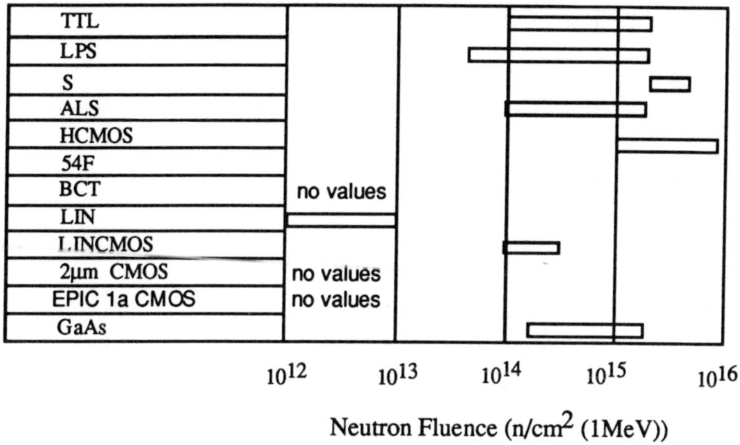


Fig. 20 Failure range due to neutron fluence of device families from Texas Instruments

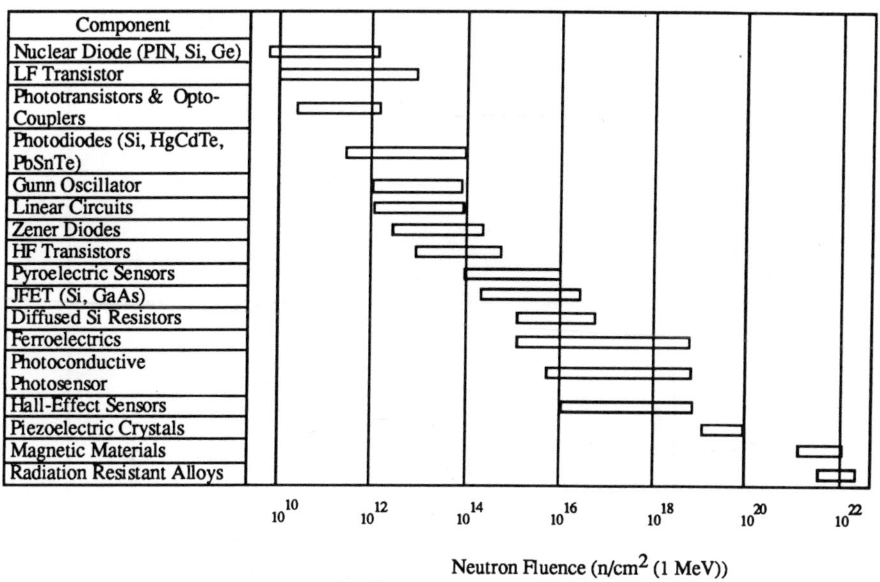


Fig. 21 Radiation sensitivity of different components due to neutron fluence

| Class | Component | Radiation Level | | Effect | Reference Page |
|---|---|---|---|---|---|
| | | n/cm² | Gy | | |
| Optical Components | **Windows, Lenses, and Prisma** | | | | |
| | Silica, fused | | $5 \times 10^7$ | Transmission in visible reduced to 89% due to gammas; unstated thickness | A, 135 |
| | Silica, vitreous | $<10^{17}$ | | Absorption of 10% near 0.6 µm and significant below 0.4 µm for 13 mm thickness irradiated in CP-5 reactor | D, 406 |
| | | $10^{19}$ | | 1% increase in specific gravity | P, 6 |
| | Quartz | | $1 \times 10^7$ | Transmission at 0.4-0.6 µm reduced to 30-35% due to gammas, unstated thickness | A, 135 |
| | | $2.65 \times 10^{17}$ | | Dilatation of $6 \times 10^{-3}$ to $1 \times 10^{-2}$ Gy | D, 406 |
| | | $3 \times 10^{19}$ | | 1% decrease in specific gravity | P, 6 |
| | Calcium Fluoride | | $>10^4$ | Slight color change due to gamma radiation | B, 11-18 |
| | Crown, protected ($CeO_2$) | | $>10^7$ | Slight transmission loss in visible; unstated thickness | B, 11-21 |
| | Flint, dense, protected ($CeO_2$) | | $>10^7$ | Slight transmission loss in visible; unstated thickness | B, 11-21 |
| | Vycor (silica), protected | | $10^6$ | Transmission loss in visible, usable; unstated thickness | B, 11-21 |
| | | | $5 \times 10^4$ | Transmission at 0.4-0.6 µm reduced to 24-36% due to gamma; unstated thickness | A,135 |
| | Glasses, various | $10^{15}$-$10^{20}$ | $10^3$-$10^6$ | Various, see reference | F, 370-375 |
| | Pyrex Glass | $5 \times 10^{17}$ | | 1% increase in specific gravity | P, 66 |
| | Lead Glass | $3 \times 10^{18}$ | | 1% decrease inspecific gravity | P, 66 |
| | Camera lens system, 10 components | | $10^3$ | 90/50/20 % transmission loss at 0.4/0.6/0.8 µm | C, 174 |
| | **Fiber Optics** | | | | |
| | Suprasil | $6 \times 10^{17}$ | $10$ - $10^7$ | Threshold range for unstated damage | W, 10 |
| | | | $10^7$ | Threshold for unstated damage due to 14 MeV neutrons | W, 10 |
| | Silica, pure | | $<3 \times 10^4$ | Transmission loss at 0.82 µm of 20 to 2000 dB/km | C, 198 |
| | **Fiber Optic Isolator System** | | | | |
| | HP 5082-4310 | | $10^4$ | Threshold level for ionization damage; used $^{60}Co$ source | K, 1 |
| | TI TIL-107 | | $10^3$ | Threshold level for ionization damage; used $^{60}Co$ source | |

Table 14-1 Summary of radiation damage effects of electronic components and materials

| Class | Component | Radiation Level | | Effect | Reference Page |
|---|---|---|---|---|---|
| | | n/cm² | Gy | | |
| Optical Components (Continued) | Quartz Crystal X-Ray Diffraction | $10^{19}$ | | Small or no effect in reflected x-ray properties; fission neutrons | E, 15 |
| | | $3 \times 10^{19}$ | | Slight shift in reflected x-ray wavelength; fission neutrons | |
| | | $8 \times 10^{19}$ | | Severely broadened and shifted reflected lines; fission neutrons | |
| Lasers | Solid State, Flash Lamp Pumped Ruby | | $10^{2}$ | Power output decrease by 40 % or more at 0.69 μm | A, 158 |
| Detectors | Zinc Sulfide Scintillator | $2 \times 10^{13}$ | | Light output loss of 10 % by neutrons for 10 μm layer | E, 3 |
| | InSb (photoconductive) | $5 \times 10^{11}$ | | Unstated threshold damage out to 5.6 μm by 14 MeV neutrons | E, 4 |
| | InSb (photovoltaic) | $10^{13}$ | | Unstated threshold damage out to 5.6 μm by 14 MeV neutrons | |
| | InSb (heat sensor) | $10^{14}$ | | Unstated resistance change at 100-2000 μm by 14 MeV neutrons | |
| | Germanium bolometer | $<2 \times 10^{12}$ | | Threshold for significant damage by 14 MeV neutrons | E, 5 |
| | Germanium (photoconductive) | $5 \times 10^{13}$ | $6 \times 10^{5}$ | Threshold for significant damage | E, 5 |
| | Germanium (photovoltaic) | $10^{12}$ | $3 \times 10^{4}$ | Threshold for significant damage | E, 5 |
| | CCD (surface channel) | $10^{11}$ | $10^{2}$ | Prohibitive dark current caused by fission neutrons | E, 8 |
| | CCD (buried channel) | $5 \times 10^{10}$ | $10^{2}$ | Prohibitive dark current caused by fission neutrons | E, 8 |
| | Silicon thermistor | $>10^{14}$ | | Resistance increases: effect depends on circuit details | E, 10 |
| | Thermistor | | $10^{7}$ | Threshold for unstated damage | A, 130 |
| | Metal-Oxyde Thermistor | | $14 \times 10^{5}$ | Reactor irradiation increases room temperature by 40 %. Three-fold increase in noise during irradiation of $5 \times 10^{3}$ rad/s | Y, 1 |
| | Platinum Sensor (x-ray absorption) | | $5 \times 10^{5}$ | Properties unchanged after irradiation | Y, 2 |
| | (Li) Si spectral distribution of charged particles | $10^{10}$ | $3 \times 10^{3}$ | Insignificant degradation of energy distribution response due to Pu-Be source neutrons | W 6,14 |
| | | $6 \times 10^{10}$ | | Loss of energy response at 100 V bias due to Pu-Be source neutrons | W 6 |
| | | $5 \times 10^{9}$ | | Probable threshold for loss of energy distribution response due to 14 MeV neutrons | W 6 |
| | (Li) Ge | $10^{10}$ | $3 \times 10^{3}$ | Threshold for unstated level of reduced performance. Radiation damage anneals at high temperature. | W 13,14 |
| | | | $2.5 \times 10^{3}$ | Probable threshold for loss of response due to gamma radiation | W 6 |
| | Si-Schottky Barrier | $10^{15}$ | $10^{6}$ | Threshold for unstated damage. Threshold depends on design details, see reference | W, 7 |
| | | $10^{14}$ | $5 \times 10^{3}$ | Threshold for unstated damage for platinum electrode device without control gate | W, 8 |
| | GaAs Schottky Barrier | $10^{14}$ | | Threshold for unstated damage | W, 8 |
| | Surface Barrier | $10^{14}$ | $10^{5}$ | Threshold for unstated damage | W, 14 |

Table 14-2 Summary of radiation damage effects of electronic components and materials

| Class | Component | Radiation Level | | Effect | Reference Page |
|---|---|---|---|---|---|
| | | n/cm$^2$ | Gy | | |
| Detectors (Continued) | HgCdTe (photovoltaic) | $5 \times 10^{11}$ | $10^4$ | Rads (Si) threshold for unstated damage | C, 233 |
| | | | | 1 MeV equivalent neutron threshold for unstated damage | C, 233 |
| | HgCdTe (photoconducting) | $10^{14}$ | | 58% loss of signal due to fission neutrons in low bias mode. 17% loss of signal due to fission neutrons in high bias mode. | W, 2, 3 |
| | Pb3aTe (photovoltaic) | $5 \times 10^{11}$ | $10^4$ | Rads (Si) threshold for unstated damage | C, 223 |
| | | | | 1 MeV equivalent neutron threshold for unstated damage | C, 233 |
| | GaAs (photovoltaic) | | $10^4$ | Rad (Si) threshold for unstated damage | C, 223 |
| | | $10^{12}$ | | Threshold for unstated level of damage | W, 14 |
| | | $10^{13}$ | | Threshold for probable loss of useful operation | W, 5 |
| | Crystal Ge (photoconductive) | | $10^4$ | Rad (Si) threshold for unstated damage | C, 223 |
| | (Cu or Hg doped) | $5 \times 10^{13}$ | | 1 MeV equivalent neutron threshold for unstated damage | C, 223 |
| | NaI - scintillator | | $10^3$ | Rads (Si) threshold for unstated damage | C, 223 |
| | | $10^{14}$ | | 1 MeV equivalent neutron threshold for unstated damage | C, 223 |
| | Pyroelectric (temperature change) | | $10^5$ | Rads (Si) threshold for unstated damage | C, 223 |
| | | $10^{14}$ | | 1 MeV equivalent neutron threshold for unstated damage | C, 223 |
| | | | | High sensitivity to gamma induced background noise limits dose rates to 10 rads/hour | V, 14 |
| Microwave Sources | Klystron | | $10^5$ | Threshold for unstated level of reduced performance | A, 130 |
| | Magnetron | | 10 | Threshold for unstated level of reduced performance | A, 130 |
| | Gunn Oscillator | | $10 - 10^3$ | Rads (Si) threshold for unstated level of damage | C, 275 |
| | | $10^{12} - 10^{14}$ | | 1 MeV equivalent neutron threshold for unstated damage | C, 275 |
| Electrical and Electronic Components | <u>Insulators</u> | | | Radiation threshold depends on application | |
| | Steatite | | $10^{10}$ | Threshold for unstated level of radiation damage | A, 128 |
| | Barium oxide | | $10^{10}$ | Threshold for unstated level of radiation damage | A, 128 |
| | Ceramic | | $10^{10}$ | Threshold for unstated level of radiation damage | A, 128 |
| | Mica | | $10^7$ | Threshold for unstated level of radiation damage | A, 128 |
| | Duroc ceramic | | $>3 \times 10^9$ | Threshold for unstated level of radiation damage | A, 131 |
| | Estane polyethylene | | $1 \times 10^7$ | Increased stiffness, temperature sensitive above 32°C | A, 131 |
| | Silicone varnish | | $1.4 \times 10^7$ | Acceptable, probably usable to $5 \times 10^7$ Gy | A, 131 |
| | Bakelite | | $10^6$ | Threshold for unstated level of radiation damage | A, 128 |
| | Mylar | | $10^5$ | Threshold for unstated level of radiation damage | A, 128 |
| | Nylon | | $10^3$ | Threshold for unstated level of radiation damage | A, 128 |
| | Teflon | | $10^2$ | Threshold for unstated level of radiation damage | A, 128 |

Table 14-3 Summary of radiation damage effects of electronic components and materials

| Class | Component | Radiation Level | | Effect | Reference Page |
|---|---|---|---|---|---|
| | | n/cm$^2$ | Gy | | |
| Electrical and Electronic Components (Continued) | **Insulators** (continued) | | | | |
| | Polyethylene | | $10^5$ | Threshold for unstated level of radiation damage | A, 128 |
| | Polyvinylchloride | | $10^3$ | Threshold for unstated level of radiation damage | A, 128 |
| | Polystyrene | | $10^6$ | Threshold for unstated level of radiation damage | A, 128 |
| | Epoxy resin | | $10^2$ | Threshold for unstated level of radiation damage | A, 128 |
| | Silicone treated mica | | $10^5$ | Threshold for unstated level of radiation damage | A, 128 |
| | Silicone varnished glass fiber | | $10^5$ | Threshold for unstated level of radiation damage | A, 128 |
| | **Fiberglass / Epoxy Composites** | | | | |
| | G-10 | $1.6 \times 10^{19}$ | $4.1 \times 10^9$ | Fast neutron fluence ($E > 0.1$ MeV) reduces electrical resistivity from $3.6 \times 10^{15}$ ohm-cm to $5.4 \times 10^6$ ohm-cm. Compressive fatigue tests on materials at room temperature show failure at compressive stress levels of 40 ksi or greater with the number of cycles-to-fall decreasing as the stress level is raised. | Z |
| | G-7 | $1.6 \times 10^{19}$ | $4.1 \times 10^9$ | Compressive fatigue tests at room temperature show failure at stress levels of 30 ksi | Z |
| | G-11 | $1.6 \times 10^9$ | $4.1 \times 10^9$ | Compressive fatigue tests at room temperature show failure at stress levels of 30 ksi | Z |
| | **Connectors** | | | | |
| | Duroc ceramic | | $> 3 \times 10^6$ | Radiation threshold depends on application | A, 131 |
| | Melamine plastic | | $> 3 \times 10^6$ | Radiation threshold depends on application | A, 131 |
| | **Transformers and Relays** | | $> 10^7$ | Radiation threshold depends on application | A, 130 |
| | Various transformers | $10^{14}$ - $10^{20}$ | $10^5$ - $10^7$ | Various, see reference | F, 638 - 639 |
| | Various relays and switches | $10^{12}$ - $10^{16}$ | $10^5$ - $10^7$ | Various, see reference | F, 659 - 660 |
| | **Magnetic Materials** | | $10^8$ | Radiation threshold depends on application | A, 130 |
| | Ferrites | | $10^{16}$ | Increases coercive force and reduces permeability | O, I-54 |

Table 14-4 Summary of radiation damage effects of electronic components and materials

| Class | Component | Radiation Level | | Effect | Reference Page |
|---|---|---|---|---|---|
| | | n/cm² | Gy | | |
| Electrical and Electronic Components (Continued) | **Linear-voltage Differential Transformer** | | > $1.4 \times 10^7$ | Slight change in null point | A, 132 |
| | **Nonmagnetic Structural Material** | | | | |
| | Filled phenolic | | $1 \times 10^7$ | Acceptable, slight dimensional changes when heated or irradiated | A, 132 |
| | **Motors and Seloyns** | | $3 \times 10^2$ - $5 \times 10^6$ | 25 % change in mechanical properties of most radiation sensitive parts | A, 136, 137 |
| | **Switches** | | $3 \times 10^4$ - > $10^7$ | 25 % change in mechanical properties of base and shell insulation; threshold depends on insulating material used | A, 137 |
| | **Coaxial Cables** | | | | |
| | Polystyrene | | > $5 \times 10^7$ | 25 % change in important mechanical properties of insulation | A, 138 |
| | Polythene | | $9 \times 10^5$ | 25 % change in important mechanical properties of insulation | A, 138 |
| | Rayon | | $1.6 \times 10^5$ | 25 % change in important mechanical properties of insulation | A, 138 |
| | Polyvinylchloride | | $1.2 \times 10^6$ | 25 % change in important mechanical properties of insulation | A, 138 |
| | Neoprene | | $7 \times 10^4$ | 25 % change in important mechanical properties of insulation | A, 138 |
| | **Printed Circuits** | | $8 \times 10^4$ - $6.4 \times 10^5$ | 25 % change in mechanical properties of base insulation | A, 138 |
| | Printed circuit boards | $10^{18}$ | | Various, see reference | F, 643 |
| | **Conductors** | | | | |
| | Copper CFHC | $1.5 \times 10^{18}$ (fast n) | | Resistivity increases by 0.2 % when exposed at 60°C and measured at 20°C | A, 133 |
| | Aluminum, 28 solution | $4 \times 10^{19}$ (fast n) | | Resistivity increases by 10 % when exposed at 50°C and measured at -190°C | A, 133 |
| | Stainless steel 304 | $4 \times 10^{19}$ (fast n) | | Resistivity increases by 5 % when exposed at 280°C and measured at 20°C | A, 133 |
| | Stainless steel 347 | $2.5 \times 10^{20}$ (fast n) | | Resistivity decreases by 1 % when exposed at 260°C and measured at 25°C | A, 133 |
| | Stainless steel 322 W | $2.5 \times 10^{20}$ (fast n) | | Resistivity decreases by 7 % when exposed at 260°C and measured at 25°C | A, 133 |

Table 14-5 Summary of radiation damage effects of electronic components and materials

| Class | Component | Radiation Level | | Effect | Reference Page |
|---|---|---|---|---|---|
| | | n/cm² | Gy | | |
| Electrical and Electronic Components (Continued) | **Pressure Transducers** | | | | |
| | Barium Titanate (Gulton Co.) (shear mode) | $7.6 \times 10^{10}$ | $9.6 \times 10^{4}$ | Threshold level for unstated damage by neutrons | K, 14 |
| | Lead Zirconate Titanate (Gulton Co.) (compression mode) | $2.6 \times 10^{18}$ | $4 \times 10^{8}$ | Threshold level for unstated damage by neutrons | K, 14 |
| | **Hall Effect Sensor** | $10^{12}$<br>$10^{15}$ | | 1 MeV equivalent neutron threshold for unstated damage to amplifier<br>1 MeV equivalent neutron threshold for unstated damage to sensor | C, 161<br>C, 161 |
| | **Oscillator Crystal** | $10^{12}$ | | Shift in frequency by one part in $10^{7}$ | C, 162 |
| | **Magnetic Materials** | | | | |
| | Permanent Magnets | | | | |
| | Alnico 2, 5, 12<br>Cunico | $10^{20}$ | | > 0.5 MeV neutrons reduce remanence and induction between 60°C and 325°C | A, 127 |
| | Cunife<br>Cr Steel<br>Co Steel<br>Barium ferrites<br>Silmeal<br>Pt-Co<br>ESD Iron | $10^{19}$ | | Neutrons with energy > 0.5 MeV cause unstated level of damage at 275°C | A, 127 |
| | Soft Magnetic Materials | | | | |
| | Ferrites<br>Garnets<br>Spinnels | $10^{12}$ | | Neutrons at energies > 1 MeV reduce remanence, and permeability, increases coercive force at 100°C | A, 127 |

Table 14-6 Summary of radiation damage effects of electronic components and materials

| Class | Component | Radiation Level | | Effect | Reference Page |
|---|---|---|---|---|---|
| | | n/cm$^2$ | Gy | | |
| Electrical and Electronic Components (Continued) | Soft Metals and Alloys | | | | |
| | Fe<br>Si-Fe<br>Al-Fe<br>Co-Fe | $10^{17}$ | | Neutrons with energies > 0.5 MeV cause unstated level of damage at temperatures lower than 100°C | A, 127 |
| | Mumetal | | | | |
| | 4-79 Mo Permalloy<br>5-79 Mo Permalloy<br>Binary Ni-Fe | $10^{15}$ | | Neutrons with energies > 0.5 MeV cause unstated level of damage at temperatures lower than 100°C | A, 127 |
| Solid State Components and Devices | **Diodes** | | | | |
| | Tunnel, Ge | | $10^7$ | Threshold for unstated level of radiation damage | A, 130 |
| | Tunnel, Si | | $10^6$ | Threshold for unstated level of radiation damage | A, 130 |
| | Tunnel, GaAs | | $10^6$ | Threshold for unstated level of radiation damage | A, 130 |
| | Microwave, Si | | $10^5$ | Threshold for unstated level of radiation damage | A, 130 |
| | Switching, Si | | $10^5$ | Threshold for unstated level of radiation damage | A, 130 |
| | Power, Se | | $10^5$ | Threshold for unstated level of radiation damage | A, 130 |
| | Power, Si | | $10^3$ | Threshold for unstated level of radiation damage | A, 130 |
| | Low power rectifier | $3 \times 10^{12}$ | | Forward voltage change by ± 5 % by reactor neutrons, reverse leakage may increase substantially (500 %) if initial leakage is extremely low | C, 58 |
| | High power rectifier | $3 \times 10^{12}$ | | Forward voltage may increase substantially (100 %) | C, 58 |
| | Zener | $3 \times 10^{12}$ | | Zener voltage may increase by 40 mV | C, 49 |
| | | | $10^4$ | Threshold for unstated level of radiation damage | A, 130 |
| | Varactors, Si, Ge | | $10^4$ | Threshold for unstated level of radiation damage | A, 130 |
| | Switching, GaAs | | $10^3$ | Threshold for unstated level of radiation damage | A, 130 |
| | Photodiodes (Si, HgCdTe, PbSnTe) | $10^{11}$ - $10^{14}$ | 1 - $10^3$ | Threshold for unstated level of radiation damage | C, 276 |
| | SCR | | | Threshold for ionising radiation pulse triggering in small devices | W, 11, 12 |
| | SCR | $10^{11}$ | | Maximum fluence below which safe operation may be achieved in high current devices | W, 12 |
| | SCR 2N1772 | $8 \times 10^{12}$ | | Trigger current increases by one order of magnitude | W, 12 |

Table 14-7 Summary of radiation damage effects of electronic components and materials

| Class | Component | Radiation Level | | Effect | Reference Page |
|---|---|---|---|---|---|
| | | n/cm² | Gy | | |
| Solid State Components and Devices (Continued) | **Diodes** (continued) | | | | |
| | SCx Unitrode GA 102 | $10^{12}$ | | Trigger voltage, gate current and gate holding current begin to increase | W, 12 |
| | | $10^{14}$ | | Trigger voltage, gate current and gate holding current increase by a factor of two | W, 12 |
| | SCR | | $10^3$ | Threshold for unstated level of radiation damage | A, 130 |
| | SCR | $10^{12}$ | | 1 MeV equivalent neutrons increase gats current and triggering point over ten times, forward voltage in conduction may double | C, 169-170 |
| | Varistor, Si | | $10^7$ | Threshold for unstated level of radiation damage | A, 130 |
| | Varistor, Se and Cu | | $10^3$ | Threshold for unstated level of radiation damage | A, 130 |
| | GaAs LED (Si or Zn dc ed) | $10^{11}$ | | Threshold for unstated damage by 14 MeV neutrons | E, 2 |
| | PIN diodes | $5 \times 10^{10}$ | | Threshold for unstated damage by 14 MeV neutrons | E, 2 |
| | | $10^{11}$ | | Threshold for unstated damage by fission neutrons | |
| | | | $10^3$ | Threshold for unstated damage by $^{60}Co$ | E, 2 |
| | **Transistors** | | | | |
| | High level switching | | $10^5$ | Threshold for unstated level of radiation damage | A, 130 |
| | Low level switching | | $10^5$ | Threshold for unstated level of radiation damage | A, 130 |
| | High frequency | | $10^5$ | Threshold for unstated level of radiation damage | A, 130 |
| | FET | | $10^5$ | Threshold for unstated level of radiation damage | A, 130 |
| | JFET | $10^{14}$ - $10^{15}$ | | Threshold for unstated level of radiation damage | C, 150 |
| | Audio | | $10^3$ | Threshold for unstated level of radiation damage | C, 150 |
| | Unijunction | $5 \times 10^{11}$ | $10^3$ | Threshold for unstated level of radiation damage | A, 130 |
| | | | | | C, 51 |
| | MOSFET | | 10 - 30 | Threshold for unstated level of radiation damage | C, 50 |
| | Bipolar transistors | $10^{11}$ to $10^{16}$ | | Neutrons degrade $\beta/\beta o$ from 0.99 to < 0.01 | C, 84 |
| | Power bipolar transistors | $1 \times 10^{14}$ | | Typical reduction of $I_m/I_{mo}$ and $\beta/\beta o$, $0.83 < I_m/I_{mo} < 0.95$, $0.72 < \beta/\beta o < 0.75$ | C, 69 |
| | | $3 \times 10^{14}$ | | Typical reduction of $I_m/I_{mo}$ and $\beta/\beta o$, $0.83 < I_m/I_{mo} < 0.90$, $0.57 < \beta/\beta o < 0.59$ | C, 69 |
| | | $1 \times 10^{15}$ | | Typical reduction of $I_m/I_{mo}$ and $\beta/\beta o$, $0.34 < I_m/I_{mc} < 0.44$, $0.31 < \beta/\beta o < 0.32$ | C, 69 |
| | | $2 \times 10^{15}$ | | Typical reduction of $I_m/I_{mo}$ and $\beta/\beta o$, $0.17 < I_m/I_{me} < 0.19$, $0.14 < \beta/\beta o < 0.15$ | C, 69 |
| | Various transistors | $10^{11}$ - $10^{14}$ | | Various, see references | F, 484-486 |

Table 14-8 Summary of radiation damage effects of electronic components and materials

| Class | Component | Radiation Level | | Effect | Reference Page |
|---|---|---|---|---|---|
| | | n/cm² | Gy | | |
| Solid State Components and Devices (Continued) | **Integrated Circuits** | | | | |
| | Emitter coupled logic | $10^4$ | $10^6$ (Si) | 1 MeV equivalent neutrons cause rate functional failures | C, 101 |
| | MOS | | $8 - 3 \times 10^2$ (Si) | See Ref. for details | C, 117-120 |
| | CMOS technology | | > 10 (Si) | Unselected components, see reference | C, 143 |
| | | | > $10^4$ (Si) | Custom made components, see reference | C, 143 |
| | Microprocessor, Memory (NMOS) | | 8 - 40 | Range for failure threshold | O, 2-22 |
| | Digital Logic (CMOS) | | $40 - 2 \times 10^2$ | Range for failure threshold | O, 2-22 |
| | Digital Logic (CMOS hardened gate) | | $10^3 - 10^4$ | Range for failure threshold | O, 2-22 |
| | Analog Switch (CMOS) | | $10^2 - 10^3$ | Range for failure threshold | O, 2-22 |
| | Operational Amplifier (Bipolar) | | $10^2 - 10^4$ | Range for failure threshold | O, 2-22 |
| | Comparator (Bipolar) | | $10^2 - 10^5$ | Range for failure threshold | O, 2-22 |
| | Digital Logic ($I^2L$) | | $5 \times 10^2 - 10^4$ | Range for failure threshold | O, 2-22 |
| | Digital Logic (Schottky TTL) | | > $10^5$ | Range for failure threshold | O, 2-22 |
| | **Microprocessor Memory Chips** | | | | |
| | RCA Memory CDP1821 | $10^{15}$ | 50 | Threshold for data loss | E, 13 |
| | RCA Memory CD 4061 | $10^{15}$ | $10^3$ | Threshold for data loss | E, 20 |
| | Harris Memory HMI6508 | $10^{15}$ | 10 | Threshold for data loss | E, 20 |
| | HMOS Memory - 4 kilobit | $10^{15}$ | 10 | Threshold for data loss | E, 20 |
| Mechanical, Vacuum, and Fluids | **Pumps** | | | | |
| | Most Sensitive Parts | | | | |
| | Packings | | | | |
| | Seals | | | | |
| | Gaskets | | | | |
| | Materials | | | | |
| | Natural rubber | | $2.6 \times 10^5$ | Dosage gives 25 % change in important mechanical properties, see reference | A, 140 |
| | Buna-N | | $4 \times 10^4$ | | |
| | Thiolcol | | $1 \times 10^4$ | | |
| | Silicone rubber | | $4 \times 10^4$ | | |
| | Teflon | | $3 \times 10^2$ | | |
| | Cotton | | $5 \times 10^4$ | | |
| | Rayon | | $1.6 \times 10^5$ | | |
| | Leather | | $4 \times 10^4$ | | |
| | Kel-F | | $1.7 \times 10^5$ | | |
| | Butyl rubber | | $4 \times 10^4$ | | |

Table 14-9 Summary of radiation damage effects of electronic components and materials

| Class | Component | Radiation Level | | Effect | Reference Page |
|---|---|---|---|---|---|
| | | n/cm² | Gy | | |
| Mechanical, Vacuum, and Fluids (Continued) | **Valves** | | | | |
| | Most Sensitive Parts | | | | |
| | Packings<br>Seals<br>Gaskets | | | See "Pumps" | A, 140 |
| | **Fluid Disconnects** | | | | |
| | Most Sensitive Parts | | | | |
| | O-rings (Buna-S)<br>Gaskets (Neoprene, thiolcol, silicone rubber) | | $10^4$ - $1.3 \times 10^5$ | Dosage gives 25 % change in important mechanical properties, see reference for details | A, 139 |
| | **Hydraulic System** | | | | |
| | Most Sensitive Parts - | | | | |
| | Hydraulic fluid<br>Di (2-ethylhexyl)-sebacate | | $7.5 \times 10^6$ | Dosage gives 25 % change in viscosity | A, 139 |
| | Seals, O-rings (Buna-S, neoprene, nylon)<br>Moses (Buna-S, neoprene, nylon)<br>Cylinder cups (Leather, neoprene)<br>Couplings (Leather, neoprene) | | $4 \times 10^4$ to $1.3 \times 10^5$ | Dosage gives 25 % change in important mechanical properties, see reference for details | A, 139 |
| | **Pneumatic System** | | | Similar to hydraulic system except hydraulic fluid is replaced by air | A, 140 |
| | **Pressure switches** | | | | |
| | Most Sensitive Part | | | | |
| | Diaphragm | | | | |
| | Neoprene<br>Nylon<br>Leather | | $6 \times 10^4$<br>$4 \times 10^4$<br>$6 \times 10^4$ | Dosage gives 25 % change in important mechanical properties, see reference for details | A, 140 |
| | **Seals and gaskets** | | $10^4$ - $10^7$ | Various, see reference | F, 159 |
| | **Lubricants and hydraulic fluid** | | $10^6$ - $10^7$ | Various, see reference | F, 179-193 |

Table 14-10 Summary of radiation damage effects of electronic components and materials

| Class | Component | Radiation Level | | Effect | Reference Page |
|---|---|---|---|---|---|
| | | n/cm² | Gy | | |
| Structural Materials, Metals and Alloys | **Aluminum Alloys** | | | | |
| | Type 61SO | $10^{20}$ (fast) | | Fast neutrons increase yield strength by 290 % and tensile strength by 150 %; exposed at 70°C | A, 141 |
| | Type A54S | $10^{20}$ (fast) | | Fast neutrons increase both yield and tensile strength by 6 %; exposed at 70°C | A, 141 |
| | Types 52SO and 61ST6 | $10^{20}$ (fast) | | Fast neutrons increase both yield and tensile strength by 50 % to 100 %; exposed at 70°C | A, 141 |
| | **High Purity Aluminum** | | | | |
| | Annealed | $2 \times 10^{20}$ (thermal) $5 \times 10^{19}$ (fast) | | Increased yield strength by 100 %, ultimate strength increased by 60 %, fractional elongation decreased by 67 %; exposed at 50°C | A, 141 |
| | Half hard | $2 \times 10^{20}$ (thermal) $5 \times 10^{19}$ (fast) | | Yield strength increased by 10 %; exposed at 50°C | C, 141 |
| | Beryllium | $3 \times 10^{20}$ | | Fast neutrons cause negligible change to various mechanical properties when exposed at 30°C | A, 141 |
| | **Copper** | | | | |
| | | $4 \times 10^{20}$ | | Volumetric swelling of 0.33 % at 360°C | M, T |
| | O²HC annealed | $2 \times 10^{18}$ (fast) | | Hardness increase from 41 to 57 DPH when exposed at 60°C | A, 141 |
| | | $6 \times 10^{19}$ (fast) | | Hardness increase by 53 Brinell numbers when exposed at 80°C | A, 141 |

Table 14-11 Summary of radiation damage effects of electronic components and materials

| Class | Component | Radiation Level | | Effect | Reference Page |
|---|---|---|---|---|---|
| | | n/cm$^2$ | Gy | | |
| Structural Materials, Metals and Alloys (Continued) | **Copper Alloys** | | | | |
| | 2 % Beryllium alloy quenched from 800°C | $9 \times 10^{19}$ (thermal) $2 \times 10^{19}$ (fast) | | Rockwell G hardness increases from 18 to 30 when alloy originally is in solution annealed state and exposed at 30°C | A, 142 |
| | 2 % Beryllium alloy quenched 2 hours at 285°C | $9 \times 10^{19}$ (thermal) | | Rockwell G number increases from 100 to 103 when exposed at 30°C | A, 142 |
| | See Ref. A, 142 for additional alloys | $2 \times 10^{19}$ (fast) | | | |
| | **Iron** | $2 \times 10^{17}$ (fast) | | Increases Rockwell F hardness number from 84 to 90 when exposed at 40°C | A, 142 |
| | SA-212 Pressure Vessel Steel | $10^{19}$ (fast) | | Fatigue endurance limit decreases from 48 ksi to 42 ksi for sample without notch exposed at 40°C | A, 143 |
| | SA-285 Pressure Vessel Steel | $4 \times 10^{19}$ (fast) | | Tensile strength (2 % offset) increases by 60 % when exposed under load at 1000 kg/cm$^2$ | A, 144 |
| | Various Pressure Vessel Steels | $10^{18}$ - $10^{22}$ | | Various, see reference | Q, R, S |
| | **Stainless Steel** | | | | |
| | Type 304 | $1.3 \times 10^{19}$ (thermal) $10^{18}$ (fast) | | Negligible increase in hardness when exposed at 200°C | A, 143 |
| | | $2 \times 10^{19}$ (fast) | | Insignificant change in fatigue strength when irradiated at 100°C and tested at 1,700 cpm | A, 143 |
| | Type 316 | $5 \times 10^{19}$ (fast) | | Rockwell B hardness increases from 83 to 98, ultimate strength increases by 90 %, elongation at rupture decreases by 15 % when exposed at 25°C | A, 143 |
| | Type 316 + 20 % CW | $2 \times 10^{22}$ (fast) | | Strength decreases from 675 MPa to 500 MPa and swelling of 0.2 % at 20 dpa when irradiated at 450°C. See reference for strength and swelling rates at temperatures up to 600°C | I, 63 |

Table 14-12 Summary of radiation damage effects of electronic components and materials

| Class | Component | Radiation Level | | Effect | Reference Page |
|---|---|---|---|---|---|
| | | n/cm² | Gy | | |
| Structural Materials, Metals and Alloys (Continued) | **Stainless Steel** (Continued) | | | | |
| | Type 316 + Ti modified + 20 % CW | Fluence not quoted, only dpa | | Swelling is 0.1 % at 20 dpa when irradiated at 460-475°C. See reference for swelling rates at temperatures up to 365°C | I, 41 |
| | Type 347 | $6 \times 10^{19}$ (fast) | | Rockwell B hardness increases from 75 to 93, yield point increases from 2300 kg/cm² to 5400 kg/cm², ultimate tensile strength increases from 6000 kg/cm² to 7300 kg/cm², elongation at rupture decreases from 57 % to 34 %, exposed at 160°C | A, 143 |
| | Various types | $10^{17}$ - $10^{21}$ | | Various, see reference | F, 427-433 |
| | Ferritic Steels, various | $10^{17}$ - $10^{21}$ | | Various, see reference | F, 418-425 |
| | Carbon and Low Alloy Steels | $10^{17}$ - $10^{20}$ | | Various, see reference | F, 409-414 |
| | Zirconium and Zirconium Alloys | $10^{18}$ - $10^{20}$ | | Various, see reference | F, 440-443 |
| | Molybdenum, Nickel and Zirconium | $10^{13}$ - $10^{20}$ | | Various, see reference | F, 144 |
| | Radiation Resistant Alloys | | | | |
| | Martensitic (HT-9) | $> 10^{23}$ | | 3.7 % swelling at 500°C is an order of magnitude lower than that of 20CW316. In reactor creep resistance appears better thant that of 20CW316. | H, 179 See also K, X-15 to X-30 |
| | | Fluence not stated, only dpa | | Irradiation up to 9.3 dpa increases 0.2 % yield strength from 500 MPa to 800 MPa at 300°C | I, 223 |
| | Ferritic / Bainitic | | | | |
| | 2.25 Cr - 1 Mo<br>H-11<br>EM 12<br>430 F | | | Highly resistant to irradiation induced swelling over temperature range 400-650°C | I, 188 See also J, 1725 |

Table 14-13 Summary of radiation damage effects of electronic components and materials

| Class | Component | Radiation Level | | Effect | Reference Page |
|---|---|---|---|---|---|
| | | n/cm² | Gy | | |
| Structural Materials, Metals and Alloys (Continued) | **Low Activation Materials** | | | | |
| | Graphite | | | | |
| | H-451 | $10^{21}$ (fast) | | Thermal conductivity decreases by 80 % and Young's modulus increases by 100 % | G, 3-51 |
| | | $3 \times 10^{22}$ (fast) | | Linear dimension decreases by 2 % at 400°C. Young's modulus increases from 10 GPa to 30 GPa at 400°C. Thermal conductivity decreases by 90 % at 400°C. | G, 3-48 |
| | Silicone Carbide | | | | |
| | CVC | $9 \times 10^{22}$ | | Linear expansion of 0.4 %, no evidence of structural deterioration at 550°C. | G, 3-54 |
| | Self bonded | $10^{21}$ - $10^{22}$ | | Thermal conductivity decreases by a factor of six at 800°C | G, 3-57 |
| | CVD | $10^{22}$ (fast) | | Flexual strength unchanged at 740°C | L |
| | Sintered Alpha | $10^{22}$ (fast) | | Flexual strength reduced 34 % at 740°C | L |
| | Reaction bonded | $10^{22}$ (fast) | | Flexual strength reduced 58 % at 740°C | L |
| | Aluminum | $4.6 \times 10^{20}$ | | Swelling of 0.35 % at 150°C | N, 74. T |
| | **Aluminum Alloys** | | | | |
| | Type 6061 | $10^{23}$ (fast) | | Density reduction $< 0.7$ % | M, V |
| | Type 6063 | $2 \times 10^{22}$ (fast) | | Density reduction of 0.7 % | M, U |
| | **Structural Adhesives and Organic Coatings** | | $10^6$ - $10^7$ | Various, see reference | F, 141-142 |
| | Polymers | | | | |
| | Plastics | | $10^2$ - $10^7$ | Various, see reference | F, 92-95 |
| | Rubbers | | $10^3$ - $10^7$ | Various, see reference | F, 92-95, 112-114, 125-126, 131, 135 |
| | Organic Materials | | $10^2$ - $10^7$ | Various, see reference | X, 4-5, 4-6 |

Table 14-14 Summary of radiation damage effects of electronic components and materials

References

A. Ricketts. L.W., Fundamentals of Nuclear Hardening of Electronic Equipment, Wiley-Interscience, New York, 1972

B. Rudie, Norman J., Principles and Techniques of Radiation Hardening ,Vol. I, "Interaction of Radiation with Matter and Material Effects", Western Periodicals Company, North Hollywood, CA, 1976

C. Holms-Siedle, Andrew, "Final Report. Radiation Effects in the Joint European Torus Experiment: Guidelines for Preliminary Design", Fulmer Research Laboratory Report R857/2. September 1980

D. Primek, William, "Radiation Design in Diagnostic Windows for the TFTR", Journal of Nuclear Materials 85 and 86, 405-407 (1979)

E. Brucker, George J., "TFTR Diagnostics Engineering Report No. 3", Report No. PH-I-004 (Princeton Plasma Physics Laboratory, May 1978)

F. Kircher, John F., and Bowman, Richard B., "Effects of Radiation on Materials and Components", Reinhold Publishing Corporation, New York, 1964

G. Low Activation Materials Safety Studies, General Atomic Company Report GA-A16005, UC-20de, December 1980

H. Rosewasser, S.N., et. al., "The Application of Martensitic Stainless Steels in Long Lifetime Fusion First Wall/ Blankets", Proc. First Topical Meeting on Fusion Reactor Materials, Miami Beach, Florida, Jan. 29 - 31, 1979, pp. 177-182

I. "Alloy Development for Irradiation Performance", (ADIP), Quarterly Progress Report for Period Ending, Dec.31, 1980. DOE-LR-0045/3

J. Smith, D.L., et. al., "First-Wall / Blanket Materials Selection for Starfire Tokamak Reactor". The Technology of Controlled Nuclear Fusion, Proc. of the Fourth Topical Meeting, Vol. II, pp. 1714-1733, Oct. 14-17, 1980, King of Prussia, Pennsylvania, Conf-801011

K. Witamir, I., "A Tandem Mirror Reactor Study, University of Wisconsin, UWFDM, Sept. 1980

L. Hopkins, George, General Atomic Company, data to be published, 1981

M. "Definition and Conceptual Design of a Small Fusion Reactor", EPRI ER-045, Project 922, Iteria Report, Apr. 1979

N. Cover, J.E., "Part I - Basic Radiation Effects and Electronics Technology, Radiation Effects and System Hardening", IEEE Short Course, 1981

O. Cover, J.E., "Part I - Overview of Nuclear and Space Radiation Effects Technology, Radiation Effects and System Hardening", Short Course, 1980

P. Lall, E., et al.,"Radiation Effects in Quartz, Silica and Glasses", Progress in Ceramic Sciences, 4, 3 (Ed. J.E. Burke), Pergamon Press, 1956

Q. Steele, L.E. and Serpan, C.Z., Jr., "Analysis of Reactor Vessel Radiation Effects Surveillance Programs", American Society of Testing and Materials, Special Technical Publication 481, Dec. 1970

R. Steele, L.E., "Neutron Irradiation Embrittlement of Reactor Pressure Vessel Steels", Technical Reports Series No. 163, IAKA, Vienna, 1975

S. "Irradiation Effects in Structural Alloys for Thermal and Fast Reactors", American Society for Testing and Materials, Special Technical Publication 457, Dec. 1969

T. Comolinski, M. and Brabec, C., "Models pour la croissance des cavites sous irradiation", in J. Nuclear Materials, 43, 59 (1972)

U. Sturcken, E.F., Krapp, C.W. and Alewine, G.B., "Effect of Irradiation on 6063 Aluminum Structural Components in Savannah River Reactors", Symp. on Materials Performance in Operating Nuclear Systems, CONF-730801 (1973)

V. Jostsons, A., Long. E.L., Ferguson, J.E. and Farrell, K., Quoted by J.O. Stiegler, "Radiation-Induced Voids in Metals", J.W. Corbett and L.C. Iamoiollo, Eds. CONF-710601 (1972)

W. Brucker, G.J., "TFTR Diagnostics Engineering Report No. 2", Report-No. PH-I-001/9AI (Princeton Plasma Physics Laboratory, November 1977)

X. Bruce, M.B. and Davids, M.V., "Radiation Effects on Organic Materials in Nuclear Power Plants", Georgia Institute of Technology, Nuclear Engineering Department, EPRI Report NP-2129,Project 1707-3, November 1981

Y. Schivell, J., Rends, J., Lowrance, J. and Hsuan H., "Development of Bolometers for TFTR", unpublished, Princeton Plasma Physics Laboratory

Z. Schmuck, R.E., Imel, G.R. and Harker, Y.D., "Irradiation Studies of Magnet Insulator Materials", 2nd Topical Meeting on Fusion Reactor Materials, Seattle, Washington, August 9-12, 1981


Table 14-15 Summary of radiation damage effects of electronic components and materials[106]

## 6. References


1 F. Wulf, D. Bräunig, A. Boden, *Electronic Components in a High-Dose Environment* (Proceedings of ECFA Study Week on Intrumentation Technologiy for High-Luminosity Hardron Colliders), Ed. by: E. Fernandes, G. Jarlskog, Barcelona, Spain 14-21 Sept. 1989, **Vol. 1**, p. 109-126, **Vol. 2**, p. 681-684

2 H. Schönbacher, F. Wulf, *Radiation hardness Studies for LHC Detector Materials* (Proceedings of ECFA Large Hadron Collider Workshop), Ed. by: G. Jarlskog, D. Rein, 4-9 Oct. Aachen, **Vol. 1**, p. 289-324, and Radiation hardness **Vol. 3**, p 566-749

3 T. P. Ma, P. V. Dressendorfer, *Ionizing Radiation Effects in MOS Devices and Circuits* (John Wiley & Sons, 1989)

4 V.A.J. van Lint,T.M.Flanagan,R.E.Leadon,J.A.Naber and V.C.Rodgers, *Mechamisms of Radiation Effects in Electronic Materials*, **Vol.1**,John Wiley,New York,1980

5 V.A.J. van Lint, G.Gigas and J.Barengoltz, *Correlation of displacement effects produced by electrons, protons and neutrons*, IEEE Trans. in Nucl. Sci., **NS-22**, p 2663 (1975)

6 D.Bräunig,*Wirkung hochenergetischer Strahlung auf Halbleiterbauelemente*, Springer-Verlag,1989

7 ICRU report 33, *International commission on radiation units and measurements,* Radiation Quantities and Units ,7910 Woodmont Ave., Bethesda, MD 20814, USA, Washington, DC (1980).

8 ASTM Standard E170, *Standard terminology relating to radiation measurements and dosimetry.*,1985 annual book of ASTM standards, **Vol. 12.02,** Nuclear II , Solar and Geothermal Energy, American Society for Testing and Materials, 1916 Race Street, Philadelphia, PA 19103-1187, USA, 1985, p. 25.

9 R.N. Hamm, *Dose calculations for Si-$SiO_2$ layered structures irradiated by X-rays and CO-60 Gamma Rays.*, IEEE Trans. Nucl. Sci. **NS-33**, **Vol. 6**, 1986, p. 1236.

10 G. A. Ausmann Jr. and F. B. Mc Lean, *Electron-hole pair creation energy in $SiO_2$* Appl. Phys. Lett., **26**, 1975, p. 173

11 O. L. Curtis, J. R. Scour and K. Y. Chiu, *Hole and electron transport in $SiO_2$ Films*, J. Appl. Phys. **45**, 1974, p. 4506

12 J. M. Benetto and H. E. Boesch Jr. , *The relationship between Co-60 and 10 keV X-ray damage in MOS devices*, IEEE Trans. Nucl. Sci. **NS-33**, 1986, p. 1318

13 J. R. Srour, *Basic mechanisms of radiation effects on electronic materials, devices and integrated circuits*, IEEE 1983 Nuclear and Space Radiation Effects Conference, work shop, July 17, 1983, p. 53

14 G. Jaffe, *Zur Theorie der Ionisation in Kolonnen*, Ann. Phys. (Leipzig) 42, 1913, p303; Phys. **Z. 15**, 1914, p353; **23**, 1929, p. 849

15 W. H. Bragg and R. D. Kleeman, *On the recombination of ions in air and other gases*, Phil. Mag., **12**, 1906, p 273

16 M. P. Langevin, *L´Ionization des Gaz*, Ann. Chim. Phys., [7], **28**, 1903, p. 289; *Recombinaison et Mobilities des Ions Dans Les Gaz*, Ann. Chim. Phys., [7], **28**, 1903, p. 433

17 T.R Oldham, *Charge generation and recombination in silicon dioxide from heavy charged particles*, Harry Diamond Laboratories Technical Report, HDL-TR-1985, Adelphi, MD 1982

18 T. R. Oldham and J. M. Mc Garrity, *Comparison of Co-60 response and 10 keV X-ray response in MOS capacitors*, IEEE Trans. Nucl. Sci. **NS-30**, 1983, p. 4877

19 T. R. Oldham, *Recombination along the tracks of heavy charged particles in* $SiO_2$ *films*, J. Appl. Phys. **57(8)**, 15. April 1985, p 2695

20 L. Onsager, *Initial recombination of ions*, Phys. Rev. **54**, 1938, p. 554

21 L. Onsager, *Deviation from Ohm´s law in weak Electrolytes*, J. Chem. Phys. **2**, 1934, p 599

22 G. A. Ausman, *Field dependence of geminate recombination in a dielectric medium*, Harry Diamond Laboratories Report No. 2097, Adelphi, MD 1986

23 D. B. Brown and C. M. Dozier, *Electron-hole recombination in irradiated* $SiO_2$ *from a microdosimetry viewpoint*, IEEE Trans. Nucl. Sci. **NS-28, No. 6**, 1981, p. 4142

24 C. M. Dozier and D. B. Brown, *Effect of photon energy on the response of MOS devices*, IEEE Trans. Nucl. Sci. **NS-28, No. 6**, 1981, p. 4137

25 C. M. Dozier, D. M. Fleetwood, D. B. Brown and P. S. Winokur, *An Evaluation of low-energy X-ray and Cobalt-60 irradiations of MOS Transistors*, IEEE Trans. Nucl. Sci. **NS-34, No. 6**, 1987, p. 1535

26 R. C. Hughes, *Charge-carrier transport phenomena in amorphous* $SiO_2$*: direct measurement of the drift mobility and lifetime*, Phys. Rev. Lett., **30**, 1973, p. 1333

27 R. C. Hughes, *Hot electrons in* $SiO_2$, Phys. Rev. Lett., **35**, 1975, p. 449

28 R. C. Hughes, *High field electronic properties of* $SiO_2$, Solid-State Electron. **21**, 1978, p. 251

29 S. Othmer and J. R. Srour, *Electron transport in* $SiO_2$ *films at low temperature*, in The Physics of MOS Insulators, G. Lucovsky, S. T. Pantelides, and F. L. Galeener, Eds. Pergamon Press, Elmsford, NY, 1980, p. 49

30 J. R. Srour, O. L Curtis and K. Y. Chiu, *Charge transport studies in* $SiO_2$*: Processing effects and Implication for radiation hardening*, IEEE Trans. Nucl. Sci. **NS-21, No. 6**, 1974, p. 73

31 D. K. Ferry, *Electron transport and breakdown in* $SiO_2$, J. Appl. Phys. **50**, 1979, p. 1422

32 R. C. Hughes, *Hole mobility and transport in thin $SiO_2$ films*, Appl. Phys. Lett., **26**, 8, 1975, p. 436

33 R. C. Hughes, E. P. Eernisse, H. J. Stein, *Hole transport in MOS oxides*, IEEE Trans. Nucl. Sci. **NS-22, No. 6**, 1975, p. 2227

34 R. C. Hughes, *Time-resoved hole transport in Si-$SiO_2$*, Phys. Rev. **B15**, 1977, p. 2012

35 M. H. Wood and R. Williams, *Hole traps in silicon dioxide*, J. Appl. Phys., **47,3**, 1976, p. 1082

36 L. Manchanda, J. Vasi, and A. B. Btattacharyya , *The nature of instrinsic hole traps in thermal silicon dioxide*, J. Appl. Phys., **52,7**, 1981, p. 4690

37 R. Freeman, A. Holmes-Siedle, *A simple model for the predicting radiation effects in MOS devices*, IEEE Trans. Nucl. Sci., **NS-25**, **Vol. 6**, 1978, p. 1216

38 C. R. Viswanathan, J. Maserjian, *Model of thickness dependence of radiation charging in MOS-structures*, IEEE Trans. Nucl. Sci., **NS-23**, **Vol. 6**, 1976, p. 1540

39 G. W. Huges, R. J. Powel, *MOS hardness characterization and its dependence upon some process and measurement variables*, IEEE Trans. Nucl. Sci., **NS-23**, **Vol. 6**, 1976, p. 1569

40 F. J. Grunthaner, P. J. Grunthaner, R. P. Vasquez, B. F. Lewis, J. Maserjian and A. Madhukar, *Local atomic and electronic structure of oxide/GaAs and $SiO_2$/Si interfaces using high-resolution XPS*, J. Vac. Sci. Technol. **16**, 1979, p. 1443

41 F. J. Grunthaner, P. J. Grunthaner, R. P. Vasquez, B. F. Lewis, J. Maserjian and A. Madhukar, *High-resolution X-ray photoelectron spectroscopy as a probe of local atomic structure:* , Application to amorphous $SiO_2$ and the Si-$SiO_2$ interface, Phys. Rev. Lett. **43**, 22, 1979, p. 1683

42 S. T. Chang and S. A. Lyon, *Location of the positive charge trapped near the Si-$SiO_2$ interface at low temperature*, Appl. Phys. Lett. **48**, 2, 1986,p. 136

43 D. J. Di Maria, F. J. Feigl and S. R. Butler, *Capture and emission of electrons at 2.4 eV-deep trap level in $SiO_2$ films*, Phys. Rev. B, **Vol. 11**, No. 12, 15 June 1975, p. 5023

44 D. J. Di Maria, *Determination of insulator bulk trapped charges and centroids from photocurrent-voltage characteristics of MOS structures*, J. Appl. Phys. **47**, 9, 1976, p. 4073

45 R. J. Powel and C. N. Berglund, *Photoinjection studies of charge distribution in oxide of MOS structures*, J. Appl. Phys. **42**, 11, 1971, p. 4390

46 G. Hu, *Study of the generation of interface states in the Si-$SiO_2$ system after high field stress and after X-irradiation*, Ph. D. Dissertation 1980, Princeton University

47 L. Lipkin, A. Reisman, C.K. Williams, *Hole trapping phenomena in the gate insulator of As-fabricated insuled gate field effect transitors*, J. Appl. Phys. **68, 9**, 1. Nov. 1990, p. 4620

48 D. R. Young, E. A. Erene, D. J. Di Maria and R. F. De Keersmaeker, H. Z. Massond, *Electron trapping in $SiO_2$ at 295 and 77° K*, J. Appl. Phys. **50**, 10, 1979, p. 6366

49 J. M. Aitken, D. R. Young, *Avalanche injection of the holes into $SiO_2$*, IEEE Trans. Nucl. Sci., **NS-24**, **Vol. 6**, 1977, p. 2128

50 L. Lipkin, A. Reisman, C.K. Williams, *Conservation and filling of neutral traps in $SiO_2$ during ionzing radiation exposure*, Appl. Phys. Lett., **57**,21, 19. Nov. 1990, p. 2237

51 R. Freeman, A. Holmes Siedle, *A simple model for predicting radiation effects in MOS devices*, IEEE Trans. Nucl. Sci., **NS-25**, **Vol. 6**, 1978, p. 1216

52 J. J. Tzou, J. Y. C. Sun, and C. T. Sah, *Field dependence of two large hole capture cross sections in thermal oxide on silicon*, Appl. Phys. Lett. **43**, 1983, p. 861

53 J. R. Schwank, P. S. Winokur, P. J. McWhorter, F. W. Sexton, P. V. Dressendorfer and D. C. Turpin, *Physical mechanisms contributing to device rebound*, IEEE Trans. Nucl. Sci., **NS-31**, **Vol. 6**, 1984, p. 1434

54 A. J. Lelis, T. R. Oldham, H. E. Boesch Jr. and F. B. Mc Lean, *The nature of the trapped-hole annealing process*, IEEE Trans. Nucl. Sci., **NS-36**, **Vol. 6**, 1989, p. 1808

55 M. Simons and H. L. Huges, *Short term charge annealing in electron-irradiated silicon dioxide*, IEEE Trans. Nucl. Sci., **NS-18**, **Vol. 6**, 1971, p. 106

56 M. Simons and H. L. Huges, *Determining the energy distribution of puls-radiation-induced charge in MOS structures from rapid annealing measurements*, IEEE Trans. Nucl. Sci., **NS-19**, **Vol. 6**, 1972, p. 282

57 F. B. McLean, *A direct tunneling of charge transfer at the insulator-semiconductor interface in MIS devices*, U. S. Government Report, HDL-TR-1765, Oct. 1976

58 P. J. McWhorter, S. L. Miller and W. M. Miller, *Modeling the anneal of radiation-induced trapped holes in a varying thermal environment*, IEEE Trans. Nucl. Sci., **NS-37**, **Vol. 6**, 1990, p. 1682

59 A. H. Johnston, *Super recovery of total dose damage in MOS devices*, IEEE Trans. Nucl. Sci., **NS-31**, **Vol. 6**, 1984, p. 1427

60 S.C. Sun and J.D. Plummer, *Electron mobility in inversion and accumuluation layers on thermally oxidized silicon surfaces.* , IEEE Electr. Dev., **ED-27**, 1980, p. 1497.

61 K. F. Galloway, M. Gaitan, and T. J. Russel, *A simple model for seperating interface and oxide charge effects in MOS device characteristics*, IEEE Trans. Nucl. Sci., **NS-31**, **Vol. 6**, 1984, p. 1497

62 K. F. Galloway, C. L. Wilson and L. C. Witte, *Charge-sheet model to extract radiation-induced oxide and interface charge*, IEEE Trans. Nucl. Sci., **NS-32**, **Vol. 6**, 1985, p. 4461

63 F. W. Sexton and J. R. Schwank, *Correlation of radiation effects in transistors and integrated circuits*, IEEE Trans. Nucl. Sci., **NS-32**, **Vol. 6**, 1985, p. 3975

64 W. M. Soppa and H. G. Wagemann, *Investigation and modelling of the surface mobility of MOSFET's from -25° to +25°C*, IEEE Trans. on Electron. Dev., **Vol. 35**, No. 7, 1980, p. 970

65 D. Zupac, K. F. Galloway, R. D. Schrimpf and P. Augier, *Effects of radiation-induced oxide-trapped charge on inversion-layer hole mobility at 300 and 77 K*, Appl. Phys. Lett., **60**, 25. June 1992, p. 3156

66 W. D. Eades and R. M. Swanson, *Calculation of surface generation and recombination velocities at the Si-$SiO_2$ interface*, J. Appl. Phys., **58**, 11, 1985, p. 4267

67 P. S. Winokur, J. R. Schwank, P. J. McWhorter, P. V. Dressendorfer and D. C. Turpin, *Correlating the radiation response of MOS capacitors and transistors*, IEEE Trans. Nucl. Sci., **NS-31**, **Vol. 6**, 1984, p. 1453

68 P. S. Winokur, F. W. Sexton, J. R. Schwank, D. M. Fleetwood, P. V. Dressendorfer, F. Wrobel and D. C. Turpin, *Total-dose radiation and annealing studies: Implication for hardness assurance testing*, IEEE Trans. Nucl. Sci., **NS-33**, **Vol. 6**, 1986, p. 1343

69 T.R. Oldham, A. J. Lelis, H. E. Boesch Jr., J. M. Benedetto, F. B. McLean and J. M. Mc Garrity, *Post-irradiation effects in field-oxide isolation structures*, IEEE Trans. Nucl. Sci., **NS-34**, **Vol. 6**, 1987, p. 1184

70 R. L. Pease and R. M Turfler, D. Platteter, D. Emily and R. Blice, *Total dose in recessed oxide digital bipolar microcircuits*, IEEE Trans. Nucl. Sci., **NS-30**, **Vol. 6**, 1983, p. 4216

71 M. L. Buschbom, E. N. Jeffry, L. E. Rhine and D. B. Spratt, *Gamma total dose effects on ALS biplolar oxide sidewall isolated devices*, IEEE Trans. Nucl. Sci., **NS-30**, **Vol. 6**, 1983, p. 4105

72 R. Pease, D. Emily, H. E. Boesch Jr., *Total dose induced hole trapping and interface state generation in bipolar recessed field oxides*, IEEE Trans. Nucl. Sci., **NS-32**, **Vol. 6**, 1985, p. 3946

73 F. Wulf, A. Boden, D. Bräunig, *Total dose study on LS,- ALS,- and F-technology integrated circuits*, Proceedings ESA Electronic Components Conference, ESTEC Noordwijk, 12-16 Nov. 1990, ESA SP-313, p. 407

74 D.M. Long, 1981, *Hardness of MOS and Bipolar Integrated Circuits*, IEEE Trans. on Nucl. Sci., **NS 27**, **Vol. 6**, pp 1674-1679

75 J.P. Raymond, E.L. Petersen, 1987, *Comparison of Neutron, Proton and Gamma Ray Effects in Semiconductor Devices*, IEEE Trans. on Nucl. Sci., **NS 34**, **Vol. 6**, pp 1622-1628

76 J.L. Titus and D.G. Platteter, 1987, *Wafer Mapping of Dose Failure Thresholds in a Bipolar Recessed Field Oxide Technology*, IEEE Trans. on Nucl. Sci., **NS-34**, **Vol. 6**, pp 1751-1756

77 D. Bräunig, F. Wulf, W. Gaebler, A. Boden, 1982, GfW Handbook, Irradiation *Test Guidelines for Radiation Hardness of Electronic Components*, HMI-Report B-380, TN53/10, Dec. 1982, 1. edition

78 A. Boden,F. Wulf, D. Bräunig, *GfW Handbook for Datacompilation of Irradiation Tested Electronic Components*, HMI-Report B-353, TN 53/08, **Vol. 7,** Dec. 92

79 A. Boden, F. Wulf, D. Bräunig, 1990, *Irradiation Tests and Radiation Hardness Prediction Considerations on Electronic Components for Space Application*, ESA Electronic Components Conference, ESTEC, Noordwijk, 12-16 Nov. 1990

80 A. Boden, F. Wulf, D. Bräunig, *Irradiation tests and radiation hardness prediction considerations on electronic components for space application* , Proceedings ESA Electronic Components Conference, ESTEC Noordwijk, 12-16 Nov. 1990, ESA SP-313, p. 413

81 A. Bennemann, A. Boden, D. Bräunig, D. Brumbi, J. W. Klein, J. U. Schott, C. Ch. Seifert, H.-G. Spillekothen and F. Wulf, *The effects on radiation an electronic devices and circuits*, Kerntechnik, **55**, 5, 1990, p. 261

82 R. V. Schrimpf, P. J. Wahle, R. C. Andrews, D. B. Cooper and K. F. Galloway, *Dose-rate effects on the total-dose threshold-voltage shift of power MOSFETS*, IEEE Trans. Nucl. Sci., **NS-35**, **Vol. 6**, 1988, p. 1536

83 P. S. Winokur, F. W. Sexton, G. L. and D. C. Turpin, *Total dose failure mechanisms of integrated circuits in laboratory and space environments*, IEEE Trans. Nucl. Sci., **NS-34, Vol. 6**, 1987, p. 1448

84 D. M Fleetwood, P. S. Winokur and J. R. Schwank, *Using laboratory X-ray and Co 60 irradiation to predict CMOS device response in strategic and space environments*, IEEE Trans. Nucl. Sci., **NS-35**, **Vol. 6**, 1988, p. 1497

85 P. V. Dressendorfer, J. M. Soden, J. J. Harrington, T. V. Nordstrom, *The effects of test conditions on radiation hardness results*, IEEE Trans. Nucl. Sci., **NS-28, Vol. 6**, 1981, p. 4281

86 J. R. Schwank, F. W. Sexton, D. M. Fleetwood, R. V. Jones, R. S. Flores, M. S. Rodgers and K. L. Hughes, *Temperature effects on the radiation response of MOS devices*, IEEE Trans. Nucl. Sci., **NS-35**, **Vol. 6**, 1988, p. 1432

87 R. A. Kohler, R. A. Kushner, K. H. Lee, *Total dose radiation hardness of MOS-devices in hermetic ceramic packages*, IEEE Trans. Nucl. Sci., **NS-35**, **Vol. 6**, 1988, p. 1492

88 P. V. Dressendorfer, *Radiation-hardening technology*, Ed. by T.P. Ma and P. V. Dressendorfer, Ionizing radiation effects in MOS devices and circuits, John Wiley & Sons, 1989, p. 333

89 P. S. Winokur, E. B. Errett, D. M. Fleetwood, P. V. Dressendorfer and D. C. Turpin, *Optimizing and controlling the radiation hardness of a Si-gate CMOS process*, IEEE Trans. Nucl. Sci., **NS-32**, **Vol. 6**, 1985, p. 3954

90 P. V. Dressendorfer, *Radiation effects on MOS devices and circuites*, Ed. by T.P. Ma and P. V. Dressendorfer, Ionizing radiation effects in MOS devices and circuits, John Wiley & Sons, 1989, p. 256

91 MIL HDBK 279, *Total dose hardness assurance guidelines for semiconductor devices and microcircuits*, 25 Jan. 1985

92 MIL-STD-883C, Method 1019.4, *Steady state total dose irradiation procedure*

93 ESA/SCC Basic Specification No. 22900, *Total dose steady-state irradiation test method*

94 D. L. Fleetwood, P. S. Winokur, L. C. Riewe and R. L. Pease, *An improved standard total dose test for CMOS space electronics*, IEEE Trans. Nucl. Sci., **NS-36, Vol. 6**, 1989, p. 1963

95 D. B. Brown and A. H. Johnston, *A framework for an integrated set of standards for ionizing radiation testing of microelectronics*, IEEE Trans. Nucl. Sci., **NS-34**, **Vol. 6**, 1987, p. 1720

96 P. S. Winokur, F. W. Sexton, D. M. Fleetwood, M. D. Terry, M. R. Shaneyfelt, P. V. Dressendorfer and J. R. Schwank, *Implementing QML for radiation hardness assurance*, IEEE Trans. Nucl. Sci., **NS-37**, **Vol. 6**, 1990, p. 1794

97 D. R. Alexander, *Implication of qualified manufacturers list reliability procedures for radiation hardness assurance*, Nucl. and Space Rad. Effects Conference, Short Course, Reno, Nevada July 1990

98 F. Wulf. D. Bräunig, A. Boden, *GfW-Handbook for data compilation of irradiation tested electronic components*, Hahn-Meitner- Institut Berlin, test report HMI-B248, **Vol. 1** to **Vol. 6**, 1981-90,

99 J. L. Leray, E. Dupont-Nivet, J. F. Péré, Y. M. Coic, M. Raffaelli and A. J. Auberton-Hervé, M. Bruel, B. Giffard, J. Margail, *CMOS/SOI hardening at 100 Mrad($SiO_2$)*, IEEE Trans. Nucl. Sci., **NS-37**, **Vol. 6**, 1990, p. 2013

100 J. R. Schwank, F. W. Sexton, D. M. Fleetwood, M. R. Shaneyfelt, K. L. Hughes, and M. S. Rodgers, *Strategies for Lot-Accaptance Testing using CMOS transistors and IC's*, IEEE Trans. Nucl. Sci., **NS-36**, **Vol. 6**, 1989, p. 1971

101 D. M. Fleetwood, P. S. Winokur and T. L. Meisenheimer, *Hardness assurance for low-dose space applications*, IEEE Trans. Nucl. Sci., **NS-38**, **Vol. 6**, 1991, p. 1552

102 E. W. Enlow, R. L. Pease, W. Combs, R. D. Schrimpf and R. N. Nowlin, *Response of advanced bipolar processes to ionizing radiation*, IEEE Trans. Nucl. Sci., **NS-38**, **Vol. 6**, 1991, p. 1342

103 D. M. Fleetwood, P. V. Dressendorfer and D. C. Turpin, *A reevaluationof worst-case postirradiation response for hardened MOS transistors*, IEEE Trans. Nucl. Sci., **NS-34**, **Vol. 6**, 1987, p. 1178

104 D. M. Fleetwood, P. S. Winokur, and L. C. Riewe, *Prediction switched-bias response from steady-state irradiation*, IEEE Trans. Nucl. Sci., **NS-37**, **Vol. 6**, 1990, p. 1806

105 IEEE *Annual Conference on Nuclear and Space Radiation Effects* IEEE Trans. Nucl. Sci., NS-17 to NS-37, 1970-1990

106 J. F. Baur, B. A. Engholm, M. P. Hecker, I. Maya, P. H. Müller, W. E. Teffele, and S. S. Wajtowlez, *Radiation Hardnening of Diagnostics for Fusion Reactors*, General Atomic Company, GA-A-16614 - UC-20

# RECOMMENDATIONS OF THE DETECTOR GROUP

To satisfy the requirements for a general purpose detector with tracking in its inner cavity, we recommend some changes of the machine characteristics to achieve the modified parameters of colliding beams as listed in the last ELN- column of table 1. In particular the beam crossing angle should be reduced from 250 μrad (150 μrad for the high luminosity version ELN-35) to 50 μrad. Consequently, the lengths of luminous regions will increase to 0.1 m (ELN-34) or 0.12 m (ELN-35). The corresponding calculations are detailed in the contribution of C. D'Ambrosio (these proceedings).

Bunch separation should be kept between 15 ns and 20 ns. Shorter spacings will become incompatible with the detectors busy times, as outlined in the contribution of T. Gys (these proceedings). Longer bunch separations yield to many events per bunch crossing, which makes the associations of decay signatures to one common vertex in most cases impossible (see contribution of U. Gensch, these proceedings).

Comparing the average number of vertices per mm for different colliders (C. D'Ambrosio, these proceedings), we notice from table 3, that for the ELN-34 version the vertex density will be about 21 times higher than for LHC. For the ELN-35 version, the vertices are about 76 times denser than for LHC, which presents already a serious handicap for the vertex resolution, which then should be as high as about 70 μm to safely resolve two vertices, taking into account the statistical fluctuations.

The annual dose rates for ELN-35 will lead, apart from the high vertex density, to unpleasant radiation hardness problems for the central tracking components (see contributions of H. Schönbacher and F. Wulf, these proceedings). The corresponding dose rates for ELN-36 are certainly not tolerable for an omni purpose detector, and considering a vertex density of about 68 $mm^{-1}$ this luminosity will be not very useful for any physics at all, since it will become impossible to guarantee, that specific particle patterns originate from the same vertex.

Particle cross-sections increase rapidly with energy : from table 5, of the contribution from H. Leutz, these proceedings, we learn, that the Higgs-cross-section (assuming 1000 GeV mass) is 104 times higher for ELN than for LHC at the same luminosity. On the other hand, the min. biased background increases much less : between ELN and LHC we find in the same table only a ratio of 3.5. This means, to improve the quality of physics with respect to signal over background ratios, priority should be given to an energy upgrade rather than a luminosity increase. Therefore, the machine design specifications should foresee a 350 km circumference of the ELN-tunnel rather than the originally proposed 300 km. In addition the development of 13 T dipole magnets should be serioulsy considered to envisage the possibility of an energy upgrade (see contribution of Q. Lin, these proceedings).

# Parameters of colliding beams at future hadron colliders

| BEAM PARAMETERS | | LHC | SSC | ELOISATRON* | | ELOISATRON** | |
|---|---|---|---|---|---|---|---|
| Centre of mass energy | TeV | 16 | 40 | 200 | | 200 | |
| Full beam crossing angle | μrad | 200 | 0–150(a) | 250 | 150 | 50 | |
| Beam diameter (4 σ = 95% of particles)(b) | μm | 42 | 19.2 | 4 | 6 | 4 | 6 |
| Bunch length (4 σ = 95% of particles)(b) | m | 0.31 | 0.24 | 0.24 | | 0.24 | |
| Depth of focus (β*) | m | 0.5 | 0.5 | 1.0 | 2.0 | 1.0 | 2.0 |
| Length of luminous region (95% of interactions)(b) | m | 0.19 | 0.15(c) | 0.022 | 0.054 | 0.1 | 0.12 |
| Bunch crossing time (95% of interactions)(b) | ns | 0.52 | 0.4 | 0.4 | | 0.4 | |
| Bunch separation | ns | 15 | 16.7 | 22 | 20 | 15 to 20 | |
| Bunch collision frequency | $\times 10^6 s^{-1}$ | 66.7 | 60 | 45 | 50 | 67 to 50 | |
| Luminosity | $cm^{-2}s^{-1}$ | $10^{34}$ | $10^{33}$ | $10^{34}$ | $10^{35}$ | $10^{34}$ | $10^{35}$ |
| Average number of vertices per bunch crossing | | 13 | 1.7 | 30 | 270 | 22 to 30 | 225 to 300 |
| Avr. Nr of vertices per mm (68% of interactions) | $mm^{-1}$ | 0.09 | 0.014 | 1.9 | 6.8 | 0.3 to 0.4 | 2.6 to 3.4 |

(a) adjustable (b) for 68% divide by two (c) crossing angle = 75 μrad

* original version

** recommended version